Advanced Series in Agricultural Sciences 20

Advanced Series in Agricultural Sciences

Volumes already published in the series

Volume 1: *A.P.A. Vink*
Land Use in Advancing Agriculture

Volume 2: *H. Wheeler*
Plant Pathogenesis

Volume 3: *R.A. Robinson*
Plant Pathosystems

Volume 4: *H.C. Coppel, J.W. Mertins*
Biological Insect Pest Suppression

Volume 5: *J.J. Hanan, W.D. Holley, K.L. Goldsberry*
Greenhouse Management

Volume 6: *J.E. Vanderplank*
Genetic and Molecular Basis of Plant Pathogenesis

Volume 7: *K.J. Matsushima*
Feeding Beef Cattle

Volume 8: *R.J. Hanks, G.L. Ashcroft*
Applied Soil Physics

Volume 9: *J. Palti*
Cultural Practices and Infectious Crop Diseases

Volume 10: *E. Bresler, B.L. McNeal, D.L. Carter*
Saline and Sodic Soils

Volume 11: *J.R. Parks*
A Theory of Feeding and Growth of Animals

Volume 12: *J. Hagin, B. Tucker*
Fertilization of Dryland and Irrigated Soils

Volume 13: *A.J. Koolen, H. Kuipers*
Agricultural Soil Mechanics

Volume 14: *G. Stanhill*
Energy and Agriculture

Volume 15: *E.A. Curl, B. Truelove*
The Rhizosphere

Volume 16: *D.P. Doolittle*
Population Genetics: Basic Principles

Volume 17: *A. Feigin, I. Ravina, J. Shalhevet*
Irrigation with Treated Sewage Effluent

Volume 18: *D. Gianola, K. Hammond*
Advances in Statistical Methods for Genetic Improvement of Livestock

Volume 19: *I. Rosenthal*
Electromagnetic Radiations in Food Science

Volume 20: *D. Russo, G. Dagan*
Water Flow and Solute Transport in Soils

D. Russo G. Dagan (Eds.)

Water Flow and Solute Transport in Soils

Developments and Applications
In Memoriam Eshel Bresler (1930 – 1991)

With 103 Figures

Springer-Verlag
Berlin Heidelberg New York
London Paris Tokyo Hong Kong
Barcelona Budapest

Professor DAVID RUSSO
Department of Soil Physics
Agricultural Research Organization
Ministry of Agriculture
Bet Dagan 50250, Israel

Professor GEDEON DAGAN
Faculty of Engineering
Tel Aviv University
Ramat Aviv 69978, Israel

ISBN-13: 978-3-642-77949-7 e-ISBN-13: 978-3-642-77947-3
DOI:10.1007/978-3-642-77947-3

Library of Congress Cataloging-in-Publication Data. Water flow and solute transport in soils: developments and applications / D. Russo, G. Dagan (eds.). p. cm. – (Advanced series in agricultural sciences; 20) Includes bibliographical references and index.
1. Soil moisture. 2. Soils – Solute movement. 3. Soil percolation. 4. Groundwater flow. I. Russo, D. (David), 1949–. II. Dagan, G. (Gedeon), 1932–. III. Series. S594.W24 1993 631.4'32–dc20 92-47096

Softcover reprint of the hardcover 1st edition 1993

Typesetting: Thomson Press, New Delhi
31/3145-5 4 3 2 1 0 –

Foreword

A year has passed since Eshel Bresler, my good friend and colleague, and a member of the editorial board of the *Advanced Series in Agricultural Sciences*, died suddenly while on a visit to the Chinese Academy of Sciences in Beijing.

We had worked together for almost 30 years at the Institute of Soils and Water, ARO, The Volcani Center at Bet Dagan. At the very beginning of our scientific careers we cooperated directly and as a result one of our first publications was coauthored (Soil Sci. 101:205–209, 1966). Thereafter, our specific research interests diversified, but we continued to work together, with similar approaches to research, and to strive towards the development of Israel soil science and its integration into general worldwide scientific progress.

I don't need to emphasize Eshel's contribution to the understanding of the processes governing water flow and solute transport processes in soils and unsaturated zones. The contributions to this Volume by such a body of outstanding scientists shows the appreciation of the international scientific community to his research achievements.

This Volume of *Advanced Series in Agricultural Sciences* is dedicated to Eshel Bresler's memory. In the name of the editorial board of our series, I would like to thank the contributors who honor Eshel's memory by answering my call without any hesitation, to Gedeon Dagan and David Russo, Eshel's close research partners, who agreed to serve as editors, and finally to Dieter Czeschlik and Springer-Verlag for agreeing to include the memorial volume in our Series.

I am convinced that the present collection of research works on water flow and solute transport in soils will be highly appreciated and together with Eshel Bresler's publications in the past will constitute landmarks in soil science for many years to come.

Bet Dagan, Israel BRUNO YARON

Preface

The present volume consists of a series of articles on water flow and solute transport in soils to honor the memory of Eshel Bresler. The authors have interacted with Eshel in various stages of his rich activity and are leaders in the field of soil physics. Since the choice of the subjects was left to the discretion of the authors, the volume somewhat reflects the main trends of contemporary research in this field and its applications. The various contributions fall into three main areas which make up the three Parts of the book.

The first area is that of stochastic modeling of solute transport through heterogeneous soil in the uper layer of the unsaturated zone. The relatively large number of articles in this area reflects, on the one hand, the deep interest of the community in the subject and, on the other hand, Eshel's important contribution to it. This is underscored by the first chapter, by G. Dagan, which reviews some recent works related to the model of transport through heterogeneous fields and developed by Eshel. The chapters by G. Destouni, D. Russo and G. Sposito are original contributions on the same topic, while that by R. J. Luxmoore and L. A. Ferrand focuses on pore-scale mechanisms contributing to solute dynamics at larger scales using the percolation theory. We believe that readers may get from this Part a fair picture of the latest developments in this area.

The second area is within the more traditional scope of analysis of flow through homogeneous or layered formations. Still, the complexity of unsaturated flow poses open and fascinating unsolved problems. This is underscored by the contribution by J. R. Philip on hillslopes, part of his extensive work on two-dimensional flows near the soil-atmosphere boundary. The two chapters by D. Hillel and H. J. Morel-Seytoux, dealing with flow through a two-layered soil of different permeabilities and arriving at somewhat contradictory conclusions, illustrate the complex behavior of unsaturated media. The chapter by C. Dirksen and his colleagues deals with numerical models to simulate one-dimensional vertical simultaneous water and solute transport in the unsaturated zone considering water uptake by roots and hysteresis in the water saturation-capillary pressure relationships, while the contribution by C. W. Rose concentrates on the transport of sorbed chemicals due to erosion by water. It is worthwhile mentioning here Eshel's contributions in the area, e.g., his work on the simultaneous water and solute transport in the unsaturated zone considering one- and two-dimensional flow domains.

Finally, we received a few interesting and diverse contributions related to applications. These span several subjects, like new devices for field measurement of water flow and solute transport parameters (D. E. Elrick and colleagues), estimation of effective properties of actual fields subjected to spatially and time-varying boundary conditions (R. A. Feddes and colleagues), calculation of solute movement through a soil cover (C. T. Kincaid and G. W. Gee), analysis of impacts of vertical heterogeneity on water flow in Hawaiian basaltic saprolite during rainfall (T. R. Green and colleagues), irrigation scheduling considering soil variability and climate uncertainty (D. Or and R. J. Hanks), and use of geostatistical methods to quantify solute concentration in spatially variable soils (S. R. Yates and colleagues). These various works reflect the deep interest Eshel had in field applications and real-life cases.

Thus, we hope this book offers a broad perspective of areas of research in soil physics and of the wide professional interest of our late friend Eshel Bresler.

GEDEON DAGAN and DAVID RUSSO

Contents

Introduction

The Contributions of Eshel Bresler to Soil Science
W. D. KEMPER 3

E. Bresler: List of Publications 6

Part I Stochastic Modeling of Flow and Transport in Unsaturated Soil at Field Scale

1 The Bresler-Dagan Model of Flow and Transport: Recent Theoretical Developments
G. DAGAN

1.1 Introduction 14
1.2 The Assumptions and the Main Results of the Original BD Model 15
1.3 The Effect of Local Dispersion 18
1.4 Unsteady Flow (Infiltration and Redistribution) and Transport 20
1.5 Transport of Reactive Solutes in Heterogeneous Fields . 22
1.6 Sensitivity Analysis of Crop Yield in a Heterogeneous Field 24
1.7 Solute Flux in Heterogeneous Soils 25
1.8 Mass Arrival of Sorptive Solute into the Groundwater . 28
1.9 Validity of One-Dimensional Approximation for Infiltration in Heterogeneous Soils 29
References 30

2 Field-Scale Solute Flux Through Macroporous Soils
G. DESTOUNI

2.1 Introduction 33
2.2 Transport Model 34
2.3 Travel Time PDF 36
2.4 Illustration of Results 38
2.5 Discussion and Conclusions 42
References 43

3 Towards Pore-Scale Analysis of Preferential Flow and Chemical Transport

R. J. LUXMOORE and L. A. FERRAND

3.1 Scales of Preferential Flow 45
3.2 Immobile Water 46
3.3 Old Water – New Water 46
3.4 Subsurface Transport Investigations at Oak Ridge, Tennessee 47
3.5 Pathlength-Supply Hypothesis 49
3.6 Percolation Theory 51
3.7 Percolation Modeling 52
3.7.1 Hydraulic Conductivity and Dispersion Coefficient .. 54
3.7.2 Mass Transfer Coefficient 56
3.8 Stochastic Methods (Latin Hypercube Sampling) 57
3.9 Synopsis 57
References 59

4 Analysis of Solute Transport in Partially Saturated Heterogeneous Soils
D. RUSSO

4.1 Introduction 61
4.2 Basic Concepts and Definitions 63
4.3 Modeling of Solute Transport in Heterogeneous Porous Media 65
4.3.1 Transport in Saturated Porous Media 65
4.3.2 Transport in Unsaturated Porous Media 67
4.3.2.1 Simplified Stochastic Approach 67
4.3.2.2 Simulation of Solute Transport 69
4.3.2.3 General Stochastic Approach 73
4.4 Summary and Conclusions 78
References 79

5 Solute Lifetime Correlations in Chemical Transport Through Field Soils
G. SPOSITO

5.1 Introduction 82
5.2 Fundamental Statistical Concepts 83
5.3 Perfect Correlations Among Solute Lifetimes 86
5.4 Less Than Perfect Correlations Among Solute Lifetimes 90
5.5 Concluding Remarks 93
References 94

Part II Solutions of Flow and Transport in Unsaturated Media by Deterministic Models

6 HYSWASOR – Simulation Model of Hysteretic Water and Solute Transport in the Root Zone
C. DIRKSEN, J.B. KOOL, P. KOOREVAAR, and M.TH. VAN GENUCHTEN

6.1 Introduction 99
6.2 Governing Equations 102
6.2.1 Water Transport 102
6.2.2 Closed Scanning Loop Hysteresis Algorithm 102
6.2.3 Solute Transport 104
6.2.4 Root Water Uptake 104
6.3 Input 106
6.3.1 General Data 106
6.3.2 Soil Parameters 109
6.3.3 Boundary Conditions 109
6.3.4 Initial Conditions 110
6.4 Output 111
6.4.1 Screen Output 111
6.5 Simulations 112
6.5.1 Daily Irrigation 112
6.5.2 Six-Day-Interval Irrigation 116
6.6 Concluding Remarks 116
References 120

7 Unstable Flow: A Potentially Significant Mechanism of Water and Solute Transport to Groundwater
D. HILLEL

7.1 Nature of the Problem 123
7.2 Review of Unstable Flow 124
7.3 Instability Between Fluids Differing in Density or Viscosity 125
7.4 Instability During Infiltration into Unsaturated Soils . 126
7.5 Preliminary Evidence from Field Experiments 131
7.6 Final Comment 133
References 133

8 Capillary Barrier at the Interface of Two Layers
H. J. MOREL-SEYTOUX

8.1 The Particular Physical Problem Considered 136
8.2 Conditions at the Interface 136
8.3 Basic Equations 139
8.3.1 Two-Phase Flow Background and Notations 139

8.3.2 Soil Characteristics Representation 140
8.3.2.1 Capillary Pressure 140
8.3.2.2 Relative Permeability 140
8.3.2.3 Effective Capillary Drive 140
8.3.3 Propagation Velocities of Fronts 141
8.3.4 Total Velocity Expression for Particular Problem ... 141
8.4 System of Equations to Be Solved 141
8.5 Solution 142
8.5.1 Procedure 142
8.5.2 Result of Integration 143
8.5.3 Numerical Procedures 144
8.5.3.1 Initialization 144
8.5.3.2 Repetitive Steps 144
8.6 Applications 144
8.6.1 Values of Parameters for Reference Run 144
8.6.2 Reference Run Results 144
8.6.3 Numerical Scheme Parameter Sensitivity 147
8.6.4 Influence of Effective Capillary Drive Magnitude .. 147
8.6.5 Influence of Supply Rate 147
8.7 Discussion 148
8.8 Conclusions 149
References 150
List of Symbols 151

9 Constant-Rainfall Infiltration on Hillslopes and Slope Crests
J. R. PHILIP

9.1 Introduction 152
9.2 Constant-Rainfall Slope-Crest Infiltration: Nonlinear Formulation 153
9.2.1 Flow Equation and Conditions 153
9.2.2 Time-to-Ponding: The Two Cases 153
9.2.2.1 Case 1: $R_* \leqslant K_1$ 154
9.2.2.2 Case 2: $R_* > K_1$ 154
9.2.3 Remark 154
9.3 Constant-Rainfall Slope-Crest Infiltration: Linearized Formulation 154
9.3.1 Flow Equation and Conditions 154
9.3.2 Dimensional Forms 155
9.3.3 Time-to-Ponding: The Two Cases 156
9.3.4 Remark 156
9.4 Long-Slope Constant-Rainfall Infiltration: Nonlinear Formulation 156
9.4.1 Flow Equation and Conditions 157
9.4.2 Remark 157

9.5 Long-Slope Constant-Rainfall Infiltration: Linearized Formulation 157
9.5.1 Flow Equation and Conditions, Dimensionless Forms 157
9.5.2 Solution 158
9.6 Physical Implications of Long-Slope Solutions 158
9.6.1 Distribution of Potential and Moisture Content 159
9.6.2 Standard and Rotated Flow Components 161
9.6.2.1 Horizontal and Vertical Components 161
9.6.2.2 Horizontal Inslope Flow Velocity 162
9.6.2.3 Vertical Flow Velocity 163
9.6.2.4 Downslope and Normal Components 163
9.6.3 Integrated Horizontal and Downslope Components 163
9.6.3.1 Integrated Inslope Horizontal Flow 163
9.6.3.2 Integrated Downslope Flow 164
9.6.4 Time Dependence of Surface Flow Velocity Vector . 164
9.6.5 Long-Slope Time-to-Ponding 165
9.7 Constant-Rainfall Slope-Crest Solution for $\gamma = 45$ .. 166
9.8 Physical Implications of Slope-Crest Solution 168
9.8.1 Evolution of Moisture Content and Potential Distributions 168
9.8.2 The Time Course of Surface Moisture Content 170
9.8.3 Slope-Crest Time-to-Ponding 171
9.8.4 Criterion for Validity of Long-Slope Solution 172
9.8.5 Limits on Long-Slope Solution for Arbitrary γ 173
9.9 Concluding Discussion 174
9.9.1 Inslope Horizontal Flow 174
9.9.2 Downslope Flow 174
9.9.3 The Surface Flow Velocity Vector 175
9.9.3.1 Ponded Infiltration 175
9.9.3.2 Constant-Rainfall Infiltration 175
9.9.4 Comparing the Dynamics of Ponded and Constant-Rainfall Long-Slope Infiltration 175
9.9.5 The Slope-Crest Effect 176
9.9.6 The Downslope Propagation of Ponding from a Slope-Crest 177
Appendix: Properties of $G(X, Y)$ 177
References 178

10 The Transport of Sorbed Chemicals in Eroded Sediment
C.W. Rose

10.1 Introduction 180
10.1.1 Overview of Chemical Enrichment Mechanisms 180
10.2 Sediment Transport with Rainfall Impact the Dominant Erosive Agent 182

10.2.1 Deposition ... 182
10.2.2 Rainfall Detachment and Redetachment ... 182
10.2.3 Changes in Time of Sediment Settling-Velocity Distributions During Erosion in the Presence of a Water Layer ... 183
10.2.4 Sediment Detachment and Transport Under Rainfall with Small Water Depth on the Soil ... 184
10.3 Sorbed Chemical Enrichment with a Significant Water Depth ... 186
10.3.1 The Effect of Erosion Process and Surface Contact Cover on the Settling-Velocity Distribution of Eroded Sediment ... 186
10.3.2 The Effect of Erosion Process and Surface Contact Cover on the Enrichment Ratio of Nitrogen ... 187
10.3.3 Theoretical Framework for Interpreting Enrichment Effects with a Significant Depth of Overland Flow . 188
10.4 Sorbed Chemical Enrichment with Shallow Water Depths ... 191
10.4.1 Enrichment due to Differential Sorbed Chemical Concentration Within Soil Aggregates ... 191
10.4.2 Relationship Between Enrichment Ratio and Cumulative Loss of Eroded Sediment ... 193
10.5 General Discussion and Conclusions ... 194
References ... 195

Part III Experimental Techniques and Application of Statistical Methods to Field Problems

11 Field Measurement of Water and Solute Transport Parameters in Soils
D. E. Elrick, W. D. Reynolds, R. G. Kachanoski, and G. W. Parkin

11.1 The Use of Permeameters and Infiltrometers to Determine Hydraulic Properties ... 199
11.2 Hydraulic Properties ... 199
11.3 Flow Equations and Analyses ... 202
11.4 Results ... 204
11.5 Discussion ... 204
11.6 The Use of TDR to Determine Hydraulic and Solute Transport Properties ... 205
11.6.1 Hydraulic Properties ... 205
11.6.2 Solute Transport Parameters ... 207
References ... 208

12 Estimation of Regional Effective Soil Hydraulic Parameters by Inverse Modeling
R. A. FEDDES, G. H. DE ROOIJ, J. C. VAN DAM, P. KABAT, P. DROOGERS, and J. N. M. STRICKER

12.1 Introduction 211
12.2 Methods of Regionalization 212
12.2.1 Scaling Approaches 212
12.2.2 Non-Scaling Approaches 212
12.2.3 Inverse Modeling Approaches of Transient Flow 213
12.2.4 Outline of the Present Study 214
12.3 Simulation Experiments 216
12.3.1 Numerical Infiltration-Evaporation Experiment on 32 Bare-Soil Columns 216
12.3.2 Inverse Modeling Approach on Bare-Soil Columns 217
12.4 Validation 222
12.5 Validation of the Water Balance of Grass 226
12.6 Conclusions 227
References 229

13 Impacts of Vertical Heterogeneity on Simulated Water Flow in Hawaiian Basaltic Saprolite: Relation to Recharge
T. R. GREEN, R. E. GREEN, and M. E. MILLER

13.1 Introduction 232
13.2 Physical Setting of Study Area 233
13.3 Methodology 234
13.3.1 Estimates of $K(h)$ from Water Characteristic Data 234
13.3.2 Rainfall and Evaporation Data 235
13.3.3 Variability Saturated Zone Simulation Model 235
13.3.4 Hydraulic Properties and Stratigraphy of the Soil/Saprolite 238
13.3.5 Simulation Scenarios 238
13.4 Results 239
13.4.1 Time Series: Infiltration, Drainage and Water Content 239
13.4.2 Pressure Head Profiles for Extreme Periods 241
13.4.3 Profiles of Pressure Exceedence 241
13.5 Discussion 243
13.6 Conclusions 244
References 244

14 Estimating Infiltration at Waste Sites: Methodology Development
C. T. KINCAID and G. W. GEE

14.1 Introduction 246

14.2 Methodology and Assumptions 247
14.2.1 Estimation of the Parameters 249
14.2.2 Calculations for Steady-State Flow 250
14.2.3 Calculations for Nonsteady Flow 251
14.3 Results 253
14.3.1 Estimating Surface Recharge Flux by Calibration with Measured Values of Saturation 253
14.3.2 Cumulative Probability of Saturation 254
14.3.3 Probability of Ponding and of Fluxes for $q<R$ 255
14.3.4 Expected Values of Flux 257
14.3.5 Expected Values of Saturation and Flux at $z = 1$ m for Non Steady Infiltration 257
14.4 Summary 259
References 260

15 Irrigation Scheduling Considering Soil Variability and Climatic Uncertainty: Simulation and Field Studies
D. OR and R. J. HANKS

15 Introduction 262
15.2 Theoretical Considerations 263
15.2.1 Crop Yield Model 263
15.2.2 Soil-Water Balance 264
15.2.3 Stochastic *ETp* Model 264
15.2.4 Mathematical Programming (Optimization) 265
15.2.4.1 Irrigation Scheduling in a Homogeneous Field 266
15.2.4.2 Irrigation Scheduling in a Heterogeneous Field 267
15.3 Materials and Methods 268
15.3.1 Simulation Study 268
15.3.2 Field Study 268
15.3.3 Stochastic *ETp* Model for the Study Site 270
15.3.4 Irrigation Scheduling Scheme 271
15.4 Results and Discussion 273
15.4.1 Simulation Study – The Effect of Seasonal Irrigation Amount on Crop Yield 273
15.4.2 Field Experiment 275
15.4.2.1 Water Balance and Crop Yield 275
15.4.2.2 *ETp* Predictions 276
15.4.2.3 The Irrigation Schedule 278
15.5 Summary and Conclusions 280
References 281

16 Use of Geostatistics in the Description of Salt-Affected Lands
S.R. YATES, R. ZHANG, P.J. SHOUSE, and M.TH. VAN GENUCHTEN

16.1 Introduction ... 283
16.2 Geostatistical Analyses ... 285
16.2.1 Pseudo Cokriging ... 285
16.2.2 Disjunctive Kriging ... 287
16.3 Examples ... 289
16.3.1 Estimation of NO_3 Using Cokriging with EC and Ca as Auxiliary Variables ... 289
16.3.2 Managing Soil EC in Agricultural Fields ... 295
16.3.2.1 The Sample Correlation Function ... 298
16.3.2.2 Estimation ... 299
16.3.2.3 Conditional Probability ... 300
16.4 Summary and Conclusions ... 302
References ... 303

Subject Index ... 305

Contributors

DAGAN, G.	Faculty of Engineering, Tel Aviv University, Ramat Aviv 69978, Israel
DE ROOIJ, G.H.	Department of Water Resources, Wageningen Agricultural University, Wageningen, The Netherlands
DESTOUNI, G.	Water Resources Engineering, Royal Institute of Technology, 10044 Stockholm, Sweden
DIRKSEN, C.	Department Water Resources, Wageningen Agricultural University, Wageningen, The Netherlands
DROOGERS, P.	Department of Water Resources, Wageningen Agricultural University, Wageningen, The Netherlands
ELRICK, D.E.	Department of Land Resource Science, University of Guelph, Guelph, Ontario N1G 2W1, Canada
FEDDES, R.A.	Department of Water Resources, Wageningen Agricultural University, Wageningen, The Netherlands
FERRAND, L.A.	Department of Civil Engineering, City University of New York, New York 10031, USA
GEE, G.W.	Hydrology Section, Geosciences Department, Battelle, Pacific Northwest Laboratory, Richland, Washington, USA
GREEN, T.R.	Stanford University, Department of Civil Engineering, Stanford, California 94305, USA and US Geological Survey, Water Resources Division, Mento Park, California, USA

Green, R.E.	University of Hawaii, Department of Agronomy and Soil Science, Honolulu, Hawaii 96822, USA
Hanks, R.J.	Department of Plants, Soil, and Biometeorology, Utah State University, Logan, Utah, USA
Hillel, D.	University of Massachusetts, Amherst, MA 10003, USA
Kabat, P.	DLO The Winand Staling Centre for Integrated Land, Soil and Water Research (SC-DLO), Wageningen, The Netherlands
Kachanoski, R.G.	Department of Land Resource Science, University of Guelph, Guelph, Ontario N1G 2W1, Canada
Kemper, W.D.	USDA-ARS, National Agricultural Library Building, Beltsville, Maryland, USA
Kincaid, C.T.	Hydrology Section, Geosciences Department, Battelle, Pacific Northwest Laboratory, Richland, Washington, USA
Kool, J.B.	Hydro Geologic Inc., Herndon, Virginia, USA
Koorevaar, P.	Department of Water Resources, Wageningen Agricultural University, Wageningen, The Netherlands
Luxmoore, R.J.	Environmental Sciences Division, Oak Ridge National Laboratory, Oak Ridge, Tennessee 37831-6038, USA
Miller, M.E.	Daniel B. Stephens and Associates, 6020 Academy NE, Albuquerque, New Mexico 87109, USA
Morel-Seytoux, H.J.	57 Selby Lane, Atherton, California 94027-3926, USA

OR, D.	Department of Civil Engineering, University of California, Berkeley, California, USA
PARKIN, G.W.	Department of Land Resource Science, University of Guelph, Guelph, Ontario N1G 2W1, Canada
PHILIP, J.R.	CSIRO Centre for Environmental Mechanics, Canberra, Australian Capitel Territory, 2601, Australia
REYNOLDS, W.D.	Centre for Land and Biological Resources Research, Research Branch, Agriculture Canada, Ottawa, Ontario K1A 0CA, Canada
ROSE, C.W.	Centre for Catchment and In-Stream Research, Faculty of Environmental Sciences, Griffith University, Nathan Campus, Brisbane, Queensland 4111, Australia
RUSSO, D.	Department of Soil Physics, Agricultural Research Organization, Ministry of Agriculture, Bet-Dagan 50250, Israel
SHOUSE, P.J.	USDA-ARS, US Salinity Laboratory, 4500 Glenwood Drive, Riverside, California 92501, USA
SPOSITO, G.	Department of Soil Science University of California, 108 Hilgard Hall, Berkeley, California 94720, USA
STRICKER, J.N.M.	Department of Water Resources, Wageningen Agricultural University, Wageningen, The Netherlands
VAN DAM, J.C.	Department of Water Resources, Wageningen Agricultural University, Wageningen, The Netherlands
VAN GENUCHTEN, M.TH.	USDA-ARS US Salinity Laboratory, 4500 Glenwood Drive, Riverside, California 92501, USA
YARON, B.	Institute of Soils and Water, Agricultural Research Organization, The Volcani Center, Bet-Dagan, Israel

YATES, S. R. — USDA-ARS, US Salinity Laboratory, 4500 Glenwood Drive, Riverside, California 92501, USA

ZHANG, R. — USDA-ARS, US Salinity Laboratory, 4500 Glenwood Drive, Riverside, California 92501, USA

Introduction

Prof. Eshel Bresler
1930-1991

The Contribution of Eshel Bresler to Soil Science

W. Doral Kemper[1]

Born in Kibbutz Givat Hashlosha, Eshel gained an early appreciation of soil and water and what they can do for us. During his professional career he directed his keen intellect to gaining a knowledge of the processes by which water and nutrients find their way to plants, how accumulations of salt and sodium degrade soils and how we can enhance our soil and water resources.

During his early years with the extension service he interacted closely with innovative farmers and field managers in nahals and kibbutzim and gained an understanding of their ability to grasp new ideas and integrate them into solutions of practical problems. During these years he became acutely aware of the limited extent of Israel's water supply, the vital role of water in crop production systems, and the efficiencies with which water was being used in crop production. Based on these facts he concluded that one of Israel's greatest needs was to improve the efficiency of water use. This led to his early recognition of the potential of drip irrigation and the enthusiastic encouragement which he expressed to Zohar and other pioneer practitioners of this type of irrigation.

Early in the history of trickle irrigation he and his coworkers (references 13 and 14 from his technical publications list) showed that flow into soils from trickle irrigation emitters could be described with satisfactory accuracy by theory derived from basic principles. He was the first to publish a method for simulating two-dimensional water and solute flow in soils in response to trickle irrigation. His continuing contributions and growing knowledge of the theory and practice involved in trickle irrigation led to an invitation from the American Society of Agronomy to write a definitive monograph on trickle irrigation. This monograph was published in *Advances in Agronomy* (Volume 29, 1977) and provided many irrigation scientists with a comprehensive understanding of the factors involved in trickle irrigation, including its potential advantages, problems and practice, and the water and solute regimes involved.

Eshel provided major impetus to the new journal *Irrigation Science* in 1978 and helped it establish high standards with its first article which he titled "Analysis of Trickle Irrigation with Application to Design Problems." In 1985 he was asked to coauthor the *Drip Irrigation Manual* whcih was published by the International Information Center, and serves as a primary

---

[1] USDA-ARS, National Agricultural Library Building, Beltsville, Maryland, USA

guide for individuals, corporations and nations contemplating the use of "trickle" or 'drip" irrigation.

With his broad interest and experience in water and solute movement in the distribution system, through the soil and their disposition as evaporation, plant uptake or drainage, and his exposure to developments in America and Israel, Eshel was often the integrator. Working with many colleagues he played a major role in the development of mathematical approaches and finite difference models which simulated movement of both water and solutes moving through soil. His willingness to do more than his share of the work and to take less than his share of the credit set a standard that attracted many cooperators to his good causes.

Working with multidimensional models with the objective of solving practical problems Eshel often found that the systems became so complicated that solutions were possible only if some of the processes which he knew were taking place in the soil could be disregarded. Some of his most significant contributions to the advancement of practical solutions were in determining conditions under which the possible complicating factors could be considered practically negligible. For instance, he was concerned about the interaction of solute ions with the electrical charge on clay minerals. Under certain conditions these interactions can result in retardation of the anions in water passing through the soil. He demonstrated experimentally and showed by calculation that while this effect was significant under some conditions, it was sufficiently small to be disregarded in most of the irrigation and drainage models which he and his coworkers were developing (e.g. publications 20–25, 31, and 32).

Needing good estimates of hydraulic conductivity in his models, it is not surprising that Eshel and his coworkers developed some of the simplest and quickest methods for estimating unsaturated hydraulic conductivity (e.g. publications 35–38).

Eshel's early experience in extension and his continuing interest in the real world of water management gave him an understanding of the large variability of soils and their water-holding and transmission characteristics within fields. Many scientists, upon recognizing the complications introduced by this variability, have either given up this type of research and development, or have become so permanently and completely immersed in documenting the variability that they have not gotten back to the development of solutions. Eshel and his coworkers documented the variability of the soil hydraulic conductivity, evaluated factors that caused the variability and developed methods for estimating how this variability interacted with phenomena of interest to affect crop yields and subsequent soil properties (publications 46–50, 53, 54, 57–63, and 65). This research laid the foundation for guidelines on how to optimize utilization of our spatially variable resources. The fact that all our soil resources are spatially variable makes this contribution particularly significant (e.g. publications 72, 73, 75, 76, and 78).

Working with his models and becoming more aware of the spatial variability inherent in soils and yields gave Eshel an insight into the accuracy needed in measuring hydraulic conductivity, matric potential and associated parameters. This insight enabled him to play a vital role in the development of measurement techniques to provide good estimates of unsaturated hydraulic conductivity without undue expenditure of time and resources (e.g. publications 68 and 69).

Some of Eshel's earliest experimental studies (ref. 5) showed us that when water percolates down through large pores in soils, its rate of movement is so fast that its solutes do not have time to come to equilibrium with solutes and adsorbed ions in the soil. The practical significance of this bypass phenomenon was not fully recognized until we became interested in groundwater quality and the role which preferential flow and soil management play therein.

Another of Eshel's early experimental findings, the importance of which has not been fully recognized, is the effects of rainfall on subsequent evaporation from undisturbed soils. Eshel measured the hydraulic potential gradient in the top few centimeters of soil during drying and found that in the early stages of drying the layer (2–3 mm thick) disrupted by raindrop impact retained its water and was more conductive than the layer immediately below it (i.e. 3–10 mm in depth), which desaturated and became the primary resistance to water flow. This subsurface resistance to water flow results in quicker drying of soil surfaces and less total evaporation after being wetted by rainfall (or sprinkler irrigation) than of soils wetted by flooding. Subsequent data, which were never published, showed that air (primarily oxygen and nitrogen) absorbed in the cold falling raindrops is liberated as the water enters the soil and is warmed. This liberated air accumulates in the largest pores where its pressure may reach several centimeters of head when the gas phase in that pore is entrapped by surrounding water and soil. These pressurized bubbles expand the large pores which continue to grow during successive irrigation events if the soil is not cultivated and form a highly vesicular layer that substantially reduces unsaturated water conductivity and evaporation. These phenomena will become major factors in our evaluation of the relative merits of different types of irrigation and of till and no-till agriculture.

In addition to these direct contributions to soil science, Eshel played a major role in educating and motivating others to do research in soil-water-solute-crop production processes. He was the major professor and mentor for five successful PhD and eight MSc candidates. In addition, he gave support, encouragement, and insight to a host of coworkers. His desire to learn made him an excellent listener and many of us were drawn to him because of this characteristic. Having observed his excellent example of listening it was easier for us to listen and learn from him.

His desire to hear all viewpoints before making a decision made him a surprisingly effective supervisor and administrator.

In spite of his quiet self-effacing manner, his experties was recognized internationally and he was invited to give lectures and/or be a consultant in more than a dozen countries. A common response from those with whom he consulted is "He listened to us and understood our problems which enabled him to help us!"

To help us fully benefit from Eshel's contributions to soil and water science, a list of his publications follows. In them are the roots of many solutions to the problems of the future.

Eshel Bresler: List of Publications

I. Books

1a. Bresler E, McNeal BL, Carter DL (1982) Saline and sodic soils: principles – dynamics – modeling. Springer, Berlin Heidelberg New York, 236 pp
1b. Bresler E, McNeal BL, Carter DL (1982) Saline and sodic soils: principles – dynamics – modeling. Gidrometeoizdat, Leningrad, 296 pp, (Russian translation, 1987)
2. Dasberg S, Bresler E (1985) Drip irrigation manual. International Information Irrigation Center (IIIC), Bet Dagan, Israel

II. Technical, Refereed Journal Articles

1. Yaron B, Bresler E, Shalhevet J (1966) A method for uniform packing of soil columns. Soil Sci 101: 205–209
2. Bresler E, Cohen OP, Broom M, Sharabani N (1966) Uniformity of irrigation water in citrus groves and its effect on water content, consumptive use of water, and irrigation efficiency. Ktavim 17: 181–190 (in Hebrew)
3. Bresler E (1967) A model for tracing salt distribution in the soil profile and estimating the efficient combination of water quality and quantity under varying field conditions. Soil Sci 104: 227–233
4. Cohen OP, Bresler E (1987) The effect of non-uniform water application on soil moisture content, moisture depletion and irrigation efficiency. Soil Sci Soc Am Proc 31: 117–121
5. Mokady RS, Bresler E (1968) Reduced sodium exchange capacity in unsaturated flow. Soil Sci Soc Am Proc 32: 463–467
6. Bresler E, Kemper WD, Hanks RJ (1969) Infiltration, redistribution and subsequent evaporation of water from soil as affected by wetting rate and hysteresis. Soil Sci Soc Am Proc 33: 832–840
7. Bresler E, Hanks RJ (1969) Numerical method for estimating simultaneous flow of salt and water in unsaturated soil. Soil Sci Soc Am Proc 33: 827–832
8. Hanks RJ, Klute A, Bresler E (1969) A numeric method for estimating infiltration, redistribution, drainage and evaporation of water from soil. Water Resour Res 5: 1064–1069
9. Bresler E, Kemper WD (1970) Soil water evaporation as affected by wetting method and crust formation. Soil Sci Soc Am Proc 34: 3–8
10. Yaron D, Bresler E (1970) A model for economic evaluation of water quality in irrigation. Aust J Agric Econ 14: 53–62
11. Bresler E (1970) Numerical solution of the equation for interacting diffuse layers in mixed ionic system with nonsymmetrical electrolytes. J Colloid Interface Sci 33: 278–283
12. Shainberg I, Bresler E, Klausner Y (1971) Studies on Na/Ca montmorillonite system. I. Swelling pressure. Soil Sci 111: 214–219
13. Brandt A, Bresler E, Diner N, Ben-Asher I, Heller J (1971) Infiltration from a trickle source. I. Mathematical model. Soil Sci Soc Am Proc 35: 675–682
14. Bresler E, Heller J, Diner N, Ben-Asher I, Brandt A, Goldberg D (1971) Infiltration from

a trickle source: experimental data and theoretical predictions. Soil Sci Soc Am Proc 35: 683–689

15. Bresler E (1971) Effect of the irrigation method on soil water regime. Bull Rech Agron Gembloux. pp. 233–254 in the Volume Extraordinaire Edite A L'Occasion De La Semaine D'Etude Des Problemes Mediterarraneens, Hors Ser, pp 233–254
16. Bresler E, Yaron D (1972) Soil water regime in economic evaluation of salinity in irrigation. Water Resour Res 8: 791–800
17. Bar-Yosef B, Kafkafi U, Bresler E (1972) Uptake of P by plants growing under field conditions. I. Theoretical model and experimental determination of its parameters. Soil Sci Soc Am Proc 36: 783–788
18. Bar-Yosef B, Bresler E, Kafkafi U (1972) Uptake of P by plants growing under field conditions. II. Computed and experimental results for corn plants. Soil Sci Soc Am Proc 36: 788–794
19. Stevenson DS, Bresler E (1972) Infiltration from a trickle source. Part II. Comments and letters to the editor. Soil Sci Soc Am Proc 36: 382–283
20. Bresler E (1972) Interacting diffuse layers in mixed mono-divalent ionic systems. Soil Sci Soc Am Proc 36: 891–896
21. Bresler E (1973) Simultaneous transport of solutes and water under transient unsaturated flow conditions. Water Resour Res 9: 975–986
22. Lahav N, Bresler E (1973) Exchangeable cations-structure parameter relationships in montmorillonite. Clays Clay Miner 21: 249–255
23. Bresler E (1973) Anion exclusion and coupling effects in nonsteady transport through unsaturated soils. I. Theory. Soil Sci Soc Am Proc 37: 663–669
24. Bresler E, Laufer A (1974) Anion exclusion and coupling effects in nonsteady transport through unsaturated soils. II. Laboratory and numerical experiments. Soil Sci Soc Am Proc 38: 213–218
25. Bresler E, Laufer A (1974) Simulation of nitrate movement in soils under transient unsaturated flow conditions. Isr J Agric Res 23: 141–153
26. Feddes RE, Bresler E, Neuman SP (1974) Field test of a modified numerical model for water uptake by root systems. Water Resour Res 9: 1199–1206
27. Neuman SP, Feddes RA, Bresler E (1975) Finite element analysis of two-dimensional flow in soils considering water uptake by roots. I. Theory. Soil Sci Soc Am Proc 39: 224–230
28. Feddes RA, Neuman SP, Bresler E (1975) Finite element analysis of two-dimensional flow in soils considering water uptake by roots. II. Field applications. Soil Sci Soc Am Proc 39: 231–237
29. Shimshi D, Yaron D, Bresler E, Weisbrod M, Strateener G (1975) Simulation model for evapotranspiration of wheat: empirical approach. J Irrig Drainage Div ASCE, vol 101, No IR1, Proc Pap 11170: 1–12
30. Strateener G, Yaron D, Bresler E, Shimshi D (1975) Simulation model for evapotranspiration of wheat: effect of evaporative conditions. J Irrig Drainage Div ASCE, vol 101, No IR1, Proc Pap 11169: 13–19
31. Bresler E, Russo D (1975) Two-dimensional solute transfer during nonsteady infiltration: laboratory test of mathematical model. Soil Sci Soc Am Proc 39: 585–587
32. Bresler E (1975) Two-dimensional transport of solutes during nonsteady infiltration from a trickle source. Soil Sci Soc Am Proc 39: 604–613
33. Miller RD, Loch JPG, Bresler E (1975) Transport of water and heat in a frozen permeameter. Soil Sci Soc Am Proc 39: 1029–1036
34. Feddes RA, Kowalik P, Neuman SP, Bresler E (1976) Finite difference and finite element simulation of field water uptake by plants. Hydrol Sci 21: 81–98
35. Russo D, Bresler E (1977) Effect of mixed Na/Ca solutions on the hydraulic properties of unsaturated soils. Soil Sci Soc Am J 41: 713–717
36. Russo D, Bresler E (1977) Analysis of the saturated-unsaturated hydraulic conductivity in a mixed Na/Ca soil system. Soil Sci Soc Am J 41: 706–710
37. Miller RD, Bresler E (1977) A quick method for estimating soil water diffusivity functions. Soil Sci Soc Am J 41: 1020–1022
38. Bresler E, Russo D, Miller RD (1978) Quick estimate of unsaturated hydraulic conductivity function. Soil Sci Soc Am J 42: 170–172

39. Bresler E (1978) Analysis of trickle irrigation with application to design problems. Irrig Sci 1: 3–17
40. Bresler E (1978) Theoretical modeling of mixed-electrolyte solution flows for unsaturated soils. Soil Sci 125: 196–203
41. Bresler E, Bielorai H, Laufer A (1979) Field test of solution flow models in a heterogeneous irrigated cropped soil. Water Resour Res 15: 645–652
42. Dagan G, Bresler E (1979) Solute dispersion in unsaturated heterogeneous soil at field scale. I. Theory. Soil Sci Soc Am J 43: 461–467
43. Bresler E, Dagan G (1979) Solute dispersion in unsaturated heterogeneous soil at field scale. II. Application. Soil Sci Soc Am J 43: 467–472
44. Dasberg S, Bresler E (1979) L'irrigation. Recherche 10(104): 968–975
45. Yaron D, Bresler E, Bielorai H (1980) A model for optimal irrigation scheduling with saline water. Water Resour Res 16: 257–262
46. Russo D, Bresler E (1980) Field determinations of soil hydraulic properties for statistical analysis. Soil Sci Soc Am J 44: 697–702
47. Russo D, Bresler E (1980) Scaling soil hydraulic properties of a heterogeneous field. Soil Sci Soc Am 44(4): 681–684
48. Bresler E, Dasberg S, Russo D, Dagan S (1981) Spatial variability of crop yield as a stochastic soil process. Soil Sci Soc Am J 45: 600–605
49. Russo D, Bresler E (1981) Soil hydraulic properties as stochastic processes. I. An analysis of field spatial variability. Soil Sci Soc Am J 45: 682–687
50. Russo D, Bresler E (1981) Effect of field variability in soil hydraulic properties on solutions of unsaturated water and salt flows. Soil Sci Soc Am J 45: 675–681
51. Bresler E (1981) Transport of salts in soils and subsoils. Agric Water Manage 4: 35–62
52. Bresler E, Dagan G (1981) Convective and pore-scale dispersive solute transport in an unsaturated heterogeneous field. Water Resour Res 17: 1683–1693
53. Russo D, Bresler E (1982) Soil hydraulic properties as stochastic processes. II. Errors of estimates in a heterogeneous field. Soil Sci Soc Am J 46: 20–26
54. Russo D, Bresler E (1982) A univariate versus a multivariate parameter distribution in a stochastic conceptual analysis of unsaturated flow. Water Resour Res 18: 483–488
55. Bresler E, Dagan G, Hanks RJ (1982) Statistical analysis of crop yield under controlled line-source irrigation. Soil Sci Soc Am J 46: 841–847
56. Stern J, Bresler E (1983) Nonuniform sprinkler irrigation and crop yield. Irrig 4: 17–29
57. Dagan G, Bresler E (1983) Unsaturated flow in spatially variable fields. I. Derivation of models of infiltration and redistribution. Water Resour Res 19: 413–420
58. Bresler E, Dagan G (1983) Unsaturated flow in spatially variable fields. II. Application of water flow models to various fields. Water Resour Res 19: 421–428
59. Bresler E, Dagan G (1983) Unsaturated flow in spatially variable fields. III. Solute transport models and their application to two fields. Water Resour Res 19: 429–435
60. Bresler E, Yaron D, Segev A (1983) Evaluation of irrigation water quality for a spatially variable field. Water Resour Res 19: 1382–1391
61. Bresler E, Dagan G, Wagenet RJ, Laufer A (1984) Statistical analysis of salinity and texture effects on spatial variability of soil hydraulic conductivity. Soil Sci Soc Am J 48: 16–25
62. Wagenet RJ, Knighton RE, Bresler E (1984) Soil chemical and physical effects on spatial variability of hydraulic conductivity. Soil Sci 137: 252–262
63. Feinerman E, Bresler E, Dagan G (1985) Optimization of spatially variable resource: an illustration for irrigated crops. Water Resour Res 21: 793–800
64. Dagan G, Bresler E (1985) Comment on "flux-averaged concentration in continuum approaches to solute transport" by Parker and van Genuchten. Water Resour Res 21: 1299–1300
65. Feinerman E, Dagan G, Bresler E (1986) Statistical inference of spatial random functions. Water Resour Res 22: 935–942
66. Bresler E, Hoffman GJ (1986) Irrigation management for soil salinity control: theories and tests. Soil Sci Soc Am J 50: 1552–1560
67. Bresler E, Green RE (1987) Transport of degradable substance and its metabolites under drip irrigation. Agric Water Manage 12: 195–206

68. Shani U, Hanks RJ, Bresler E, Olivera CAS (1987) A simple field method for estimating the hydraulic conductivity and matric potential-water content relations in soils. Soil Sci Soc Am J 51: 298–302
69. Bresler E, Naor A (1987) Estimating transport parameters in soils by a maximum likelihood approach. Soil Sci Soc Am J 51: 870–875
70. Bresler E (1987) Application of conceptual model to irrigation water requirement and salt tolerance of crops. Soil Sci Soc Am J 51: 788–793
71. Bresler E, Dagan G (1988) Variability of yield of an irrigated crop and its causes. I. Statement of the problem and methodology. Water Resour Res 24: 381–388
72. Bresler E, Dagan G (1988) Variability of yield of an irrigated crop and its causes. 2. Input data and illustration of results. Water Resour Res 24: 389–396
73. Dagan G, Bresler E (1988) Variability of yield of an irrigated crop and its causes. 3. Numerical simulation and field results. Water Resour Res 24: 397–406
74. Bresler E, Laufer A (1988) Statistical inferences of soil properties and crop yields as spatial random functions. Soil Sci Soc Am J 52: 1234–1244
75. Feinerman E, Bresler E, Dagan G (1989) Optimization of input in a spatially variable natural resources: unconditional vs. conditional analysis. J Environ Econ Manage 17: (2) 140–154
76. Feinerman E, Shani Y, Bresler E (1989) Economic optimization of sprinkler irrigation considering uncertainty of spatial water distribution. Aust J Agric Econ 33: 88–107
77. Bresler E (1989) Estimation of statistical moments of spatial field averages for soil properties and crop yields. Soil Sci Soc Am J 53: 1645–1653
78. Feinerman E, Bresler E, Achrish H (1989) Economics of irrigation tecnology under conditions of spatially variable soils and nonuniform water distribution. Agronomie 9: 819–826

III. Invited Reviews, Proceedings, and Book Chapters

1. Mokady R, Shainberg I, Hagin J, Bresler E (1962) Cation exchange phenomena as affected by the soil moisture content. I. Cation exchange capacity. In: Radiosiotopes in soil-plant nutrition studies. IAEA, Vienna pp. 31–41
2. Monselise SP, Bresler E (1964) Irrigation of citrus trees. In: Hamadrich Hahaklaii 43: 1–11 (in Hebrew)
3. Aharoni A, Bresler E, Karmeli S (1968) Irrigation of crops in Israel: citrus. In: Irrigated soils. Hakibbutz Hameuchad Publ, pp 342–350 (in Hebrew)
4. Yaron D, Bresler E (1968) Toward economic evaluation of water quality in irrigation. Proc Water for Peace Conference 6: 749–757. Superintendent of Documents, Washington, DC
5. Bresler E (1972) Control of soil salinity. In: Hillel D (ed) Optimizing the soil physical environment toward greater crop yields. Academic Press, New York, pp 101–132
6. Bresler E (1973) Solute movement in soils. In: Yaron B, Danfors E, Vaadia Y (eds) Arid zone irrigation. Ecological studies 5. Springer, Berlin Heidelberg New York, pp 166–175
7. Heller J, Bresler E (1973) Trickle irrigation. In: Yaron B, Danfors E, Vaadia Y (eds) Arid zone irrigation. Ecological Studies 5. Springer, Berlin Heidelberg New York, pp 339–351
8. Magduff F, Bresler E (1973) Evaluation of method for reclaiming sodic soils with $CaCl_2$. In: Hadas A, Swatzendruber D, Rijtema PE, Fuchs M, Yaron B (eds) Physical aspects of soil water and salt in ecosystems. Ecological Studies 4. Springer, Berlin Heidelberg New York, pp 441–452
9. Zur B, Bresler E (1973) A model for the water and salt economy in irrigation agriculture. In: Hadas A, Swatzendruber D, Rijtema PE, Fuchs M, Yaron B (eds) Physical aspects of soil water and salt in ecosystems. Ecological Studies 4. Springer Berlin Heidelberg New York, pp 395–407
10. Neuman SP, Feddes RA, Bresler E (1975) Galerkin method of simulating water uptake by plants. In: Vansteenkiste GC (ed) Computer simulation of water resources systems. North Holland/American Elsevier, Amsterdam pp 325–346

11. Neuman SP, Narasimhan TN, Feddes RA, Bresler E, Witherspoon PA (1975) Simulation of flow in soil-plant system using integral methods. Proc Summer Computer Simulation Conf (SCSC) FIPS Press, Reston, VA, pp 822–830
12. Bresler E, Miller RD (1975) Estimation of pore blockage induced by freezing of unsaturated soil. Proc Conf Soil-water problems in cold regions. Calgary, Alberta, Canada pp 161–175
13. Bresler E (1976) Models for predicting the impact of irrigation on soil salinity. Proc Int Conf on Managing saline water for irrigation. Texas Tech Univ, Lubbock, TX, pp 229–316
14. Bresler E (1977) Water utilization and soil salinity: control in arid zone agriculture. In: Mundlak Y, Singer SF (eds) Arid zone development: prospectives and problems. Ballinger Cambridge, MA, pp 109–128
15. Bresler E (1977) Trickle-drip irrigation: principles and application to soil water management. In: Brady NC (ed) Advances in Agronomy, vol 29. Academic Press, New York, pp 343–393
16. Mualem Y, Bresler E (1979) Soil water processes: general report. In: Morel-Seytoux HJ, Salas JD, Sanders TG, Smith RE (eds) Surface and subsurface hydrology. Proc Ft Collins 3rd Int Hydrology Symp Theoretical and applied hydrology, July 27–29, 1977, III, Water Resour Publ, Ft Collins, CO, pp 297–313
17. Russo D, Bresler E (1980) Soil-water-suction relationships as affected by soil solution concentration and composition. In: Kafkafi U, Banin A (eds) Agrochemicals in soils. Intercongress of commissions II, IV, & VII of the Int Soc Soil Sci, Jerusalem, Israel pp 287–296
18. Bresler E (1980) Transport of salts in soils and subsoils. In: Holmes JW, Talsma T (eds) Land and stream salinity Elsevier, Amsterdam, pp 35–62
19. Bresler E (1980) Irrigation of crops: trickle-drip irrigation. In: Parker SP (ed) Yearbook of science and technology. McGraw-Hill, New York, pp 246–250
20. Bresler E (1981) Irrigation and soil salinity. In: Yaron D (ed) Salinity irrigation, and water resources. Marcel Dekker, New York, pp 65–102
21. Bresler E, Dagan G (1982) Modeling of water and solute transport in unsaturated heterogeneous fields. Proc Symp Unsaturated flow and transport modelling. US Nuclear Regulatory Commission, NUREG/CP-0030, PNL – SA – 10325, Seattle, Washington, pp 159–178
22. Yaron D, Bresler E (1983) Economic analysis of on-farm irrigation using response functions of crops. In: Hillel D (ed) Advances in irrigation, vol 2. Academic Press, New York, pp 234–255
23. Dagan G, Bresler E (1984) Solute transport in soil at field scale. In: Yaron B, Dagan G, Goldshmid J (eds) Pollution in porous media. Ecological Studies 47. Springer, Berlin Heidelberg New York, pp 17–26
24. Bresler E (1987) Modeling of flow, transport, and crop yield in spatially variable fields. In: Advances in soil sciences, vol. 7. Springer, Berlin Heidelberg New York, pp 1–51
25. Bresler E (1987) State-of-the are of drip irrigation technology. Proc Natl Worksh Drip irrigation for crop production. Workshop held in Bridgetown, Barbados. FAO of the UN, pp 5–23
26. Bresler E (1987) Chemigation in relation to drip irrigation technology. Proc Natl Worksh Drip irrigation for crop production. FAO of the UN, pp 62–73
27. Dagan G, Russo D, Bresler E (1990) Effect of spatial variability upon subsurface transport of solutes from nonpoint sources. In: De Coursey Donn G (ed) Proc Int Symp Water quality modeling of agricultural non-point sources. USDA ARS-81: 523–548
28. Bresler E (1991) Soil spatial variability. In: Hanks J, Ritchie JT (eds) Modeling plant and soil systems. Agronomy Monograph 31. Am Soc Agron, Madison, WI, pp 145–164

Part I
Stochastic Modeling of Flow and Transport in Unsaturated Soil at Field Scale

1 The Bresler-Dagan Model of Flow and Transport: Recent Theoretical Developments

G. Dagan[1]

Prologue

Towards the end of 1976 I moved from the Technion, Israel Institute of Technology, located in Haifa, to the newly created Faculty of Engineering at Tel-Aviv University. The geographic proximity between Tel-Aviv and the Volcani campus of the Agriculture Research Organization (ARO) raised the possibility of my collaboration with the ARO Institute of Soils and Water. In this spirit, I was offered a part-time position by Bruno Yaron, the director of the Institute at this time. A year later, Eshel Bresler, the head of the Soil Physics department, returned from his sabbatical and we started to collaborate within this framework. The main drive of our collaboration was the fusion between soil physics, the area of expertise of Eshel Bresler, and the new field of stochastic modeling of flow and transport through heterogeneous formations, in which I had started to work by that time.

Eshel and I quickly found many common grounds and our collaboration expanded during the years to come. As with any true joint work, it was not only a meeting of minds but also a chemistry of personalities that evolved into a deep friendship. The scope of our collaboration can be better understood by referring to the body of journal papers we coauthored: starting with the two-part series Dagan and Bresler (1979) and Bresler and Dagan (1979) and ending with Feinerman et al. (1989), we produced 15 articles. We have also participated in and presented together joint works at various meetings and have influenced the formation of young and promising scientists like David Russo, by now a foremost researcher in soil physics.

The present contribution is related to our aforementioned first joint work, in which we developed a model of flow and transport in the unsaturated zone of heterogeneous soils. This was one of the early works on the subject and a few of our subsequent studies relied on these articles. Furthermore, several other workers in the field have developed and expanded this model and a review of these developments is the main aim here. The impact of this first work is underlined by the extent of its citation: the Citation Index of ISI shows that by 1990 it had been quoted in 65 publications. In retrospect, I believe that this was our major contribution and associating the model with the name of Eshel Bresler, as suggested by the title, is a well-deserved tribute to his contribution. As for the presence of my name, I will be fully

[1] Faculty of Engineering, Tel-Aviv University, Ramat Aviv 69978, Israel

rewarded if it will serve as a reminder of the deep friendship and affection I felt for Eshel.

1.1 Introduction

The model of flow and transport developed by Dagan and Bresler (1979) and Bresler and Dagan (1979), denoted as the BD model hereafter, was aimed at describing water and solute movement in the upper soil layer (in the root zone) at the field scale. In line with field measurements of soil properties available at that time (Nielsen et al. 1973; Biggar and Nielsen 1976), and confirmed by later works, we were concerned with the impact of the soil heterogeneity, as well as with that associated with the non-uniformity of water application, upon flow and transport. Since the spatial distribution of the various parameters (hydraulic conductivity at saturation, rate of recharge, solute initial concentration etc.) is affected by uncertainty, we regarded them as random space functions (RSFs) or regionalized variables in the geostatistical terminology. Thus, these variables are characterized by the joint probability density functions (pdfs) of their values at an arbitrary set of points in the flow domain. Consequently, the flow and transport variables of interest are also RSFs and they can be determined in terms of their statistical moments only. All statistical moments (mean, variance) are based on ensemble averaging, i.e., averaging over all possible realizations of the heterogeneous field and boundary conditions. Since, in practice, there is only one existing field, the only way to relate the ensemble statistics to the available information is to adopt some stationarity assumption. We have adopted the simplest one, namely that the one-point pdf of various properties is independent of the coordinate. Hence, these pdfs can be identified in the usual manner, by constructing histograms of the measured values. Next, the correlation structure is characterized at the lowest order by the covariance of the properties values at two points and the stationarity implies that it depends only on the lag, the distance between the points. We have adopted a convenient correlation scale, the integral scale, which we assumed to be finite and much smaller than the field size. A generalization of the concept is that of a stationary variogram of finite range.

As mentioned before, we were concerned with processes occurring in the upper soil layer of large fields. Hence, the depth is of the order of 1 m, whereas in the horizontal plane we were interested in the behaviour of fields of lengths from hundreds to thousands of meters. This scale disparity motivated us to adopt an approximation similar to that widely used in different branches of fluid mechanics and known as the shallow-water approximation (Dupuit approximation in groundwater flow). In this approach the flow is viewed as vertical and the soil properties are averaged over the depth, such that all properties and flow variables become RSFs of the horizontal coordinates only. This is the distinctive feature of the BD model and herein is the reasoning adopted in the 1979 series to justify it. We viewed heterogeneity as of two different scales: a vertical local scale, and a horizontal field scale. Vertical heterogeneity is associated mainly with the layering present in most soils and has scales in the order of centimeters. We neglected its effect upon flow, which is characterized by the mean water velocity over the profile. Its influence was accounted for in transport by assuming an increased effective dispersion coefficient, as measured in the field in individual profiles. In any case, it is clear that for processes taking place in the upper soil layer the only heterogeneity of consequence along the vertical is that of scales smaller than the depth of, say, 1 m. In contrast, heterogeneity in the horizontal plane is generally characterized by much larger scales, in the order of tens of hundreds of meters. It is precisely this two-dimensional spatial variability in the horizontal plane that we modeled as random.

The above assumption of one-dimensional flow in the vertical direction, i.e., regarding the medium as a collection of vertical columns, considerably simplifies the modeling of flow and transport under quite general conditions of unsteady boundaries and complex solute behavior. The advantage is underscored by the comparison between this approach with other existing ones. Thus, the first-order analysis in the variance of the saturated log-conductivity developed by Yeh et al. (1985) is underlain by quite restrictive conditions of average uniform flow and small variability on the one hand, and leads to complex calculations on the other hand. The other approach, of fully two- to three-dimensional numerical solution of flow and transport (Ababou 1988; Russo 1991), is both complex and time-consuming on the computer. The cost is increased enormously if a Monte-Carlo approach is adopted, i.e., if various realizations of the medium are generated.

Under these conditions, one should not wonder that the BD model has been explored subsequent to its publication, either to check its validity or to extend it to various flow and transport scenarios. The aim of the present review is to examine and summarize a few such studies and to show the reader the various options available in the literature.

1.2 The Assumptions and the Main Results of the Original BD Model (Bresler and Dagan 1979; Dagan and Bresler 1979)

Since our main aim is to examine the various extensions of the BD model, it is worthwhile to recapitulate here the assumptions underlying it and its main results.

The Flow Model

Vertical flow takes place from the soil surface $z > 0$. The flow is steady and gravitational and it is caused by a uniform recharge of constant magnitude R applied on the surface. The hydraulic conductivity $K = K_s$ at saturation $\theta = \theta_s$, where θ is the moisture content, is regarded as a RSF of log-normal distribution. Its moments are related to those of a scaling parameter, δ, (Warrick et al. 1977) which is the basic random quantity of a given log-normal distribution. The moments (mean m_y and variance σ_y^2 of $y = \ln\delta$ and correspondingly those of $\ln K_s$) are supposed to be inferred from measurements. The saturation S is defined by $S = (\theta - \theta_{ir})/(\theta_s - \theta_{ir})$, where θ_{ir} is the irreducible saturation, and it is related to K/K_s by the empirical and deterministic formula $K/K_s = S^{1/\beta}$, similar to the Brooks and Corey (1964) relationship. The solution of the gravitational flow problem is simply $q_z = K$, where q is the specific discharge vector. Furthermore, for $R < K_s$ and $\theta < \theta_s$ one has $q_z = R$, whereas for $R > K_s$ and $\theta = \theta_s$, $q_z = K_s$. The latter relationship is based on the assumption that ponding occurs. It accounts for the excess water flux $R - K_s$ and no runoff is assumed to occur, so that the ponded areas of the field do not influence the remaining part of the field. Finally, the variable of interest for transport, the vertical pore-velocity V, is equal to $q_z/(\theta - \theta_r)$. As for the recharge R, stemming from rainfall or irrigation, we have assumed that it is affected by uncertainty as well but that it is not spatially variable. Hence, we regarded it as a random variable and assumed that it has a rectangular pdf which, in turn, is characterized by two parameters, the mean $\bar{R}$ and the band width $2d_R$. It is seen that for any given R and K_s, one can derive the values of K and θ in the profile and, in turn, that of V. The pdf of V, however, has different expressions for $K_s \langle R$ and $K_s \rangle R$. In our model V is a function of the two random variables y and R and it is useful to visualize it by considering the R, y plane and the strip $\bar{R} - d_R < R < \bar{R} + d_R$, $-\infty < y < +\infty$. The probability distribution function (PDF) of V, $P(V) = \int f(V)dV$, is obtained from the joint pdf $f(y, R)$ by integration over the strip from $y = -\infty$ to a curve $y(R)$ defined by the value of V (Dagan and Bresler 1979, Fig. 1).

Summarizing the flow model, for the given parameters θ_r (taken as equal to zero in BD if $\theta_r \ll \theta_s$), θ_s, β, m_y, σ_y^2, $\bar{R}$ and d_R, the pdf $f(V)$ and the PDF $P(V) = \int f(V)dV$, of the velocity are completely defined and can be effectively determined.

The Transport Model

The BD model dealt with a conservative solute, applied at the soil surface $z = 0$ at $t = 0$. Pore-scale dispersion was neglected as it was assumed that the effect of field-scale heterogeneity upon the spreading process is much stronger than the local one. The solute particles are therefore convected downward at the constant velocity V which is solely a function of x and y. Thus, the considered boundary and initial conditions were $C(x, y, 0, t) = C_0$ and $C(x, y, z, 0) = 0$, respectively. Here C is the *residence* concentration and C_0 is constant. Then, the concentration field is simply given by the "piston-flow" solution:

$$C(x, y, z, t) = C_0 H(Vt - z), \tag{1}$$

where $H(a)$ is the Heaviside step function ($H = 1$ for $a > 0$ and $H = 0$ for

$a < 0$). C is a RSF since V is such, and can be characterized only in statistical terms. In BD the aim was to calculate the expected value $\langle C \rangle = \int Cf(C)dC = \int Cf(V)dV$. This was regarded as the most important result of the analysis since ergodic conditions were assumed to prevail, i.e., $\langle C \rangle$ was assumed to be equal to the field space average $\bar{C} = (1/A)\iint_A C\,dx\,dy$, where A is the field area. This is the case if V has a finite correlation scale in the plane and A is very large at this scale. As for the recharge R, it is regarded as a random variable and averaging over its pdf renders $\langle C \rangle$ as an expected value. In Fig. 3 of Dagan and Bresler (1979), we presented a schematic vertical crosssection showing the solute body at time t for a given R. It is seen that $\bar{C}(z,t)$ can be viewed in this case as the ratio between the area of the field for which the front has moved beyond z, i.e., for which $V > z/t$, and the total area A.

Hence, $\langle C \rangle$ is obtained by multiplying C [Eq. (1)] by $f(V)$ and integrating, as follows:

$$\langle C(z,t|R) \rangle = C_0 \int_0^{\infty} H(Vt - z) f(V|R)dV$$

$$= C_0 \int_{z/t}^{\infty} f(V|R)dV = C_0[1 - P(z/t|R)], \tag{2}$$

where $P(V|R)$ is the PDF (probability distribution function, cumulative) of V conditioned on a fixed R. An additional integration over $f(R)$ provides the expected value (unconditional) of C as follows:

$$\langle C(z,t) \rangle = \int_{\bar{R}-d_R}^{\bar{R}+d_R} \langle C(z,t|R) \rangle f(R)dR = \int_{\bar{R}-d_R}^{\bar{R}+d_R} [1 - P(z/t|R)] f(R)dR. \tag{3}$$

The solution is illustrated in Bresler and Dagan (1979) for the field data of the Panoche soil (Warrick et al. 1977) of relatively low conductivity and large variability.

One of the important consequences of the BD model is that $\langle C \rangle$ is a function of z/t rather than of z and t. Consequently, the profiles of $\langle C \rangle$ at different t could be represented (Bresler and Dagan 1979, Fig. 1) as unique curves of $\xi = zV^*/t$, where $V^* = K_s^*/\theta_s$ and $K_s^* = \langle (K_s)^{1/2} \rangle^2$. For a given soil, each $\langle C(\xi) \rangle$ depends only on the two dimensionless parameters $\bar{R}/V^*$ and $d_R/\bar{R}$. Another feature of the $\langle C \rangle$ profiles is that they are of discontinuous slope due to the fact that for $R > K_s$ (ponding) the velocity is assumed to be independent of R.

The variance of the concentration σ_C^2 is easily determined from $\langle C \rangle$ since the random variable C has a bimodal distribution ($C = 0$ or $C = C_0$) in the absence of pore-scale dispersion. Thus, $\sigma_C^2 = \langle C \rangle (C_0 - \langle C \rangle)$ and reaches the maximal value of $C_0^2/4$ for $\langle C \rangle = C_0/2$.

Finally, by regarding $\langle(C, z, t)\rangle$ as the outcome of a dispersion process one may attempt to derive a front, say, for $\langle C\rangle/C_0 = 0.5$ and and effective (field) dispersion coefficient (EDC). Since the profiles shown in Fig. 1 of Bresler and Dagan (1979) do not appear to be Gaussian, the definition of the EDC is quite arbitrary and is related to $z_{0.9} - z_{0.1}$. The latter are the depths for which $\langle C\rangle/C_0 = 0.9$ and $\langle C\rangle/C_0 = 0.1$, respectively. The major point, however, is that both the front depth and an EDC grow linearly with time, due to the dependence of $\langle C\rangle$ upon z/t. The linear growth of EDC with time is typical for stratified media in which transvers diffusion is neglected (see, e.g. Dagan 1989). Furthermore, such a growth prevails in any transport process by random velocity fields, provided that the transport time is small compared with the velocity microscale (Taylor 1921; see also analysis in Dagan 1989). In other words, a more general heterogeneous system, of a three-dimensional structure, will behave as a stratified one at sufficiently short distances from the inlet. Hence, the BD model of independent vertical columns can be an accurate representation of the *transport* process under two different circumstances: (1) in the shallow-water approximation in which soil properties are averaged over the profile and this average varies in the horizontal plane with a correlation scale larger than the depth and (2) for sufficiently shallow depths beneath the soil surface to render the transient, linearly growing stage of the EDC a valid regime.

A methodology similar to BD has been developed by Jury et al. (1986) (see also the recent book by Jury and Roth 1990) and was called the transfer function approach. In common with the BD model, it assumes that transport is one-dimensional and vertical and that parcels of solute are convected at constant velocity in each profile. Consequently, C is a function of z/Vt and its expected value is related in a simple way to the pdf of V. The main difference in principle is the proposition that the derivation of the pdf of z/Vt is achieved by carrying out a field transport experiment and identifying indirectly the pdf of V from concentration measurements. In contrast, the BD model is underlain by a flow model which implies identification of soil properties and of flow boundary conditions as a prerequisite to predicting transport. We will now examine various generalizations of the BD model.

1.3 The Effect of Local Dispersion (Bresler and Dagan, 1981)

As a further step, a dispersive term has been incorporated in the transport equation but other assumptions are as in Section 1.2. Assuming that in each vertical profile the dispersivity is constant, the change is in Eq. (2) in which the step function representing a piston-flow type of transport is replaced by the standard solution for a semi-infinite column:

$$C(z, t) = \frac{1}{2}\left[1 - \operatorname{erf}\frac{z - Vt}{2(Dt)^{1/2}}\right]. \tag{4}$$

With $D = D_m + \lambda V$, where D_m is the molecular diffusion coefficient of the medium and λ the dispersivity, the salient question is: what is the magnitude of λ? If the usual value of λ, as determined in laboratory columns, is adopted (say 0.1 cm), the influence upon $\langle C \rangle$ [Eq. (3)] is negligible and the piston-flow solution of Eq(1) is accurate. Instead, we have considered a local value of λ. The latter is defined as the one characterizing the transport in any vertical profile in the field. In other words, λ incorporates the effect of the heterogeneity in the vertical profile which was neglected in the BD model. In this manner the local three-dimensional effect if partly incorporated in the model.

The values of λ adopted to illustrate the results were taken from Biggar and Nielsen (1976). They have conducted extensive field transport measurements in the Panoche soil, mentioned in Section 1.2, and found that λ is spatially variable. Assuming it is log-normal, Bresler and Dagan (1981) derived a mean value of $\bar{\lambda} = 3$ cm and $\sigma^2_{\ln\lambda} \simeq 1.4$.

A complete calculation of $\langle C \rangle$ from Eqs. (4) and (3) implies integration over the pdf of V and λ, i.e., over those of R and K_s on the one hand and of λ on the other hand. To simplify matters, and after a preliminary check, it was decided to neglect the variability of λ and that of R and to adopt their mean values. Therefore, we concentrated, on the impact of V variability due to that of K_s, i.e., on the scaling factor δ. Thus, simple results could be obtained by a single quadrature over C[Eq. (3)] multiplied by the pdf of $y = \ln\delta$. However, due to the large variability of K_s of the Panoche soil with $\sigma_{\ln K_s} \simeq 2.3$, the effect of the local dispersion of constant $\bar{\lambda} = 3$ cm upon $\langle C \rangle$ was negligible at times of interest. In contrast, the variance of C was greatly reduced due to the smooting effect of the local dispersion.

The results were somewhat different for the Bet Dagan soil when transport with neglect of local dispersion was compared with transport with $\bar{\lambda} = 3$ cm. It should be said, however, that the latter value was assumed on a somewhat speculative basis from the Panoche soil field. The $\langle C(\xi) \rangle$ curve (Bresler and Dagan 1981, Fig. 5c) was smoother and closer to a Gaussian sigmoid when λ was accounted for. This is due, of course, to the much smaller variability of K_s, $\sigma_{\ln K_s} = 0.8$. Still, for a sufficiently large period of time the results were close to those obtained in the absence of λ.

An additional issue examined in Bresler and Dagan (1981) was the statistics of C space-averaged over the depth. For instance, this is a relevant entity when the effect of solutes upon roots is of interests, since the root zone extends over an interval.

Summarizing this extension of the original BD model, the following conclusions can be drawn:

1. The effect of local three-dimensional heterogeneity was incorporated in the model in an approximate and simple manner by introducing a dispersive term in the one-dimensional transport equation. The longitudinal dispersivity was taken as constant and its augmented value, with respect to the laboratory one, lumps in one parameter the complex effect of vertical variability.
2. Unlike the original BD model, $\langle C \rangle$ is no more a function of the "self-similar" variable $\xi \sim z/t$.
3. If the variability in the horizontal plane of hydraulic properties is sufficiently large, and for depths larger than $\bar{\lambda}$, the pistonflow

approximation Eq. (1) of transport of the BD model leads to results close to those which incorporate $\bar{\lambda}$. There are some differences at the tail, i.e., for low values of $\langle C \rangle$, and for the variance of C.

1.4 Unsteady Flow (Infiltration and Redistribution) and Transport (Bresler and Dagan 1983a, b; Dagan and Bresler 1983)

In the original BD model (Sect. 1.2) and in the extension in Section1.3, the flow was supposed to be steady and gravitational. Such conditions may prevail in the upper soil layer after continuous application of water to the surface for a prolonged period. Alternatively, this may be regarded as a regime representing an average behavior in time even over a period of variable application, as happens at a sufficiently deep depth due to the buffering effect of the unsaturated zone. However, in some circumstances pulses of applied water are followed by prolonged periods of no flux or of evaporation, i.e., an infiltration–redistribution cycle is present. In the series of 1983 papers we tried to extend the BD model to cover this case as well.

A complete solution of the problem of flow during an infiltration–redistribution cycle in a spatially variable soil of a three-dimensional structure can be achieved only numerically and at the expense of enormous programs and computer time. In the case of heterogeneity affected by uncertainty, the usual procedure of Monte-Carlo simulation amplifies the difficulty tremendously, since the equations of flow have to be solved in many realizations of the medium, boundary and initial conditions. The viewpoint adopted by Dagan and Bresler (1983) was that such a detailed solution is not needed if one seeks only a few statistical moments of the flow and transport variables and requires mainly the expected values. Furthermore, simplified approximate models of flow and transport may lead to quite accurate values of these moments due to mutual cancellations of errors. Lastly, we believed that once the basic assumption of independent vertical columns had been adopted, one is justified in also simplifying the treatment of the one-dimensional problem.

Thus, the approximate solution of the flow problem was obtained by replacing the actual moisture content profile $\theta(z, t)$, for fixed x, y, by one of constant $\bar{\theta}$ from $z = 0$ to $z = L(t)$. For $z > L$, θ was constant and equal to its initial value. Hence, the θ profile is a step function and $L(t)$ represents an equivalent front. We considered two types of boundary conditions at the soil surface: constant given flux q_0 or constant given $\theta = \theta_0$. In the first case the approximate solution obeyed exactly the global conservation equation: during the infiltration stage the water volume accumulated in the profile is equal to the integrated surface flux, whereas in the redistribution stage the flux is zero, and both $\bar{\theta}$ and L change with t. In the second case $\bar{\theta}$ was assumed to be constant and equal to θ_0 and only L changed. A second equation was provided by the integrated form of Darcy's law between the soil surface and the front. These two equations, of integrated conservation of mass and Darcy's law, led to simple ordinary differential equations relating $\bar{\theta}(t)$ and $L(t)$ to the soil parameters and to the initial and boundary conditions. The latter included t_i, the infiltration time, during which $q = q_0$, while $q(0, t) = 0$ for $t > t_i$. In the case of $q_0 > K_s$, ponding could occur after $t = t_p$, and t_p was also determined with the aid of the model.

In Part 2 of the sequence (Bresler and Dagan 1983a) the results of part 1 were applied to computing the expected value and variance of the saturation $S = (\theta - \theta_r)/(\theta_s - \theta_r)$ in the two heterogeneous soils (Panoche and Bet Dagan). The only random parameter was the log-normal K_s, in turn related to the scaling parameter δ; values of its mean and variance are given in Section 1.3. The surface flux q_0 and the infiltration time t_0 were taken as deterministic. The expected value $\langle S(z, t) \rangle$ and the coefficient of variation $CV_s(z, t) = \sigma_s / \langle S \rangle$ were calculated in two different ways: (1) by solving numerically the one-dimensional equations of flow for different constant values of K_s and subsequently integrating over the pdf of K_s, i.e. $\langle S \rangle = \int S f(K_s) dK_s$, and (2) similarly by using for S the approximate step function solution described above. The results were illustrated in Fig. 1 of Bresler and Dagan (1983a) showing the profiles at $t = t_i$, the end of the infiltration period. In the first column deterministic solutions for $K_s = \langle K_s \rangle$ are first shown and it is seen that the approximate solution differs from the numerical one, though conservation of mass is obeyed exactly. In contrast the expected values $\langle S \rangle$ (second column) and CV_s, based on the two solutions, are very close, supporting our basic thesis that average flow quantities can be evaluated quite accurately by approximate flow models with in the frame of the BD model. Results for $\langle S \rangle$ and CV_s at the end of the redistribution period (Fig. 2 of Bresler and Dagan 1983a) show larger discrepancies between the numerical and approximate solutions for the deterministic solution, very good agreement for $\langle S \rangle$ and CV_s for the Panoche soil and fair agreement for the Bet Dagan soil. This can be easily understood if we remember that the Panoche soil is much more heterogeneous that the Bet Dagan soil (see Sect. 1.3) and, therefore, the "mixing effect" of spatial variability upon mean quantities is much stronger.

In the third part of the series (Bresler and Dagan 1983b) solute transport was modeled for the infiltration–redistribution regime by an approach similar to that used in Section 1.3. Two additional features are present: (1) the advective velocity in the transport equation is time dependent and (2) the local dispersivity, λ, grows with distance from zero to its constant value along an interval which was taken to be equal to 13λ. The latter feature stems from the fact that λ lumps the effect of heterogeneity in the profile and, as is well known from studies of saturated flow (Dagan 1984) or from the recent study on unsaturated flow (Russo 1991), the effective dispersivity grows and reaches its asymptotic value only after a distance of a few heterogeneity scales. This improvement is particularly important for flow which develops from the soil surface to a front at a short distance below.

Again, in this part, the main objective was to derive the expected value and coefficient of variation of the resident concentration of a conservative solute under the boundary condition of constant $C = C_0$ and under the initial condition of constant C. This time, too, the solutions for $\langle C \rangle$ and CV_C, based on the approximate model of flow of parts 1 and 2, and the numerical solutions, are very close (Fig. 2 in Bresler and Dagan 1983b).

Another point examined in this series was the suitability of deriving the mean flow and transport variables by using a fictitious homogeneous medium of effective properties. This was found to be unfeasible for the unsteady regime prevailing under infiltration–redistribution conditions.

Summarizing this extension of the BD model to include unsteady flow for infiltration and redistribution, the following conclusions are drawn:

1. Even after adopting the basic assumption of the BD model, namely 1-D vertical flow in homogeneous columns, the solution of the flow and transport problem under transient conditions may require a large number of numerical simulations followed by statistical averaging.
2. Accurate results for the expected value and the coefficient of variation of the flow variables and concentration can be obtained by using simple flow models which lend themselves to analytical solutions. This is based on the smoothing effect of averaging over many columns and on mutual cancellations of model errors.
3. It does not seem that an effective medium approach is applicable to transient flow and transport, and stochastic modeling is needed in order to derive the mean variables over the field.

1.5 Transport of Reactive Solutes in Heterogeneous Fields (Van der Zee and Van Riemsdijk 1986, 1987)

This is a first attempt to extend the BD model to transport of a reactive solute. The original study (Van der Zee and Van Riemsdijk 1986) deals with transport of phosphate in the upper soil layer. In its first part it proposes an approximate model for F, the fixation of phosphate on the solid matrix expressed as mass of solute per mass of solid phase, in a homogeneous vertical column. The 1-D water flow is assumed to be of constant velocity and at constant moisture content, whereas the solute is applied periodically to the surface. After various considerations, related to the nature of the chemical processes occurring in the soil column, the approximate expression of $F(z, t)$ is:

$$F(z, t) = F(C_0)H(\mathrm{Vat}/r - z), \tag{5}$$

where C_0 is the concentration applied to the surface $z = 0$ during time $a.t$ (a is the ratio between the time of phosphate application and the total time t), $F(C_0)$ is the sorption capacity corresponding to C_0, V is the water velocity, r is the retardation factor and H is the Heaviside function. Thus, Eq. (5) describes a front at depth $z = \mathrm{Vat}/r$ above which F is constant and equal to its equilibrium value. Hence, the approach adopted is similar to that used in Section 1.4: in a heterogeneous field simplified solutions for the homogeneous profile are appropriate. The retardation factor is assumed to be much larger than unity and is given by $r = \rho_s F(C_0)/\theta C_0$, where ρ_s is the solid matrix density. Hence, r is equal to the ratio between the adsorbed mass and the total mass in the solution. By substituting r in Eq. (5) and defining the dimensionless concentration in the solid phase of the sorbed solute by $\Gamma = F/F(C_0)$, Eq. (5) is rewritten as follows:

$$\Gamma(z,t) = H(A_T/F_T - \zeta), \tag{6}$$

where $A_T = V\theta C_0$ at is the total amount of applied phosphorus and F_T is the total fixation capacity of a column of length L. The variable ζ is equal to z/L if the fixation capacity and the solid density ρ_s are constant, otherwise ζ is a weighted relative average.

Equation (6) is the starting point for modeling the phosphate fixation. Van der Zee and Van Riemsdijk (1976) assume that in a heterogeneous field A_T and F_T are spatially variable, whereas the flow variables V and θ are constant. They adopt the basic structure of the BD model, i.e., of vertical independent columns, and regard A_T and F_T as independent random variables of log-normal distribution. Subsequently, they derive the expected value $\langle\Gamma(z,t)\rangle$ by integrating Eq. (6) over the pdf of A_T and F_T. The values of the parameters determining these pdf are found by analyzing measurements in a field, after detrending some data.

Although heterogeneity manifests in the sorptive capacity and the rate of application rather than in the flow and transport variables of the original BD model, the effect upon $\langle\Gamma\rangle$ is similar: the sharp front of Γ[Eq. (6)] in an individual column results in a gradual $\langle\Gamma\rangle$. By regarding the latter as the solution of a transport equation with constant V and a constant dispersivity, one can translate the effect of heterogeneity into an effective field-scale dispersivity.

In their conclusions the authors stress that the field-scale dispersivity increases with increasing coefficients of variation of the two variables, as one would expect, and also with a negative correlation between the two. An important aspect is that prediction may require detection of possible trends of the random variables.

A similar study was presented later (Van der Zee and Van Riemsdijk 1987), and this time the equilibrium sorption was given by the Freundlich isotherm $F = kC^n$. For $0 < n < 1$ a traveling wave solution is assumed to take place in a homogeneous column and the relative adsorbed solute is approximately represented by a sharp front, Eq. (6), but this time with $R = 1 + (\rho_s k/\theta)\ [(C_0^n - C_i^n)/(C_0 - C_i)]$, where $C_i < C_0$ is the initial concentration. Again, heterogeneity manifests in the spatial variability of the time of application of the solute to the surface and in the retardation coefficient, precisely like in Van der Zee and Van Riemsdijk (1986). However, this time the authors also allow for V, the water downward velocity, to be random and of log-normal distribution. Various $\langle\Gamma(z,t)\rangle$ profiles are computed by adopting a few values for the statistical parameters of the three variables and the correlation between them. For data of k and n pertinent to heavy metals (Cu and Cd) and for soils and water velocities typical for Holland, it was found that the $\langle\Gamma\rangle$ profiles are influenced mostly by the variability of the retardation coefficient and of the cumulative duration of solute input. However, this conclusion stems from the low value of 0.0193 for the log-velocity variance while the value of the log-retardation variance was much larger, namely 1.16.

1.6 Sensitivity Analysis of Crop Yield in a Heterogeneous Field (Bresler and Dagan 1988a, b; Dagan and Bresler 1988)

In this series the main aim was to establish the influence upon crop yield of spatial variability of various factors in an irrigated field. The basic tenets of the BD model were adopted here as well: the field was regarded as a collection of vertical columns in which 1-D flow and transport takes place. The yield Y in a homogeneous column was computed by using deterministic models developed over the last decades (see, for instance, Bresler et al. 1982). In these models relative yield is related to seasonal transpiration from a given crop. In turn, the transpiration is determined by solving the flow and transport equations in the presence of roots. Such a solution can be achieved only numerically, but the required effort is moderate because of the assumed 1-D nature of the flow. However, the number of parameters influencing the yield is extremely large. They belong to several groups, summarized in Table 1 of Bresler and Dagan (1988a): plant (distance between roots, root distribution with depth, critical suction and crop factor), soil (scaling factor δ, air entry value, saturated θ_s and residual θ_r, exponents in Brooks and Corey relationships, and dispersivity) and boundary and initial conditions (potential evapotranspiration, applied water during the irrigation period, irrigation water concentration C_0, initial concentration in the soil and initial water content). All these parameters are affected by uncertainty which manifests in that of the yield. The relative yield can be computed by a numerical solution of the equations of flow and transport for a given set of parameters. Hence, the statistics of the yield can be determined in principle by a Monte-Carlo simulation. This is, however, prohibitive due to the huge number of realizations needed in order to account for the uncertainty of all the parameters. Furthermore, the required input is the pdfs of parameters, including the correlations between them. Such information has to be obtained from extremely elaborate field measurements which are neither available nor justified. Consequently, we have used the BD model for a more limited but useful task: we studied the sensitivity of the yield expected value $\langle Y \rangle$ and its variance σ_Y^2 by a first-order Taylor expansion around $\langle Y \rangle$. Thus, if Y depends on a vector of n parameters a of expected values $\langle a_k \rangle$ and variances $\sigma_k^2 (k = 1, 2, \dots, n)$, one obtains for the first order:

$$\langle Y \rangle = Y(\langle a \rangle) + \frac{1}{2} \sum_{k=1}^{k=n} \sigma_k^2 \frac{\partial^2 Y(\langle a \rangle)}{\partial a_k^2}$$

$$\sigma_Y^2 = \sum_{k=1}^{k=n} \sigma_k^2 \left[\frac{\partial Y(\langle a \rangle)}{\partial a_k} \right]^2 . \tag{7}$$

Our main aim was to evaluate the sensitivity coefficients in Eq. (7) and subsequently to establish the parameters whose variability has the largest impact upon Y. This was done by using some reasonable values, based on those in the literature, for the parameters mean and variance and again for the two aforementioned Panoche and Bet-Dagan fields. Thus, the numerical computations have been limited mostly to $Y(\langle a \rangle)$ and its first two derivatives, as needed in Eq. (7). It is beyond the scope of this review to describe the results as the main aim is to point out various applications of the BD model.

Another point of interest discussed by Bresler and Dagan (1988a, b) relates to the ponding boundary condition. We have mentioned in Section 1.2 that in the original BD model it was assumed that in the portion of the field in which the rate of water application R exceeds the saturated conductivity K_s, the downward velocity is $V = K_s$. Furthermore, this assumption had a large impact on the rate of spreading of the solute in the highly heterogeneous field (Panoche) of low average conductivity, for which a significant portion was simulated as ponded. In contrast, two different possibilities were discussed in Bresler and Dagan (1988a, b) for the case of application of pulses of water of finite duration. The first one, of ponding, was identical with the previous one and the infiltration period extended until the downward flow at velocity $V = K_s$ exhausted the volume of applied water. The second case was termed "complete runoff" and implied that the infiltration period is equal to the application time and that the excess water is drained instantaneously from the field. These two cases were regarded as bounding ones. In reality, as it well known from watershed hydrology, the process of runoff is much more complex. An excess of water in areas of low permeability will move out of the field but will also redistribute and infiltrate through areas of higher conductivity. It seems that until more thorough field investigations have been carried out, the two limits of "no runoff" or "complete runoff", suggested in Bresler and Dagan (1988a, b), may serve as bounding cases. If transport is not too different in the two cases, as happens for low rates of application relative to soil permeability, one may indeed conclude that the accurate knowledge of the process is not essential in this context.

1.7 Solute Flux in Heterogeneous Soils (Cvetkovic and Destouni 1989; Dagan et al, 1989, 1991; Destouni and Cvetkovic 1989)

In the works of Destouni and Cvetkovic, which were also part of the doctoral dissertation of Dr. G. Destouni (Dr. Destouni has dedicated her thesis to the memory of Eshel Bresler), the basic assumptions of the BD model served as a starting point for further developments.

In the first two studies (Cvetkovic and Destouni 1989; Destouni and Cvetkovic 1989) steady flow between the soil surface and a water-table at depth Z was considered, under the same conditions as in the BD model: the flow was gravitational and generated by a random applied recharge on the surface. One of the main contributions was modeling transport in terms of solute flux, or flux-averaged concentration, rather than resident concentration. Another new point was to incorporate the uncertainty of Z into the analysis, which, like R, was regarded as a random variable of rectangular pdf.

The solute flux s through an element of area is defined as the mass of solute per unit time and unit area. If transport is caused by advection at velocity V and local dispersion with a coefficient D, s is related to the resident concentration C by the simple relationship $s = \theta(VC - D\partial C/\partial z)$. The flux-averaged concentration is, in turn, defined by $s/(\theta V)$ and usually is very close to C for transport at large Peclet numbers. The derivation of $s(z, t)$ for 1-D flow with constant V and D has been presented by Kreft and Zuber (1978) for different injection boundary conditions at $z = 0$. Two such boundary conditions for a pulse are given mass or given flux. The first case may correspond to leaching of a solute which was applied uniformly upon the soil surface in unsaturated flow. The second mode is appropriate to ponding of a solution whose concentration is maintained constant.

We shall refer to the first mode, discussed in Cvetkovic and Destouni (1989). Thus, the solution for s (Kreft and Zuber 1978) is as follows:

$$s = m\frac{z + Vt}{2(4\pi Dt^3)^{1/2}}\exp\left[-\frac{(z - Vt)^2}{4Dt}\right], \tag{8}$$

where m is the constant mass per unit total horizontal area applied on the surface at $z = 0$, $t = 0$. The resident concentration is:

$$C = \frac{m}{\theta(4\pi Dt)^{1/2}}\exp\left[-\frac{(z - Vt)^2}{4Dt}\right]. \tag{9}$$

Cvetkovic and Destouni (1989) assumed that V is random, exactly as in the original BD model, i.e., the flow is steady and gravitational and K_s and R are random. They were also concerned with the solute mass transfer to a water-table at depth Z beneath the surface and Z was also considered to be random. Hence, with K_s, R and Z regarded as independent random variables the mean $\langle C\rangle$ and (s) are given by:

$$\langle C\rangle = \iiint C(z, t; V, Z) f(K_s) f(R) f(Z) dK_s dR dZ;$$

$$\langle s\rangle = \iiint s(z, t; V, Z) f(K_s) f(R) f(Z) dK_s dR dZ. \tag{10}$$

In turn, the mean flux-averaged concentration is given by $\langle C_f\rangle = \langle s\rangle/\langle V\theta\rangle$. Cvetkovic and Destouni (1989) computed and compared $\langle C\rangle$ and $\langle C_f\rangle$ for various combinations of statistics of K_s, R and Z, with fixed and relatively low dispersivity λ. The results are presented in the form of breakthrough curves for $\langle C \rangle / m$ and $\langle C_f \rangle / m$ as functions of t for the fixed average Z. Generally, the differences between $\langle C \rangle$ and $\langle C_f \rangle$ are quite small, except for the case of variable R and K_s and relatively high $\langle R \rangle / \langle K_s \rangle$. In the latter case the peak of $\langle C_f \rangle$ is of a lower value and is delayed with respect to that of $\langle C \rangle$, which is compensated by a tailing of $\langle C_f \rangle$ at large t.

These results can be understood in the light of the analysis carried out by Dagan et al. (1989, 1991). The investigation was aimed at interpreting the transport experiments conducted in a large caisson at Los Alamon National Laboratory. Dagan et al. (1989) analyzed saturated flow experiments in which a pulse of *fixed duration* of solute was applied at the caisson entrance, while the flux-averaged concentration was measured at the caisson outlet. In the second study, Dagan et al. (1991), the same type of transport occurred in unsaturated flow generated by sprinkling at a constant rate on the caisson upper surface.

In the saturated flow experiment both the injection and detection modes were of concentration flux, since the solution was stirred at the inlet. The concentration at the outlet displayed spreading far exceeding that that could be attributed to pore-scale dispersion, indicating the effect of medium heterogeneity. Dagan et al. (1989) applied the BD model, modified to account for flow taking place under a constant head-drop and with a velocity proportional to K_s, the latter being assumed to be log-normal. By neglecting pore-scale dispersion a close-form solution for $\langle s \rangle$ could be obtained and compared with the measurements. In the context of a comparison between resident and flux-averaged concentration, the differences are quite large, the $\langle C_f \rangle$ breakthrough curve displaying a much larger spread than that of $\langle C \rangle$. The reason is quite transparent: for C_f both the mass of pulse at the inlet and the mass flux at the outlet are proportional to V, i.e., with K_s. In contrast, C injected and detected in the resident mode is not affected by these two effects.

The situation was different in the unsaturated caisson experiment for which the velocity was controlled by the constant rate of application to the top. By ignoring ponding, the only source of velocity variability in the BD model was that of θ, the moisture content. This resulted, for the same medium, in a rate of spreading much smaller than that in saturated flow. Furthermore, in this case the injection was in the resident mode and the product $q = V\theta$ was constant and equal to the rate of application R, so there was no difference between $\langle C_f \rangle$ and $\langle C \rangle$.

In conclusion, the average solute flux and the associated flux-averaged concentration play an important role when evaluating the mass transferred to groundwater or the solution concentration at the outlet of columns. The BD model was extended toward computing the solute flux. Under common conditions of unsaturated flow it was found that the difference between the resident and flux-averaged mean concentrations is rather small, except in the case of extensive ponding which causes saturated flow in a major part of the field.

1.8 Mass Arrival of Sorptive Solute into the Groundwater (Destouni and Cvetkovic 1991)

This study addresses in principle the same problem as that in Section 1.5, namely, transport of a reactive solute through the unsaturated zone. There are, however, a few differences: the non-equilibrium sorption–desorption process is described by linear kinetics and transport is represented in terms of solute flux rather than resident concentration.

The flow model adopted by Destouni and Cvetkovic (1991) is precisely the original BD model; vertical steady flow takes place in independent columns and is caused by a constant recharge R applied to the surface. Thus, the velocity V varies in the plane due to the variability of K_s, which is assumed to be log-normal. Ponding is taken into account whenever $R > K_s$.

Again, along the basic premises of the BD model, transport is also assumed to be one-dimensional. After neglecting the pore-scale dispersion, the transport equations adopted by Destouni and Cvetkovic (1991) are:

$$\theta\frac{\partial C}{\partial t}+\frac{\partial s}{\partial x}=-\frac{\partial C^*}{\partial t}-\frac{\partial C_e^*}{\partial t};\quad \frac{\partial C^*}{\partial t}=k_1\theta C-k_2C^*;\quad C_e^*=K_e\theta C, \tag{11}$$

where $s=\theta VC$ is again the solute flux, C_e^* is the concentration of immobilized solute per total unit volume in equilibrium sorption sites, and k_1 and k_2 are sorption and desorption rate coefficients. The effect of K_e is in the retardation factor $r=1+K_e$. It is emphasized that V, θ, k_1, k_2 and K_e are assumed to be independent of z or t. A set of equations for s are obtained by replacing $C=s/\theta V$ and introducing the retardation factor, as follows:

$$\frac{\partial s}{\partial t}+\frac{V}{r}\frac{\partial s}{\partial x}=-\frac{\partial S^*}{\partial t};\quad \frac{\partial S^*}{\partial t}=(k_1/r)S-k_2S^*, \tag{12}$$

where $S^*=VC^*/r$.

The system in Eq. (12) can be solved exactly for a semi-infinite column $z>0$ and instantaneous mass injection at $z=0$. The solution is given in Destouni and Cvetkovic (1991, Eq. 6). The solution is represented as a function of time for a fixed z and is made up from two terms: the first one, s_1, is a pulse moving at the retarded velocity V/r and of attenuated mass due to kinetically controlled sorption represented by k_1; the second term, s_2, a continuous tail trailing behind the pulse, represents the delayed contribution due to the sorption–desorption mechanism and depends on both k_1 and k_2.

The next step was to compute the mean solute flux $\langle s\rangle$ and the accumulated mass function (the breakthrough curve), i.e., $\langle M\rangle=\int_0^t\langle s(z,t')\rangle dt'$ under different assumptions.

First, the sorption parameters were assumed to be constant, so that spatial variability stemmed only from K_s. The expected value was obtained along the lines of the BD model, like in Eq. (9) for K_s. The results are shown in Destouni and Cvetkovic (1991, Fig. 1) for fixed values of the log-conductivity variance and ratio R/K_{sG} between the rate of application and the conductivity geometric mean, but for a few values of k_1 and k_2. The overall effect of the presence of kinetics is to diminish the peak of $\langle s(t) \rangle$ and create a tail at large t. It is emphasized that for a sorption time scale (of the order $1/k_1$) much smaller than advection time, while the ratio k_1/k_2 is fixed, equilibrium conditions are reached. Under these circumstances the effect is retardation majored by a factor $1+(k_1/k_2)$. One would expect that under the slow flow conditions that exist in unsaturated flow this would be the case. This is not evident in the results of Destouni and Cvetkovic (1991, Fig. 1) because of the relatively high variability of K_s and the high ratio between R/K_{sG}. This manifested in ponding over about 50% of the simulated field, for which flow was saturated. One would expect that for cases of smaller R/K_{sG} the results would have been closer to those of local equilibrium.

Additional cases examined by Destouni and Cvetkovic (1991) are of variable and random k_1 and k_2, negatively, positively or uncorrelated with K_s. Interesting features like the presence of double-peak breakthrough curves are observed and the reader is advised to consult the original article for details.

1.9 Validity of One-Dimensional Approximation for Infiltration in Heterogeneous Soils (Protopapas and Bras 1991a, b)

The basic assumption of the BD flow model (Sect. 1.2) is of one-dimensional flow in each vertical profile, leading to the major simplifications which permitted various authors to arrive at the results described in the preceding sections. As mentioned in Section 1.2 this approximation may be justified by the general approach of shallow-water flow, which implies that the length scale of variability in the horizontal plane is much larger than the vertical dimension of the flow domain. Protopapas and Bras (1991a) have undertaken the task of establishing the validity of the one-dimensional approximation by solving *exactly* problems of infiltration in heterogeneous soils and comparing the outcome with that obtained by the 1-D approximation.

Protopapas and Bras (1991a) solved the problem of infiltration through a heterogeneous soil of 2-D structure. Here are the main features of the model they investigated:

1. The soil hydraulic properties are encapsulated in two constant parameters, μ in the assumed linear relationship $\partial K/\partial\theta = \mu$, and α in $K/K_s = \exp(\alpha\psi)$.
2. A semi-infinite domain $|x| > 0$, $z > 0$ is considered. The initial and boundary conditions are for dry soil $\psi \to -\infty$, except for the strip $|x| < L$ on which moisture content is at saturation $\psi(x,0) = 0$. Thus, a saturation front propagates from the surface into the medium.

Heterogeneity of a deterministic nature has been considered in two stages. In the first stage, the saturated conductivity was assumed to vary exponentially:

$$K_s = K_0 \exp[-\lambda_x|x| - \lambda_z z], \tag{13}$$

and the problem allowed an analytically exact solution, given in Protopapas and Bras (1991b). The one-dimensional solution is the one at the center $x = 0$, obtained from the equations of flow in which derivatives with respect to x are neglected. More precisely, the derivative with respect to x is neglected in the continuity equation while the equation of flow has z and t as independent variables. Hence, the variation of K_s with z and the unsteady nature of the solution are preserved.

The accuracy of the one-dimensional approximation has been checked for two types of soils (given as μ, α) and for various combinations of other parameters. The main result was that the one-dimensional approximation is an accurate one if $L \gg 2\mu t\lambda_x/\alpha$. For the two types of soil and for $t = 10$ h it was found that the 1-D approximation is valid for relatively narrow strips of say, $L > 1$ m.

In the second stage, three more complex heterogeneous structures were considered, namely:

$$K_s = K_0 \exp[\cos(2\omega_x x) + \cos(2\omega_z z)];$$
$$K_s = K_0 \exp[\cos(2\omega_x x) - \lambda_x x^2 + \cos(2\omega_z z) - \lambda_z z^2];$$
$$K_s = K_0 \exp[\cos(2\omega_x x) + \cos(\omega_x x) + \cos(2\omega_z z) - 1]. \tag{14}$$

The 2-D solution was obtained numerically in this case and the same was true for the 1-D equation, solved separately for each x. Though no quantitative criteria similar to the first ones could be obtained, in most cases the 1-D approximation yielded a remarkably accurate reconstruction of the 2-D form.

The authors' conclusion is that for the assumptions used in the study and for uniform application of water on the soil surface, the heterogeneous soil medium can be represented by a set of noninteracting columns. Since this conclusion was reached for relatively small values of L, one is entitled to assume that the BD flow model and its subsequent developments, conceived for distributed sources over large areas, provide an accurate flow solution.

References

Ababou R (1988) Three-dimensional flow in random porous media. Thesis, MIT, Cambridge
Biggar JW, Nielsen DR (1976) Spatial variability of leaching characteristics of a field soil. Water Resour Res 9: 975–986
Bresler E, Dagan G (1979) Solute dispersion in unsaturated heterogeneous soil at field scale. II: Applications. Soil Sci Soc Am J 43: 476–472

Bresler E, Dagan G (1981) Convective and pore scale dispersive solute transport in unsaturated heterogeneous fields. Water Resour Res 17: 1683–1693

Bresler E, Dagan G (1983a) Unsaturated flow in spatially variable fields. 2. Application of water models to various fields. Water Resour Res 19: 421–428

Bresler E, Dagan G (1983b) Unsaturated flow in spatially variable fields. 3. Solute transport models and their application to two fields. Water Resour Res 19: 429–435

Bresler E, Dagan G (1988a) Variability of yield of an irrigated crop and its causes. 1. Statement of the problem and methodology. Water Resour Res 24: 381–388

Bresler E, Dagan G (1988b) Variability of yield of an irrigated crop and its causes. 2. Input data and illustration of results. Water Resour Res 24: 389–394

Bresler E, McNeal BL, Carter DL (1982) Saline and sodic soils. Springer, Berlin Heidelberg New York

Brooks RH, Corey AT (1964) Hydraulic properties of porous media. Hydrol Pap 3. Colorado State University, Fort Collins

Cvetkovic VD, Destouni G (1989) Comparison between resident and flux-averaged concentration models for field-scale solute transport in the unsaturated zone. In: Kobus HE, Kinzelbach W (eds) Contaminant transport in groundwater. A.A. Balkema, Rotterdam, pp 245–250

Dagan G (1984) Solute transport in heterogeneous porous formations. J Fluid Mech 145: 151–177

Dagan G (1989) Flow and transport in porous formations. Springer, Berlin Heidelberg New York

Dagan G, Bresler G (1979) Solute dispersion in unsaturated heterogeneous soil at field scale. I: Theory. Soil Sci Soc Am J 43: 461–467

Dagan G, Bresler E (1983) Unsaturated flow in spatially variable fields. 1. Derivation of models of infiltration and redistribution. Water Resour Res 19: 413–420

Dagan G, Bresler E (1988) Variability of yield of an irrigated crop and its causes. 2. Input data and illustration of results. Water Resour Res 24: 389–394

Dagan G, Nguyen VV, Springer E (1989) Analysis of transport in the upper soil layer and interpretation of the caisson experiments. Proc 1st Int Conf Focus 89: Nuclear waste isolation in the unsaturated zone, Las Vegas, published by ANS and ASCE, pp 233–240

Dagan G, Nguyen VV, Springer E (1991) Analysis of solute transport in an intermediate-scale unsaturated flow experiment. Proc 2nd Int Conf High level radioactive waste management, Las Vegas, published by ANS and ASCE, pp 837–844

Destouni G, Cvetkovic V (1989) The effect of heterogeneity on large scale solute transport in the unsaturated zone. Nord Hydrol 20: 43–52

Destouni G, Cvetkovic V (1991) Field scale mass arrival of sorptive solute into the groundwater. Water Resour Res 27: 1315–1325

Feinerman E, Bresler E, Dagan G (1989) Optimization of inputs in a spatially variable natural resource: unconditional vs. conditional analysis. J Environ Econ Manage 17: 140–154

Jury WA, Roth K (1990) Transfer functions and solute movement through soil: theory and applications. Birkhauser, Basel

Jury WA, Sposito G, White RE (1986) A transfer function model of solute transport through soil. 1. Fundamental concepts. Water Resour Res 22: 243–247

Kreft A, Zuber A (1978) On the physical meaning of the dispersion equation and its solution for different initial and boundary conditions. Chem Eng Sci 33: 1471–1480

Mantouglu A, Gelhar LW (1987) Stochastic modeling of large-scale transient unsaturated flow systems. Water Resour Res 23: 37–46

Nielsen DR, Biggar JW, Erh KT (1973) Spatial variability of field measured soil-water properties. Hilgardia 42: 216–221

Protopapas AL, Bras RL (1991a) The one-dimensional approximation for infiltration in heterogeneous soils. Water Resour Res 27: 1019–1020

Protopapas AL, Bras RL (1991b) Analytical solutions for unsteady multidimensional infiltration in heterogeneous soils. Water Resour Res 27: 1029–1034

Russo D (1991) Stochastic analysis of simulated vadose zone solute transport in a vertical cross section of heterogeneous soil during nonsteady water flow. Water Resour Res 27: 267–283

Taylor GI (1921) Diffusion by continuous movements. Proc Lond Math Soc A20: 196–211
Van der Zee SEATM, van Riemsdijk WH (1986) Transport of phosphate in a heterogeneous field. Transp Porous Media 1: 339–359
Van der Zee SEATM, van Riemsdijk WH (1987) Transport of reactive solute in spatially variable soil systems. Water Resour Res 23: 2059–2069
Warrick AW, Mullen GJ, Nielsen DR (1977) Scaling field-measured soil hydraulic properties using a similar model concept. Water Resour Res 13: 355–362
Yeh T-C, Gelhar LW, Gutjahr AL (1985) Stochastic analysis of unsaturated flow in heterogeneous soils. 1. Statistically isotropic media. Water Resour Res 21: 447–456

2 Field-Scale Solute Flux Through Macroporous Soils

G. Destouni[1]

2.1 Introduction

In general, natural soils exhibit a spatial heterogeneity that may be important for the field-scale spreading of solute. One manifestation of spatial heterogeneity that is typical for the unsaturated zone is the existence of relatively large and more or less continuous voids. These voids are commonly referred to as macropores and may, for instance, be interaggregate pores, earthworm holes, drying cracks in clay soils, and decayed root channels. Such macropores are generally characterized by distinctly different soil hydraulic properties than the soil matrix and may result in rapid solute movement through the unsaturated zone.

Models of solute transport in macroporous soils are commonly based on a two-region or bicontinuum approach, in which the water inside the aggregates is assumed to be stagnant while the water in the macropores is considered to be highly mobile. Solute transfer between the two regions is then assumed to be diffusion controlled and described either by a first-order rate expression (van Genuchten and Wierenga 1976), or by Fick's law if the geometry of the aggregates can be specified explicitly (Rasmuson and Neretnieks 1980; van Genuchten and Dalton 1986). Generally, these models are deterministic and do not explicitly account for spatial variability within the soil matrix and between the various macropores.

In addition to the heterogeneity that is reflected in the existence of two distinctly different flow regions in the same field, it is conceivable that each region will also exhibit spatial heterogeneity within itself. The need to account for spatial variability in the macropore flow and geometry characteristics has been recognized in several studies of water flow problems (Beven and Germann 1981; Beven and Clarke 1986; Germann and Beven 1986).

The use of deterministic models for field-scale transport problems will generally require the evaluation of effective hydraulic properties for each flow region. Determination of these effective properties in the field will require experimentation on a sufficiently large scale to be relevant for the specific application. Moreover, because it is generally not possible to isolate one region from another during tracer experimentation, the determination of the effective soil properties will involve fitting excercises, which often may

[1] Water Resources Engineering, Royal Institute of Technology, 10044 Stockholm, Sweden

not lead to unique parameter sets. In many cases, it may be an easier task to obtain statistical field information about the soil hydraulic properties in the soil matrix and in the macropores. The basic requirement in these cases will be a sufficient number of point measurements to produce reliable statistics.

The objective of this chapter is to incorporate spatial variability in the soil matrix and in the macropores in a transport model for investigating the effect of various properties of the macroporous flow region on the field-scale solute flux through the unsaturated zone. The focus is on solute transport from areally distributed sources at the soil surface. Advection in both the macropores and the soil matrix, and first-order mass transfer between the two flow regions will be considered.

2.2 Transport Model

For advection-dominated transport in statistically homogeneous fields, the expected field-scale mass flux of nonreactive solute is defined from the probability density function (pdf) for solute travel time, p_T, as (Shapiro and Cvetkovic 1988; Dagan et al., 1992; Destouni 1992a):

$$\langle s \rangle = M_0 p_T(t; Z), \tag{1}$$

in which M_0 is the total, instantaneously injected solute mass and $p_T(t; Z)$ quantifies the probability that an indivisible parcel of water and solute crosses a horizontal control plane at the depth Z at time t. For areally distributed sources that are sufficiently large relative to the horizontal integral scale, the transport problem may be considered as ergodic and Eq. (1) can be used to quantify the actual field-scale breakthrough curve at the depth Z (Destouni 1992a). For simplicity, the ergodic assumption is assumed to be valid in the following. The problem of statistical homogeneity in macroporous soils will be handled by use of a bimodal distribution of the soil hydraulic properties which, in turn, leads to a bimodal travel time pdf, p_T (Destouni 1991a).

Solute spreading at the field scale may often be regarded as being primarily an effect of advection variability, as expressed by Eq. (1). Transverse local dispersion may, however, play an important role in macroporous soils by transferring solute between regions with distinctly different hydraulic characteristics. The resistances that are associated with mass transfer into relatively immobile flow regions may often be significant and are mainfested at the field scale as sorption–desorption reactions, commonly referred to as physical nonequilibrium (Brusseau and Rao 1989). The macropores will generally occupy only a small part of the field and are surrounded by the soil matrix. The mass exchange with the relatively immobile soil matrix can threfore be expected to play an important role for transport in this flow region. In contrast, the soil matrix covers the largest part of the field and only a small part of this flow region will be in contact with macropores.

The effect of transverse dispersion on the solute transport within the soil matrix will therefore be important only after long travel times and can, in general, be neglected (Dagan 1984).

Cvetkovic and Shapiro (1990) provided a general result for an instantaneous injection of solute that undergoes linear nonequilibrium sorption–desorption in a semi-infinite domain with a steady-state, spatially variable, three-dimensional velocity field. This result, in which the travel time pdf is uncoupled from the mass transfer reactions, can be used to describe transport in the macroporous region. For steady, vertical flow in vertically homogeneous soil, the general result simplifies to (Destouni and Cvetkovic 1991):

$$\langle s \rangle = M_0\left(\exp[-k_1 t]p_T(t) + \int_0^t s_2(t,T)p_T(T)dT\right)$$

$$s_2 = k_1 k_2 T \exp(-k_1 T - k_2 t + k_2 T)\hat{I}_1[k_1 k_2 T(t-T)]H(t-T) \tag{2}$$

$$T = Z/v,$$

in which $\hat{I}_1(W) \equiv I_1(2\sqrt{W})/\sqrt{W}$, with I_1 being a modified Bessel function of the first kind of order one, H is the Heaviside step function, k_1 and k_2 are the constant mass transfer rate coefficients into and out of the relatively immobile flow region, respectively, T is the travel time to Z of an indivisible parcel of water and solute, and v is the local mean pore-water velocity.

The mass flux Eq. (2) is essentially composed of two parts: the first term describes the movement of the front that is attenuated due to the loss of mass into the relatively immobile flow region, and the second term describes the delayed contribution due to the kinetically controlled exchange of solute mass between the two flow regions. Analogous expressions to Eq. (2) for spatially variable mass transfer rate coefficients k_1 and k_2, with various types of cross-correlation with the soil hydraulic properties, have been presented by Destouni and Cvetkovic (1991), who also related k_1 and k_2 to the parameters of the commonly used mobile–immobile fluid model (van Genuchten and Wierenga 1976).

In this case the travel time pdf p_T in Eq. (2) represents only the macroporous region. To be able to handle the entire field as statistically homogeneous through a bimodal p_T, a condition has to be introduced in Eq. (2), such that the mass transfer is only active for the macroporous flow region. Such a condition may, for instance, be expressed as:

$$\langle s \rangle = M_0\left(\exp[-k_1 t]p_T(t) + \int_0^t s_2(t,T)p_T(T)dT\right) \quad t < T_c \tag{3a}$$

$$\langle s \rangle = M_0\left(p_T(t) + \int_0^{T_c} s_2(t,T)p_T(T)dT\right) \quad t \geqslant T_c, \tag{3b}$$

in which T_c is the travel time associated with the fastest solute pathway through the soil matrix. Solute travel times $T < T_c$ are thus associated with pathways in the macroporous flow region. The condition of (3b) implies that there is no attenuation of the solute front in the soil matrix. The upper limits in the integrals of Eq. (3) imply that the mass exchange term $s_2 \neq 0$ only in the macroporous flow region i.e., for travel times $T < T_c$. Moreover, from Eq. (2) follows that $s_2 = 0$ for $T > t$.

Although the flow conditions that underlie Eqs. (2) and (3) will generally not prevail in the field, such simplifying assumptions may still be useful for approximating flow and solute advection in the field (Destouni 1991b, 1992b; Protopapas and Bras 1991). Moreover, analyses of relatively simple systems are often useful for gaining insight into the effect of various parameters; such insight may, in turn, aid in interpreting field data and in focusing future theoretical and experimental efforts more effectively. Thus, for the comparative and illustrative purposes of the following discussion, Eq. (3) will be assumed to be valid.

2.3 Travel Time PDF

The main problem with evaluating p_T in Eq. (3) is how to relate the statistical description of the soil hydraulic properties to the advection statistics. For the flow conditions that are relevant for Eq. (3), the solute travel time $T = Z/v$ can be related to the soil hydraulic properties along an individual streamline by the model of Dagan and Bresler (1979) and Bresler and Dagan (1981) yielding:

$$T = Z\frac{(\mathscr{R}/K_s)^{\beta}(\theta_s - \theta_{ir}) + \theta_{ir}}{\mathscr{R}} \qquad \mathscr{R}/K_s < 1 \tag{4a}$$

$$T = Z\frac{\theta_s}{K_s} \qquad \mathscr{R}/K_s \geqslant 1, \tag{4b}$$

in which $\mathscr{R}$ is the constant rate of recharge, K_s and θ_s are the hydraulic conductivity and volumetric water content at saturation, respectively, θ_{ir} is the irreducible volumetric water content, and β is a soil coefficient related to the pore-size distribution. To account for spatial heterogeneity, the parameters K_s and β will, in the following, be regarded as random space functions, bimodally distributed according to:

$$p_W(W) = \frac{\nu}{\sigma_1\sqrt{2\pi}}\exp\left[-\frac{(W-\mu_1)^2}{2\sigma_1^2}\right] + \frac{1-\nu}{\sigma_2\sqrt{2\pi}}\exp\left[-\frac{(W-\mu_2)^2}{2\sigma_2^2}\right]$$

$$K_s = \exp(W); \quad \beta = \gamma + \delta W. \tag{5}$$

In Eq. (5), the pdf of W, p_W, is a combination between two normal pdfs, and ν is a weighting parameter that quantifies the fraction of the field that is occupied by the soil matrix. Moreover, μ_1 and σ_1^2 are the mean and variance of W in the soil matrix, and μ_2 and σ_2^2 are the corresponding parameters in the macropores.

In Eq. (5), a linear relation between K_s and β is assumed. Although indications of such a relation have been found from field data (Destouni 1991a), only a limited number of field studies have addressed cross-correlations between various soil hydraulic properties; the relation suggested in Eq. (5) should therefore only be considered as one particlar example used here.

The variability in K_s, expressed through Eq. (5), implies that the condition of Eq. (4b) may be applicable in some points of the field. In that case, the excess water from these saturated zones will be diverted to zones of higher conductivity (either by local surface runoff or by subsurface flow diversion) and will result in saturated conditions also in other parts of the field. The expected rate of recharge in the highly conductive zones that remain unsaturated will finally increase to (Destouni 1991a):

$$\mathscr{R}_p \equiv \mathscr{R} + \langle q_{s1} \rangle + \langle q_{s2} \rangle + \langle q_{s3} \rangle + \cdots, \tag{6}$$

in which the subscript p refers to the rate of recharge in preferential flow paths, and $\langle q_{si} \rangle$ express the expected surplus water from the zone of low conductivity, and are defined as

$$\begin{aligned}
\langle q_{s1} \rangle &= \int_0^{\mathscr{R}} (\mathscr{R} - K_s) p_{K_s} dK_s \\
\langle q_{s2} \rangle &= \int_{\mathscr{R}}^{\mathscr{R}_1} (\mathscr{R}_1 - K_s) p_{K_s} dK_s \\
\langle q_{si} \rangle &\equiv \int_{\mathscr{R}_{i-2}}^{\mathscr{R}_{i-1}} (\mathscr{R}_{i-1} - K_s) p_{K_s} dK_s \quad i = 3, 4, \ldots \\
\mathscr{R}_1 &= \mathscr{R} + \langle q_{s1} \rangle \\
\mathscr{R}_j &\equiv \mathscr{R}_{j-1} + \langle q_{sj} \rangle \quad j = 2, 3, \ldots
\end{aligned} \tag{7}$$

In Eq. (7), p_{K_s} is the pdf of K_s, defined from $p_{K_s} dK_s = p_W dW$. The series in Eq. (6) converges rapidly and $\mathscr{R}_p$ can be evaluated explicitly.

Because the soil zones in which $\mathscr{R}_p/K_s \geqslant 1$ will be saturated, Eq. (4) can be reformulated in terms of $\mathscr{R}_p$ instead of $\mathscr{R}$, and will remain valid without any other change; the effect of this reformulation on the field-scale solute flux is discussed by Destouni (1991a) and may be important in specific field situations. For the predominantly unsaturated conditions that will be investigated in the following, however, the ratio $\mathscr{R}_p/\mathscr{R} \approx 1$ (Destouni 1991a) and there is no need to reformulate Eq. (4).

A combination of Eqs. (4) and (5) yields:

$$T = \frac{\mathscr{R}^{\gamma + \delta W}}{r} \exp(-(\gamma + \delta W) W) \quad Tr < 1 \tag{8a}$$

$$T = \frac{\mathcal{R}}{r} \exp(-W) \quad Tr \geqslant 1, \tag{8b}$$

in which $r \equiv \mathcal{R}/(Z\theta_s)$ and, for simplicity, θ_{ir} has been assumed to be zero. From Eq. (8), $W(T; Z)$ can be expressed as:

$$W = \frac{\delta \ln \mathcal{R} - \gamma}{2\delta} \pm \left(\frac{(\delta \ln \mathcal{R} - \gamma)^2}{4\delta^2} - \frac{\ln(Tr\,\mathcal{R}^{-\gamma})}{\delta} \right)^{1/2} \quad Tr < 1 \tag{9a}$$

$$W = -\ln[Tr\,\mathcal{R}^{-1}] \quad Tr \geqslant 1. \tag{9b}$$

For realistic parameter values the negative root in Eq. (9a) leads to negative values of W only and can therefore be disregarded. From Eq. (9), the travel time pdf $p_T(t; Z)$ can be evaluated as $p_T = |\delta W/\delta T| P_W$, which yields the expressions:

$$p_T(t; Z) = \frac{1}{2\delta t} \left(\frac{(\delta \ln \mathcal{R} - \gamma)^2}{4\delta^2} - \frac{\ln(tr\,\mathcal{R}^{-\gamma})}{\delta} \right)^{-1/2} p_W(W(t; Z))$$

$$tr < 1 \tag{10a}$$

$$P_T(t; Z) = \frac{1}{t} p_W(W(t; Z)) \quad tr \geqslant 1. \tag{10b}$$

The relation $W(t; Z)$ quantifies the required W value for a solute particle to reach the depth Z at the time t, and is evaluated from Eq. (9). The travel time pdf p_T is independent of β for saturated flow conditions as in Eq. (10b). Furthermore, if the spatial variability in β is neglected, the expression for unsaturated flow conditions as in Eq. (10a) simplifies to:

$$p_T(t; Z) = \frac{1}{\beta t} P_W \left(\frac{1}{\beta t} \ln(tr\,\mathcal{R}^{-\beta}) \right) \quad tr < 1. \tag{11}$$

2.4 Illustration of Results

The parameter set that has been used to illustrate the results is given in Table 2.1. The parameter T_c in Eq. (3) was chosen as the travel time that corresponds to $W_c \equiv (\mu_1 + \mu_2)/2$, as given by Eq. (8). The parameter γ was calculated from $\gamma = \mu_\beta - \delta\mu_W$, in which $\mu_W = v\mu_1 + (1 - v)\mu_2$.

The effect of the weighting parameter v in Eq. (5) on the field-scale solute flux is illustrated in Fig. 2.1. The existence of macropores (for illustrative purpose chosen here as 20% of the field, i.e., $v = 0.8$) is reflected in an early peak in the field-scale breakthrough curve. This early peak has been attenuated due to the mass exchange with the soil matrix. The delayed contribution due to this exchange is reflected in the solute arrival between the early peak and the main peak. The main peak is primarily associated with transport

Tabble 2.1. Parameter values for illustration of results

	Standard value	Varies in Fig.
$\mathscr{R}$(m/yr)	0.3	—
v	0.8	1
$\exp(\mu_1)$ (m/yr)	3.0	—
$\exp(\mu_2)$ (m/yr)	300.0	4
σ_1	1.0	—
σ_2	1.0	5
k_1/r	5.0	2
k_2/r	5.0	2,3
μ_β	0.14	—
δ	0.03	—

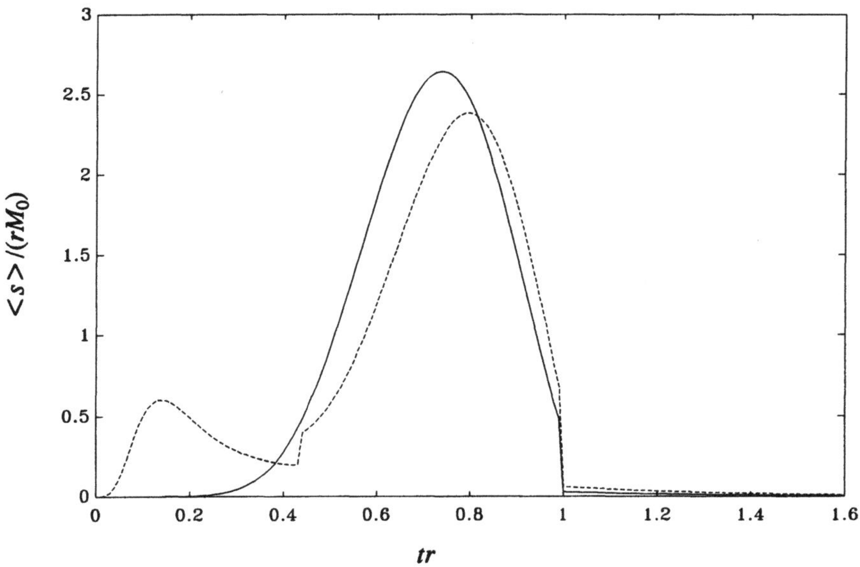

Fig. 1. Field-scale solute flux. *Solid line* $v = 1$; *dashed line* $v = 0.8$

in the soil matrix. The discontinuity at $tr \approx 0.4$ is due to the conditions in Eq. (3). The second discontinuity at $tr = 1$ is due to the conditions in Eq. (10).

Figure 2.2 illustrates that a decrease in the values of k_1 and k_2 has a similar effect on the field-scale breakthrough curve as that of decreasing v. The reason for this is that a decrease in k_1 decreases the attenuation from the solute front in the macropores and therefore increases the early peak in the breakthrough curve. Furthermore, a corresponding decrease in the rate coefficient for the mass transfer back into the macropores, k_2, implies that the ratio k_1/k_2 remains constant, whereby the arrival of the desorbed solute is unaffected by the decrease in k_1 and k_2. The main difference between

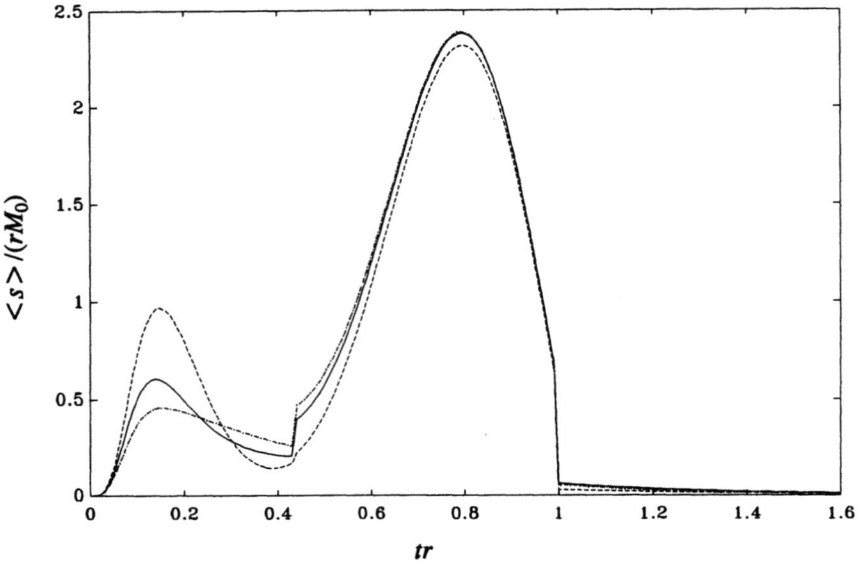

Fig. 2. Field-scale solute flux. *Dashed line* $k_1 = k_2 = 0.5$; *solid line* $k_1 = k_2 = 5$; *dashed and dotted line* $k_1 = k_2 = 10$

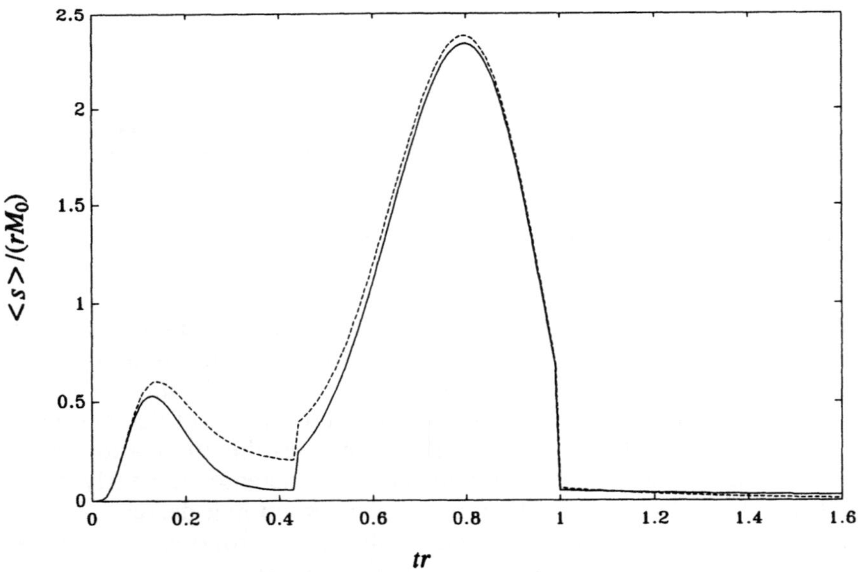

Fig. 3. Field-scale solute flux. *Solid line* $k_2 = 0.5$; *dashed line* $k_2 = 5$

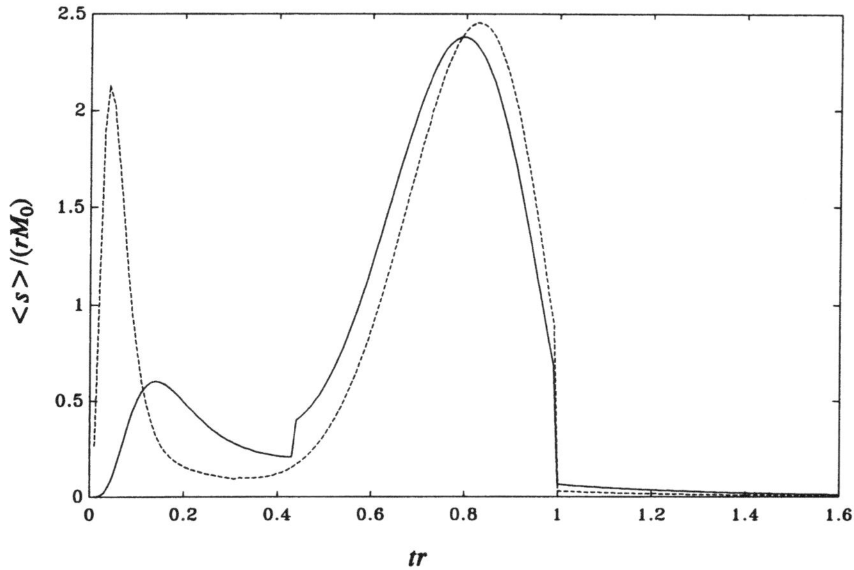

Fig. 4. Field-scale solute flux. *Solid line* $\exp(\mu_2) = 300$ m/yr; *dashed line* $\exp(\mu_2) = 3000$ m/year

the influence of v and that of k_1 and k_2 is that v has a greater effect on the main peak of the breakthrough curve.

Figure 2.3 illustrates that a decrease in k_2, without a corresponding decrease in k_1, will result in later arrival and more spreading of the solute that has been subject to exchange between the macropores and the soil matrix. This is reflected in the field-scale breakthrough curve by the fact that the solute arrival between the two peaks is smaller for the higher ratio, $k_1/k_2 = 10$; the desorbed solute mass does for these conditions instead give rise to increased tailing.

Due to the mass exchange with the soil matrix, an increase in the mean of W in the macropores, μ_2, affects not only the arrival time of the first peak in the breakthrough curve, but also the magnitude of, and the mass contained in, this peak. This effect is illustrated in Fig. 2.4 and the reason for it is that the faster transport in the macropores (through the increase in μ_2) results in decreased attenuation of the solute front because the rate of mass transfer relative to the advection rate has decreased.

Figure 2.5 illustrates the effect of the variance of W in the macropores, σ_2^2, on the field-scale breakthrough curve. In comparison to the effect of μ_2 (Fig. 2.4) the effect of σ_2^2 is smaller. Although σ_2^2 affects the magnitude and width of the early peak, it has only a small effect on the mass that is contained in that peak, or on its arrival time.

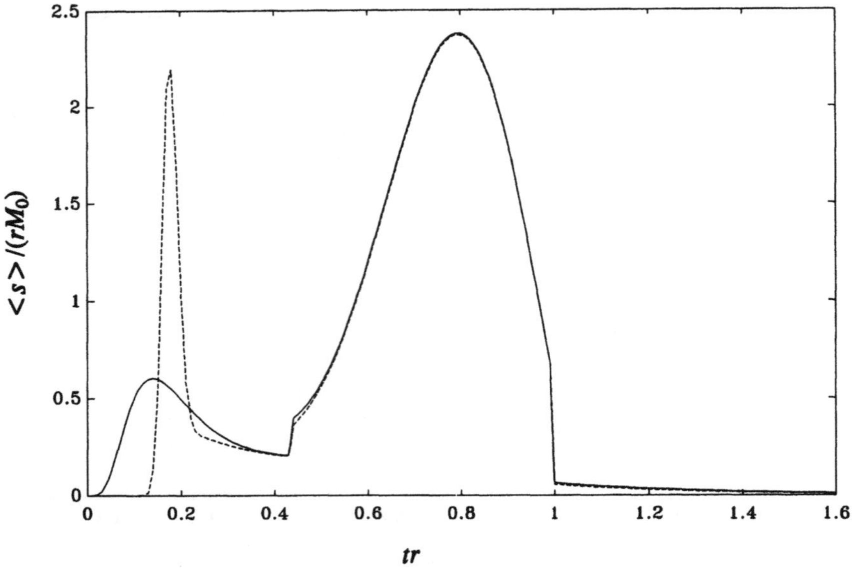

Fig. 5. Field-scale solute flux. *Solid line* $\sigma_2 = 1$; *dashed line* $\sigma_2 = 0.2$

2.5 Discussion and Conclusions

The proposed model for field-scale solute transport through macroporous soils has been used to identify similarities and differences between the influence of various soil properties on the field-scale solute flux through the unsaturated zone. Insight into the effects of different model parameters is useful for interpreting field data; if the data are compared with modelling results, such insight may assist in pointing out weaknesses in the input parameters and in the modelling approach.

The existence of macropores in the field will generally be reflected in an early peak in the field-scale breakthrough curve. However, this early peak is attenuated due to mass exchange with the soil matrix. The relative extent of the soil matrix (i.e., v) and the rate coefficient for mass transfer from the macropores to the soil matrix (i.e., k_1) have similar effects on the field-scale breakthrough curve. Decreasing v or k_1 results in increased early solute arrival through the macropores, the difference being that v also affects the main solute arrival through the soil matrix.

The mean advection rate in the macropores (given by μ_2) strongly affects the arrival time, the magnitude of and the mass contained in the early peak of the field-scale breakthrough curve. In comparison, the influence of the advection variance in the macropores (given by σ_2) is smaller because it affects only the magnitude and width of the early peak. However, the results of Destouni and Cvetkovic (1991) indicate that for fields with spatially variable mass transfer rate coefficients, and for reactive solute with spatially variable reaction parameters, advection variability within the macroporous flow region will have a more significant effect on the field-scale breakthrough curve.

Double- or multiple-peaked solute plumes or solute breakthrough curves have been observed in several experimental studies in field soils (Jury et al. 1986; Butters et al. 1989; Roth et al. 1991). However, the interpretation of these experiments has commonly been restricted to modelling the main peak behaviour only. The early peaks(s) have been neglected, partly because the rapid solute movement was unexpected and the measured data were insufficient, and partly because the models used to interpret the data could not account for this complex behaviour. The methodology presented in this chapter, and the results of Destouni and Cvetkovic (1991), may be useful for modelling early solute breakthrough in field-scale tracer experiments, and for interpreting field data where such behaviour occurs.

Acknowledgements. Financial support for this work has been provided by the Swedish Natural Science Research Council (NFR) and the Swedish Geological Survey (SGU).

References

Beven K, Clarke RT (1986) On the variation of infiltration into a homogeneous soil matrix containing a population of macropores. Water Resour Res 22: 383–388

Beven K, Germann P (1981) Water flow in soil macropores. II. A combined flow model. J. Soil Sci 32: 15–29

Bresler E, Dagan G (1981) Convective and pore scale dispersive solute transport in unsaturated heterogeneous fields. Water Resour Res 17: 1683–1693

Brusseau ML, Rao PSC (1981) Sorption nonideality during organic contaminant transport in porous media. CRC Crit Rev Environ Control 19: 33–99

Butters GL, Jury WA, Ernst FF (1989) Field scale transport of bromide in an unsaturated soil. 1. Experimental methodology and results. Water Resour Res 25: 1575–1581

Cvetkovic VD, Shapiro AM (1990) Mass arrival of sorptive solute in heterogeneous porous media. Water Resour Res 26: 2057–2067

Dagan G (1984) Solute transport in heterogeneous porous formaions. J Fluid Mech 145: 151–177

Dagan G, Bresler E (1979) Solute dispersion in unsaturated heterogeneous soil at field scale, I. Theory. Soil Sci Soc Am J 43: 461–467

Dagan G, Cvetkovic V, Shapiro A (1992) A solute flux approach to transport in heterogeneous formations, 1. The general framework. Water Resour Res 28: 1369–1376

Destouni G (1991a) Solute fluxes and travel times in heterogeneous soil. PhD Thesis, Dep Hydraulic Engineering, The Royal Institute of Technology, Stockholm

Destouni G (1991b) Applicability of the steady-state flow assumption for solute advection in field soils. Water Resour Res 27: 2129–2140

Destouni G (1992a) Prediction uncertainty in solute flux through heterogeneous soil. Water Resour Res 28: 793–801

Destouni G (1992b) The effect of vertical soil heterogeneity on field scale solute flux. Water Resour Res 28: 1303–1309

Destouni G, Cvetkovic V (1991) Field-scale mass arrival of sorptive solute into the groundwater. Water Resour Res 27: 1315–1325

German P, Beven K (1986) A distribution function approach to water flow in soil macropores based on kinematic wave theory. J Hydrol 83: 173–183

Jury WA, Elabd H, Resketo M (1986) Field study of napropamide movement through unsaturated soil. Water Resour Res 22: 749–755

Protopapas AL, Bras RL (1991) The one-dimensional approximation for infiltration in heterogeneous soils. Water Resour Res 27: 1019–1027

Rasmuson A, Neretnieks I (1980) Exact solution for diffusion in particles and longitudinal dispersion in packed beds. Am Inst Chem Eng J 26: 686–690

Roth K, Jury WA, Flühler H, Attinger W (1991) Transport of chloride through an unsaturated field soil. Water Resour Res 27: 2533–2541

Shapiro AM, Cvetkovic VD (1988) Stochastic analysis of solute arrival time in heterogeneous porous media. Water Resour Res 24: 1711–1718

Van Genuchten MTh, Dalton FN (1986) Models for simulating salt movement in aggregated field soils. Geoderma 38: 165–183

Van Genuchten MTh, Wierenga PJ (1976) Mass transfer studies in sorbing porous media. I. Analytical solutions. Soil Sci Soc Am J 40: 473–480

3 Towards Pore-Scale Analysis of Preferential Flow and Chemical Transport

R.J. Luxmoore[1] and *L.A. Ferrand*[2]

3.1 Scales of Preferential Flow

Preferential flow is a useful generic term for describing the process whereby water and chemical movement through a porous medium follow favored routes bypassing other parts of the medium. This term does not give any indication of the pore scales involved. Sometimes macropore flow is used to describe preferential flow and this term implies that large pores of some sort are conductive. There is no consensus definition of what constitutes a macropore so one may need to carefully determine what is meant when that term is used.

It is useful to distinguish between at least *two scales of preferential flow* based on the effective surface area of interaction of the preferential flow paths with the relatively immobile water in the soil matrix. In one case, referred to as *macropore channeling*, water moves through networks of holes, often biopores, that are 1 mm or more in width or diameter. These paths have a relatively low surface area and serve largely as physical conduits which may have a small influence on water quality during drainage (Luxmoore et al. 1990). In the other case, referred to as *mesopore channeling*, water moves through networks of cracks and channels that are less than 1 mm in width or diameter. We define mesopores as all pores less than 1 mm in equivalent diameter that drain by gravity, i.e., empty at matric pressure heads equivalent to field capacity (approximately -60 cm). This scale of drainable porosity has a relatively large surface area for chemical interactions with soil particles and with nondraining matrix water. Luxmoore et al. (1990) estimated that mesopores of the 5- to 14-cm suction head range had 25 times larger surface area per unit soil volume than macropores of the 0 to 2-cm suction head range. Pores that retain water by surface tensions above field capacity are referred to as micropores and these contain the "old water" in a drained field soil. "New water", introduced in precipitation events, usually has a different chemical signature from the old water and this difference can be used to estimate the old and new water contributions to preferential flow.

[1] Environmental Sciences Division, Oak Ridge National Laboratory, P.O. Box 2008, Oak Ridge, TN 37831–6038, USA

[2] Department of Civil Engineering, City University of New York (Y–120), New York, 10031, USA

Fingering flow in soils, a result of wetting front instability, is a third type of preferential flow that occurs in porous media with more or less random pore arrangements. There may not be any physically defined channels in the soil to account for this type of flow. A larger scale flow, described as *funnel flow* by Kung (1990), results from profile heterogeneity. Low permeability layers or coarse lenses in a profile may restrict vertical drainage, redirecting flow laterally through specific regions of the profile (like a funnel). Water repellency can also be a factor in the development of preferential flow (Van Dam et al. 1990).

3.2 Immobile Water

The relative immobility of water in micropores during unsaturated drainage events is due to their small size and to surface tension at air water interfaces associated with entrapped air. Smith and Browning (1942) showed entrapped air proportions of 9 to 22% in soil samples wetted under controlled laboratory conditions. Entrapped air can influence the pathway and rate of infiltration into a soil profile (Linden and Dixon 1976) and build-up of soil gas pressure is one explanation for development of fingering flows in some soils.

3.3 Old Water – New Water

Pearce et al. (1986) reviewed hypotheses of subsurface flow contributions to storm runoff generation in headwater catchments and summarized the published work in two contrasting categories. Runoff may be generated by displacement of old resident water in the catchment by new incoming rainfall which has differing water quality. The alternative mechanism of rapid throughflow of new water occurs via preferential flow channels bypassing the old water in the soil matrix. These two mechanisms represent two extremes that result in differing quality of the outflow water due to the differing qualities of the old and new water sources. Isotope and chemical analyses of the old, new, and outflow water have been used to estimate the relative contributions of old and new water to the flow process.

Several investigations have shown that old water can be a large component of outflow (Sklash and Farvolden 1979; McDonnell 1990; Wilson et al. 1993); however, explanations for the mechanism of old water displacement have not been convincing. Capillary fringe development near stream banks during rainfall has been proposed as one mechanism for rapid water-table rise, increase in the gradient for saturated flow, and the outflow of old water (Gillham 1984). However, adequate field evidence for this mechanism is lacking. McDonnell (1990) suggested that perched water-tables developed during rainfall on hillslopes could lead to (saturated) flow of old water into macropore channels. This hypothesis needs to account for how old water in the matrix converges into the preferential flow paths. An alternative hypothesis will be offered on the basis of pore-scale considerations developed from results obtained at field research facilities in Tennessee. These facilities were established to address soil heterogeneity and preferential flow effects on chemical transport (Luxmoore et al. 1990).

3.4 Subsurface Transport Investigations at Oak Ridge, Tennessee

Investigations of rainfed solute transport have been conducted at a forested hillslope site with the use of an in situ pedon and a subsurface hydrologic monitoring facility. A number of complementary laboratory studies of solute transport conducted under unsaturated and saturated flow conditions in undisturbed soil columns taken from the field sites have also been completed. Application of the two-region (mobile, immobile) form of the convection-dispersion (CD) equation to data from the soil column experiments provided reasonable descriptions of solute breakthrough when dispersion coefficients were optimized from nonreactive tracer breakthrough data and adsorption parameters were independently determined for reactive solutes (Jardine et al. 1988; Jardine and Jacobs 1992).

Field investigations of solute movement in the in situ pedon (2 by 2 m in area and 3 m deep) showed that the CD equation could be applied at selected depths to describe solute transport during ponded infiltration (Jardine et al. 1989) but could not account for the solute profiles at selected times. This was due in part to the limited spatial sampling of the heterogeneous flow paths in the soil block. Data from the same pedon site obtained during rainfed (unsaturated) transport of bromide indicated considerable exchange of Br^- contained in small pores after a drying period (old water) with the new water percolating through larger pore classes during a precipitation event (Jardine et al. 1990). A cascading process of bromide transferring from small to large pores in the upper profile was interpreted to reverse in the lower profile above a restricted flow layer (at 1.2- to 1.5-m depth) during sequences of drainage and redistribution over a 375-day study period. Solute transfer between pore classes was attributed to convection and diffusion with gradients that were inferred to change through the course of the study. A significant finding from this rainfed study was that bromide movement through large pores exhibited high concentration during the high flow periods and not a dilution effect as might be anticipated (Fig. 3.1). This behavior was similar to results obtained at two larger scales of observation. In a subwatershed of 0.47 ha, solute transport induced by rainfall was dominated by lateral flow in approximately the upper 1.2 m of soil. Subsurface flows through this “thin veneer” of the landscape had increasing chemical concentrations in some cases (S, Na, K, Mg, Ca) and decreasing

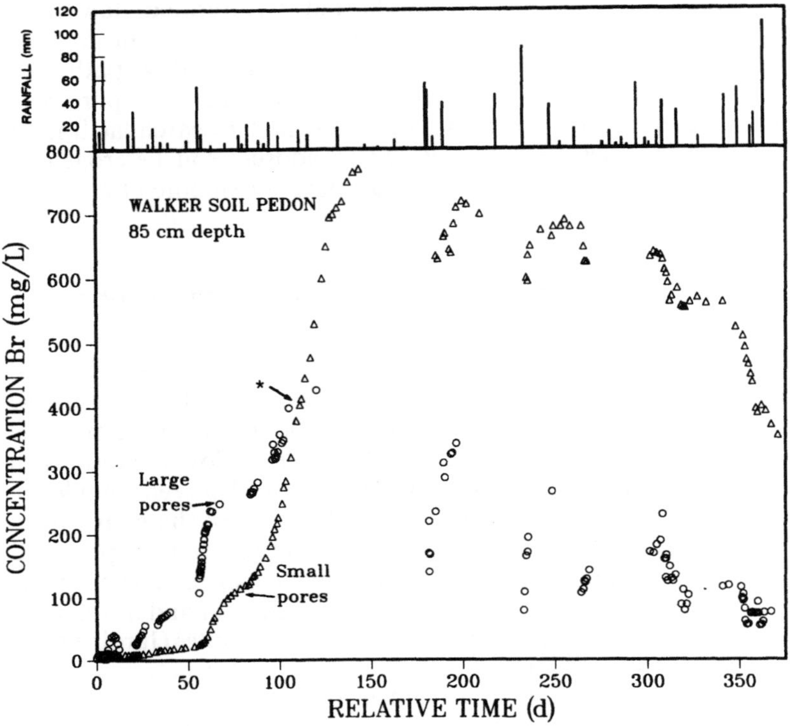

Fig. 3.1. Bromide concentration measured over a 375-day period at 85-cm depth in a pedon (2 × 2 m area) at a field site in Walker Branch watershed. The bromide was applied at the surface on day 0 and rainfall events (*upper panel*) caused downward transport in large and small pore classes. Several rainfall events are associated with rising bromide concentration. (Jardine et al. 1990)

concentrations in others (Al, Fe, Mn) during the rising limb of the subsurface hydrograph (Fig. 3.2, Luxmoore et al. 1990). At these two scales, varying in area by a factor of 1000, observations of high solute concentration occurring with increasing flow rate have been made. Similar findings have also been obtained from the perennial stream of Walker Branch watershed which contains the pedon and subwatershed sites. The west fork of Walker Branch watershed is 38 ha in area and the highest sulfate concentration has been measured at the time of peak streamflow (Fig. 3.3, Johnson and Henderson 1979). Thus, at three contrasting areal scales (4, 4700, 380 000 m^2) elevated chemical concentration has been found to co-occur with high flow rates. Our first explanation for these observations, having a common pore-scale basis, involved flow path considerations and diffusive exchange of solutes along flow paths with the matrix.

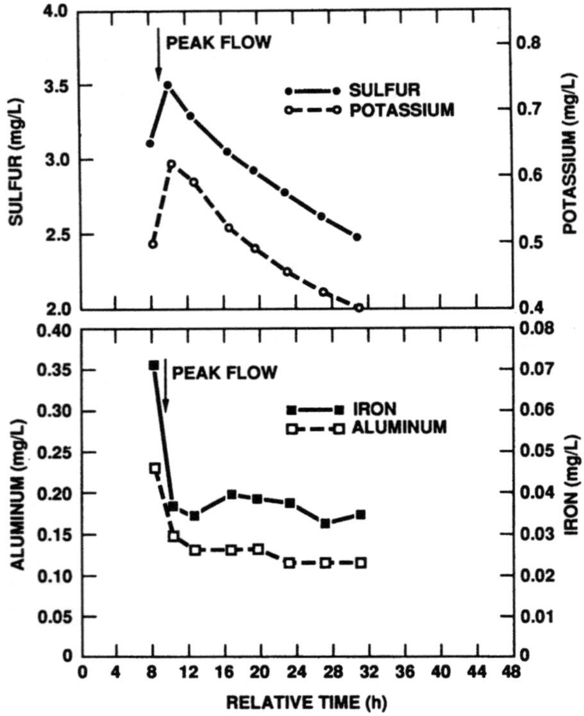

Fig. 3.2. Chemical (S, K, Al, and Fe) concentration in the outflow from a subsurface monitoring facility in a subwatershed (0.47 ha area) of Walker Branch watershed. Rising S and K concentrations were observed at peak subsurface flow, whereas Al and Fe concentrations were declining. (Luxmoore et al. 1990)

3.5 Pathlength-Supply Hypothesis

A pathlength-supply hypothesis was offered as an explanation for the above chemical transport observations (Luxmoore et al. 1990). Outflow from a landscape unit was viewed as a collection of contributing flow paths of varying lengths. At peak flow the longest and most extensive network of flow paths are hypothesized to discharge. These flow paths have the greatest area of interaction with matrix water resulting in the opportunity for diffusive transfer of solutes from the old water in micropores to the new rainfed percolating water in mesopores. This hypothesis requires that a large area of interaction exists between mobile and immobile pore regions. Interpretation of both the pedon and hillslope scales of investigation would benefit from a quantitative determination of the interaction area between draining and nondraining pore regions (Luxmoore 1991).

The chemicals that exhibit high concentration at the time of peak flow were described as "high supply" solutes and these maintained an effective diffusion gradient (strongly buffered) between the mobile and stagnant pore regions. A sustained diffusion gradient combined with the longest path lengths resulted in the most effective solute transfer between pore regions. In the case of the "low supply" chemicals there was no effective buffering

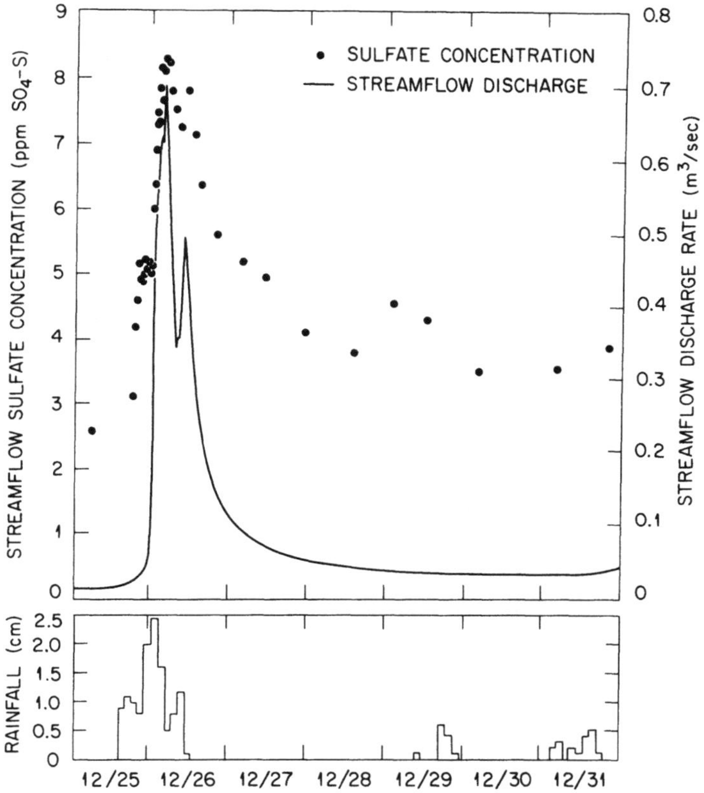

Fig. 3.3. Discharge rate and sulfate concentration in streamflow from the west fork of Walker Branch watershed showing the co-occurrence of highest sulfate concentration with peak flow rate. (Johnson and Henderson 1979)

and the initial removal of the chemical during drainage depleted the supply. For example, subsequent measurements have shown that the high initial Fe and Al concentrations (Fig. 3.2) were associated with particles that moved during early drainage (Wilson et al. 1991) and these particles were not immediately replaced at the sites of removal.

Our assessment of the role of macropores at our study sites is that they behave as physical conduits without much influence on water quality. Quantifying solute dynamics in unsaturated soil through drainable pores smaller than 1 mm in equivalent diameter (mesopores) along with the interactions with nondraining micropores were considered to be the major challenge at the Oak Ridge sites. These ideas are being modified with considerations from pore-scale analysis and the application of percolation theory. Extrapolation (scaling up) from columns to pedons and from pedons to hillslopes in being evaluated from a pore-scale basis with percolation models and a stochastic procedure called Latin hypercube sampling.

3.6 Percolation Theory

In a cubic lattice of nodes containing a random distribution of "conducting" nodes (i.e., pores), a conducting node proportion (porosity) of at least 0.31 is required to ensure the existence of a percolation path from one side of the cube to the opposite face. The threshold value of 0.31, known as the percolation threshold (Stauffer 1985), has been determined from extensive Monte-Carlo computer simulations. A typical percolation cluster of nodes consists of a main pathway (or pathways) of connected nodes called the backbone and an adjoining zone (other cluster sites) that can be viewed as a stagnant or "backwater" region. Transport out of the backwater must pass back through the backbone of the cluster (Fig. 3.4). The simulation patterns generated for percolation clusters have visual similarities to dye stain patterns obtained with rhodamine B in our column studies and as reported by Seyfried and Rao (1987).

Herrmann and Stanley (1984) describe the backbone in a regular lattice as a "string of blobs" with blob sizes selected randomly from a distribution of blob sizes. They estimated the fractal dimension for the backbone in a cubic lattice to be 1.74 ± 0.04 which means that as the size of the cube increases by the third power of length, the backbone within the cube increases by the 1.74 power. Herrmann and Stanley further describe the backbone as a "volatile fractal" due to the change in backbone shape with change in size of the observation volume. Formerly unconnected nodes can become connected to the backbone with an increase in the observation volume. Determination of the backbone and backwater proportions for random and nonrandom pore arrangements are needed and this will require extensive

Fig. 3.4. Two-dimensional percolation cluster diagram of a preferential flow path showing backbone, backwater (other cluster sites) and the non-cluster sites which are not connected to the flow process at this observation scale. The 20 × 20 lattice has a porosity of 0.6

efforts in computer analysis with percolation models. Lower percolation thresholds will be obtained with nonrandom (preferred flow path) pore arrangements. For example, field estimates of the hydrologically active macroporosity have shown that soils with low macroporosity (< 0.01) were capable of conducting large quantities of water (Wilson and Luxmoore 1988).

3.7 Percolation Modeling

In a typical percolation model for quasi-static immiscible displacement, the porous medium is conceptualized as a network or lattice of tubes (bonds) intersecting at nodes (sites) which may or may not be assigned a finite volume. Bond and site radii are assigned from specified pore-size distributions and these bond and site sizes may or may not be correlated.

The displacement of one fluid (e.g., water) by another (e.g., air) in the lattice is governed by the Young–Laplace equation which relates pressure difference (capillary pressure) and surface tension between fluids, the wettability of the solid (contact angle) and the mean radius of the fluid/fluid interface. Stable interface positions between fluids throughout the lattice are determined for a series of capillary pressure values.

In the earliest reported application of a network model to porous media Fatt (1956) used a "pencil and paper" technique to generate relationships between capillary pressure and total pore volume occupancy by a wetting fluid for a variety of two-dimensional lattices. The intervening years have seen the application of percolation-based computer simulation models to a variety of problems relating to immiscible displacement in porous media (e.g., Chatzis and Dullien 1977; Wilkinson 1984). This work has led to several relevant conclusions. First, it has been shown that with modification of a basic percolation model important aspects of immiscible displacement (such as the existence of an entry pressure, residual or irreducible saturations, and hysteresis between drainage and imbibition processes) can be captured. Second, it has become clear that a three-dimensional lattice is necessary to represent two-fluid flow in porous media (Wilkinson and Willemsen 1983; Ferrand and Celia 1990). The need to work with three-dimensional lattices has motivated our search for efficient network algorithms and the application of parallel computation for simulation of large networks.

Third, we can determine a minimum representative elementary volume (REV) for a given lattice type. Ferrand and Celia (1990) estimated the minimum REV size as approximately $9 \times 9 \times 9$ sites for drainage in a cubic lattice with randomly assigned bonds from a log-normal distribution of bond sizes. We expect REV size to increase with increasing correlation, i.e., with the degree of heterogeneity in pore sizes. Ferrand and Celia (1992) showed that fluid occupancy patterns may vary significantly at the same fluid content in media with differing patterns of lattice heterogeneity. If we wish to simulate fluid behavior at several scales, it will be necessary to determine the appropriate REV sizes for the degree of heterogeneity at each scale. Micromorphological characterization of the porous media for these REVs will be needed in deterministic and/or stochastic terms. Soils with fragmented fractal structure can be represented with fractal dimensions (Rieu and Sposito 1991) which can be readily used for scaling to increasing REVs.

The percolation model developed by Ferrand and Celia (1990, 1992) has been used to estimate the area between air-filled and water-filled pores over a range of water content values. Relationships between air/water interaction area and suction head calculated for a sand material shows maximum area between 10- and 20-cm suction heads (Fig. 3.5). At each step in the

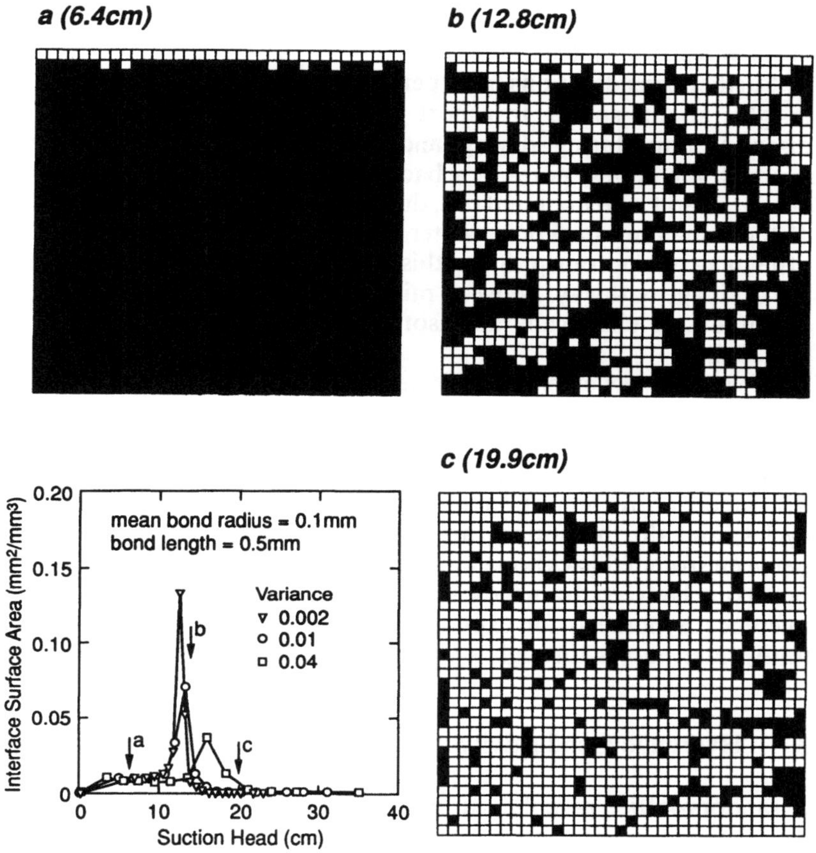

Fig. 3.5. Air/water interaction area simulated during drainage of a porous medium represented by a 3-D lattice percolation model. The calculated surface area is the sum of cross-sectional areas of pore throats intersecting the global interface between fluids at selected suction heads during simulation of quasi-static displacement. Diagrams of the water-filled pores (*dark sites*) at three suction heads (6.4, 12.8, and 19.9 cm), obtained from vertical cross sections through the 3-D percolation lattice for the 0·01 variance (bond radius) case, illustrate the stages of drainage and the development of high surface area at the intermediate suction head. The 3-D continuity of the global interface area between air and water for the 12.8-cm suction head case is not apparent in a single slice through the lattice

simulation, a global interface (i.e., connected to the boundaries of the lattice) between water and air was identified and the sum of the cross-sectional areas of pores which intersected this interface was calculated. This interaction area is less than the total air/water interface area since any isolated (discontinuous) water blobs in the medium were not included. The isolated blobs no longer contribute to the flow process. While the air/water interface in a draining medium is not identical to the old-water/new-water area in a reimbibing medium, we expect the latter to be smaller (rather than larger) than the air/water interface. Structured media would be expected to have even lower interaction areas than statistically homogeneous media due to bypass effects. These preliminary results suggest that the expectation of a large diffusive interaction area between old water and new water may not occur in random media.

Further considerations from percolation modeling have led us to another insight about solute transport in porous media, that of backbone disconnection during drying and reconnection during wetting. In this concept, the largest pores in a backbone become readily air filled during a decline in soil water content (i.e., during drainage after rainfall) causing some of the remaining backbone water to "hang up" in the medium. If there is little or no evapotranspiration this backbone water will remain stranded in the drainage pathway. Between rain events the new *stranded* water will tend to equilibrate chemically and isotopically with the old water by diffusion (Fig. 3.6).

We now expect an enhanced role for percolation modeling in the transient analysis of pore-scale effects on solute transport processes in soil. Percolation models are expected to provide (1) the surface area of interaction between backbone and backwater pore regions during drainage, (2) the dynamics of disconnection/reconnection of the backbone during drying and wetting cycles, and (3) the surface area of interaction and microscopic distance between stranded backbone water and the water in non-draining pore classes occurring between drainage events. This expanded concept provides a strong inference hypothesis for explaining how a large proportion of antecedent (old) water moves through a watershed during a storm event (Fig. 3.6).

3.7.1 Hydraulic Conductivity and Dispersion Coefficient

A random-lattice percolation model of a porous medium can also be used to estimate its transmission characteristics for fluids and associated dissolved substances. In this model, conductivity is calculated for each bond based on its size (radius and length). A pore which is not filled with the fluid of interest at a given suction has zero (or nominal zero) conductivity. Given a global pressure gradient across the lattice, we can solve for a fluid pressure at each site. Based on the discrete pressure distribution, a local steady-state flow rate can be calculated for each pore. Summing pore flow rates through a surface perpendicular to the global pressure gradient gives a total steady-state flow rate across the lattice. The flow rate can, in turn, be used

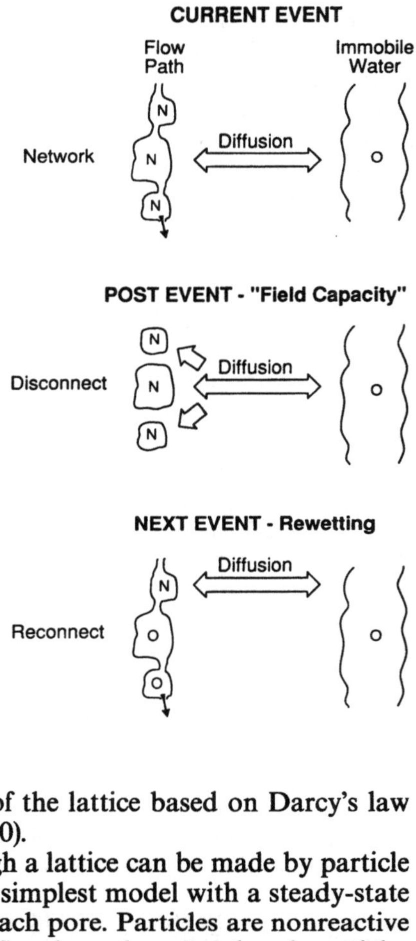

Fig. 3.6. Three conceptual stages of disconnect/reconnect of flow paths during convective-dispersive transport of solutes induced by wetting and drying cycles. New water (*N*) entering soil moves along connected flow paths and has diffusive solute exchange with immobile antecedent (old) water (*O*) within the matrix. The flow paths disconnect after rainfall ends and solutes exchange between stranded water and immobile water. Reconnection of flow paths with the next rain event allows the formerly stranded water to quickly drain with the chemical attributes of old water

to calculate the hydraulic conductivity of the lattice based on Darcy's law (Bear et al. 1987; Jerauld and Salter 1990).

Calculation of solute transport through a lattice can be made by particle tracking methods (Ferrand 1992). In the simplest model with a steady-state flow field, one assumes uniform flow in each pore. Particles are nonreactive and are introduced into the flow at the inflow boundary. In a bond, particles move with the average flow velocity. At junctions, the direction of particle movement to the next bond is chosen according to random probability weighted by the relative flow rate in the receiving bond. That is, each receiving bond has a probability of being chosen which is equal to its flow rate divided by the total outflow rate at the junction.

This simple particle tracking model can be used to estimate porous medium dispersivities as a function of pore size distribution parameters. As with the immiscible displacement model, a minimum REV size can be determined for varying degrees of heterogeneity within the lattice. This has been estimated to be approximately $25 \times 7 \times 7$ sites for the case of dispersion from the inflow face in a lattice with an uncorrelated pore-size distribution (Ferrand 1992). Again, we expect the minimum lattice size to increase with the scale of heterogeneity.

The dispersion model can be expanded to include internally distributed, as well as boundary solute sources. That is, particles may be started at arbitrary positions within the lattice and at a single boundary face. Internal sources of solute may not be "turned on" until the pore in which solutes reside becomes part of the lattice flow field. This scenario is analogous to the situation described above, in which fluid may be disconnected from continuous flow paths during drying and reconnected during rewetting. Thus, the particle tracking model can be used to estimate transient effects of solute sources added to or removed from the flow field as wetting fluid content changes.

3.7.2 Mass Transfer Coefficient

The two-region convection-dispersion model of solute transport has been applied as a means for accounting for the solute interactions between mobile and stagnant pore regions of soil (e.g. Rao et al. 1980). Controlled studies have demonstrated that diffusion is the mechanism of solute transfer between the two pore regions (see review by Brusseau and Rao 1990). Operationally, a mass transfer coefficient is empirically fitted to experimental data since the geometric complexity of a more physically based analysis of soil has been problematic. Determination of the mass transfer coefficient from diffusion calculations requires that the cross-sectional area of interaction between draining and stagnant pore regions be estimated along with the microscopic diffusion distance between pore regions and the apparent diffusion coefficient (Fig. 3.7). Percolation modeling can assist with these determinations.

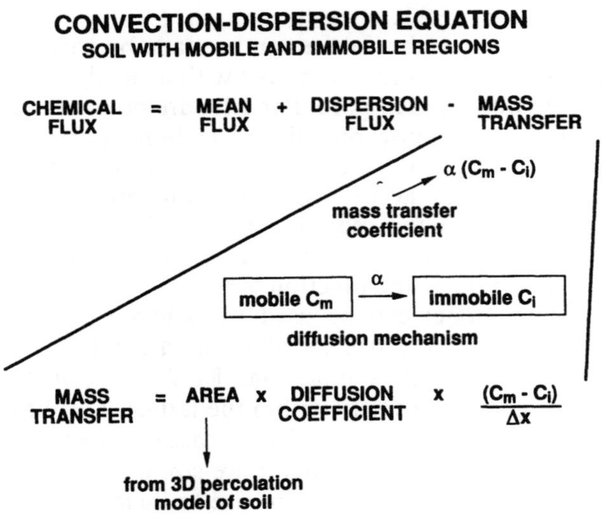

Fig. 3.7. Diffusion calculations may be used to determine mass transfer of chemical between mobile (C_m) and immobile (C_i) pore regions utilizing the area of interaction and the microscopic distance ($\Delta \times$) between these pore regions estimated with a percolation model

3.8 Stochastic Methods (Latin Hypercube Sampling)

Percolation models will be dependent on soil micromorphological characterization for representation of soil structure as well as pore-size frequency distributions. It is not practical to seek deterministic percolation models (i.e., exact pore representation) and thus stochastic approaches are needed. It may take many percolation model realizations to adequately describe the flow dynamics in soil; however, convenient Monte-Carlo procedures are available for this undertaking.

Many realizations of a percolation model derived from soil structural heterogeneity data may be used to calculate frequency distributions of hydraulic conductivity, dispersion coefficient and the area of interaction between mobile and stagnant pore-water regions. Percolation theory provides a means of identifying mobile zones (backbone) and stagnant zones (backwater) for given soil structural attributes and these can be compared with column dye tracing results. Change in backwater and backbone cluster size with a change in observation volume, and any fractal dimensions determined from percolation modeling should be valuable for extrapolating from the column scale to the pedon and to the watershed scale.

Latin hypercube sampling is an efficient stochastic method that is suitable for use with transport simulators. In the first step of this stochastic procedure, the frequency distributions of input variables (e.g., structural data) are divided into equal probability classes (e.g., 100 for each input distribution), and input values are selected by sampling without replacement so that all class intervals contribute to the input data sets of a model when the sampling has been completed. There may be random or correlation relationships between input variables, and a testing protocol identifies and rejects any spurious relationships in the input data sets not conforming to the expected behavior. Monte-Carlo methods that use sampling with replacement usually require several-fold more simulations for the same statistical precision obtained with Latin hypercube sampling. Gardner et al. (1983) describe a program called PRISM that implements the method based on the work of Iman and Conover (1980). A benefit from using a stochastic method such as Latin hypercube sampling with transport models is that outputs are in the form of frequency distributions which can be used for making statistically based comparisons between alternative modeling scenarios.

3.9 Synopsis

Rainfed chemical transport studies conducted at field facilities covering a wide range of spatial scales have all shown the occurrence of increasing solute concentration with increasing drainage rate or outflow rate. The origin of the majority of outflowing solutes is not in the new rainfall entering the soil, thus some mechanism by which increasing amounts of chemicals in antecedent water can be flushed out with rising subsurface flow rates needs to be determined.

Three hypotheses have been outlined for this mechanism during drainage:

1. Movement of old water in soil micropores into larger drainable pores.
2. Diffusion of solutes through the surface area of interaction from nondraining pores (containing old water) to larger draining pores containing new water.
3. Disconnection/reconnection of drainage water flow paths during drying and wetting cycles.

The first hypothesis cannot be a suitable explanation if entrapped air results in a high proportion of air water interfaces within the matrix. Also, a physical explanation for the movement of a high volume of micropore water into preferential flow paths is needed. The application of percolation models of pore-scale fluid relationships suggests that the surface area of interaction hypothesis (number 2) can only be a partial explanation due to insufficient surface area to allow for effective diffusive exchange *during* drainage events. Percolation modeling has shown the feasibility of the disconnection/reconnection hypothesis. A cluster of percolating pores disconnects during drying, causing some water in drainable pores to "hang up" in the percolation network. The water in these pores approaches chemical equilibration with old water in the matrix by diffusion during the period between drainage events. This equilibrated water is readily mobilized in the next drainage event. As rainfall increases the network connection of formerly stranded (drainable) old water expands in the landscape, increasing the proportion of old water in the outflow, and hence increasing the solute concentration with flow rate.

Use of the CD equation, representing the continuum scale, cannot explain the underlying mechanisms of solute transport; however, empirically derived variables can be selected to represent solute flux with the CD equation for the particular scale of calibration. Extrapolation to larger spatial scales is problematic with the CD equation since rules for scaling model variables are lacking. Percolation models offer a means for understanding pore-scale mechanisms contributing to solute dynamics, for calculation of continuum scale transport coefficients (conductivity, dispersion, mass transfer) and for identifying representative elementary volumes and fractal scaling rules that can aid in scaling up model applications to large scales of interest. Linkage of results from percolation modeling to multiregion convection-dispersion simulators is suggested as a useful way to derive dispersion and mass transfer coefficients from soil structural characteristics. Percolation-based CD models will aid our understanding of the dynamics of solute transport involving old water in immobile pores, stranded water in flow paths, and the diffusive exchange of solutes between old and new water during and between rainfall events.

Acknowledgments. This research was sponsored by the Exploratory Studies Program of Oak Ridge National Laboratory, operated for the US Department of Energy by Martin Marietta Energy Systems, Inc., under Contract No. DE-AC05-84OR21400. The research was also sponsored by the City University of New York, New York. Publication No. 3881, Environmental Sciences Division, ORNL.

References

Bear J, Braester C, Menier PD (1987) Effective and relative permeabilities of anisotropic porous media. Transp Porous Media 2: 301–316

Brusseau ML, Rao PSC (1990) Modeling solute transport in structured soils: a review. Geoderma 46: 169–192

Chatzis I, Dullien FAL (1977) Modelling pore structure by 2-D and 3-D networks with application to sandstone. J Can Petrol Technol 16: 97–108

Fatt I (1956) The network model of porous media. I. Capillary pressure characteristics. Trans AIME Petr Div 207: 144–159

Ferrand LA (1992) A pore-scale computational model for chemical transport in porous media. In: Russell TF (ed) Proc. IX International Conf. on Computational Methods in Water Resources. Computational Mechanics Publ., Southampton, UK, pp 401–408

Ferrand LA, Celia MA (1990) A percolation-based model for drainage in porous media. Water Resources Report WR-90-2, Princeton University, Princeton, NJ

Ferrand LA, Celia MA (1992) The effect of heterogeneity on the drainage capillary pressure-saturation relation. Water Resour Res 28: 859–870

Gardner RH, Röjder B, Bergström U (1983) PRISM: a systematic method for determining the effect of parameter uncertainties on model predictions. Studsvik/NW-83/555, Studsvik Energiteknik AB, Nykoping, Sweden

Gillham RW (1984) The capillary-fringe and its effects on water table response. J Hydrol 67: 307–324

Herrmann HJ, Stanley HE (1984) Building blocks of percolation clusters: volatile fractals. Phys Rev Lett 53: 1121–1124

Iman RL, Conover WJ (1980) Small sample sensitivity analysis techniques for computer models, with an application to risk assessment. Commun Stat Theor Methods A9: 1749–1842

Jardine PM, Jacobs GK (1992) Unsaturated transport of inorganic cations in undisturbed soil columns. In: Canepa JA (compiler) Proceedings of the DOE Yucca Mountain Site Characterization Project Radionuclide Adsorption Workshop at Los Alamos National Laboratory, Sept. 11–12, 1990. Rep LA 12325-C, Los Alamos National Laboratory, Los Alamos, NM, pp 177–186

Jardine PM, Wilson GV, Luxmoore RJ (1988) Modeling the transport of inorganic ions through undisturbed soil columns from two contrasting watersheds. Soil Sci Soc Am J 52: 1252–1259

Jardine PM, Wilson GV, Luxmoore RJ, McCarthy JF (1989) Transport of inorganic and natural organic tracers through an isolated pedon in a forest watershed. Soil Sci Soc Am J 53: 317–323

Jardine PM, Wilson GV, Luxmoore RJ (1990) Unsaturated solute transport through a forest soil during rain storm events. Geoderma 46: 103–118

Jerauld GR, Salter SJ (1990) The effect of pore structure on hysteresis in relative permeability and capillary pressure: pore-level modeling. Transp. Porous Media 5: 103–151

Johnson DW, Henderson GS (1979) Sulfate adsorption and sulfur fractions in a highly weathered soil under a mixed deciduous forest. Soil Sci 128: 34–40

Kung K-JS (1990) Preferential flow in a sandy vadose zone. 2. Mechanism and implication. Geoderma 46: 59–71

Linden DR, Dixon RM (1976) Soil air pressure effects on route and rate of infiltration. Soil Sci Soc Am J 40: 963–965

Luxmoore RJ (1991) On preferential flow and its measurement. In: Gish TJ, Shirmohammadi A (eds) Preferential flow: proceedings of the national symposium. Am Soc Agric Eng, St Joseph, MI, pp 113–121

Luxmoore RJ, Jardine PM, Wilson GV, Jones JR, Zelazny LW (1990) Physical and chemical controls of preferred path flow through a forested hillslope. Geoderma 46: 139–154
McDonnell JJ (1990) A rationale for old water discharge through macropores in a steep, humid catchment. Water Resour Res 26: 2821–2832
Pearce AJ, Stewart MK, Sklash MG (1986) Storm runoff generation in humid headwater catchments. 1. Where does the water come from? Water Resour Res 22: 1263–1272
Rao PSC, Rolston DE, Jessup RE, Davidson JM (1980) Solute transport in aggregated porous media: theoretical and experimental evaluation. Soil Sci Soc Am J 44: 1139–1146
Rieu M, Sposito G (1991) Fractal fragmentation, soil porosity, and soil water properties. II. Applications. Soil Sci Soc Am J 55: 1239–1244
Seyfried MS, Rao PSC (1987) Solute transport in undisturbed columns of an aggregated tropical soil: preferential flow effects. Soil Sci Soc Am J 51: 1434–1444
Sklash MG, Farvolden RN (1979) The role of groundwater in storm runoff. J Hydrol 43: 45–65
Smith RM, Browning DR (1942) Persistent water unsaturation of natural soil in relation to various soil and plant factors. Soil Sci Soc Am Proc 7: 114–118
Stauffer D (1985) Introduction to percolation theory. Taylor and Francis, London
Van Dam JC, Hendrickx JMH, van Ommen MH, van Genuchten MTh, Dekker LW (1990) Water and solute movement in a coarse-textured water-repellent field soil. J Hydrol 120: 359–379
Wilkinson D (1984) Percolation model of immiscible displacement in the presence of buoyancy forces. Phys Rev A 30: 520–531
Wilkinson D, Willemsen JF (1983) Invasion percolation, a new form of percolation theory. J Phys A (Math Gen) 16: 3365–3376
Wilson GV, Luxmoore RJ (1988) Infiltration, macroporosity, and mesoporosity distributions on two forested watersheds. Soil Sci Soc Am J 52: 329–335
Wilson GV, Jardine PM, Luxmoore RJ, Zelazny LW, Todd DE, Lietzke DA (1991) Hydrogeochemical processes controlling subsurface transport from an upper subcatchment of Walker Branch watershed during storm events. 2. Solute transport processes. J Hydrol 123: 317–336
Wilson GV, Jardine PM, O'Dell JD, Collineau M (1993) Field-scale transport from a buried line source in unsaturated soil. J Hydrol (in press)

4 Analysis of Solute Transport in Partially Saturated Heterogeneous Soils

D. Russo[1]

4.1 Introduction

Concerns over groundwater contamination by inorganic and organic chemicals in the subsurface, due to surface pills or leaks from underground storage facilities, have created a great impetus for the development of computational method of the transport process in the unsaturated (vadose) zone. Such methods are based on a conceptual framework, on subsequent theoretical developments, and, ultimately, on particular models adapted to the problem at hand. The purpose of these developments is twofold: to improve the basic understanding of the transport process, and to provide quantitative, predictive tools. In turn, these tools will be used to predict the future spread of pollutants in near-surface geologic environments, and for management purposes, i.e., to minimize the hazard of polluting underlying groundwater supplies. Our interest lies to a much larger extent in transport occurring at the field (formation) scale than at the laboratory (core) scale. This chapter focuses on some recent developments in the theory and in models of vadose-zone transport at the field scale. One of the distinctive features of a porous medium at this scale is the spatial heterogeneity of its properties which affect transport. This spatial heterogeneity is generally of an irregular fashion and occurs on a scale which is not captured by laboratory samples. These features have a distinct effect on the spatial distribution of solutes as a result of transport through the heterogeneous porous medium. To illustrate the type of heterogeneity one may encounter in the vadose zone at a field scale Fig. 4.1 shows measured soil properties from Bet Dagan, Israel (Russo and Bouton 1992). This figure shows contour lines of log transformations of two soil properties that affect transport in the vadose zone, namely, saturated conductivity, K_s, and α, a parameter related to the soil pore-size distribution. The irregular, ostensibly erratic, spatial variation of both properties is evident in this figure. It is clear that such heterogeneity in soil properties must affect flow and transport at the field scale. The complexity and irregular distribution of solutes one might encounter in the vadose zone at the field scale is illustrated in Fig. 4.2 which displays contour lines of measured chloride concentrations in the unsaturated zone, obtained from a tracer experiment conducted under

[1] Department of Soil Physics, Agricultural Research Organization, Ministry of Agriculture, Bet Dagan 50250, Israel

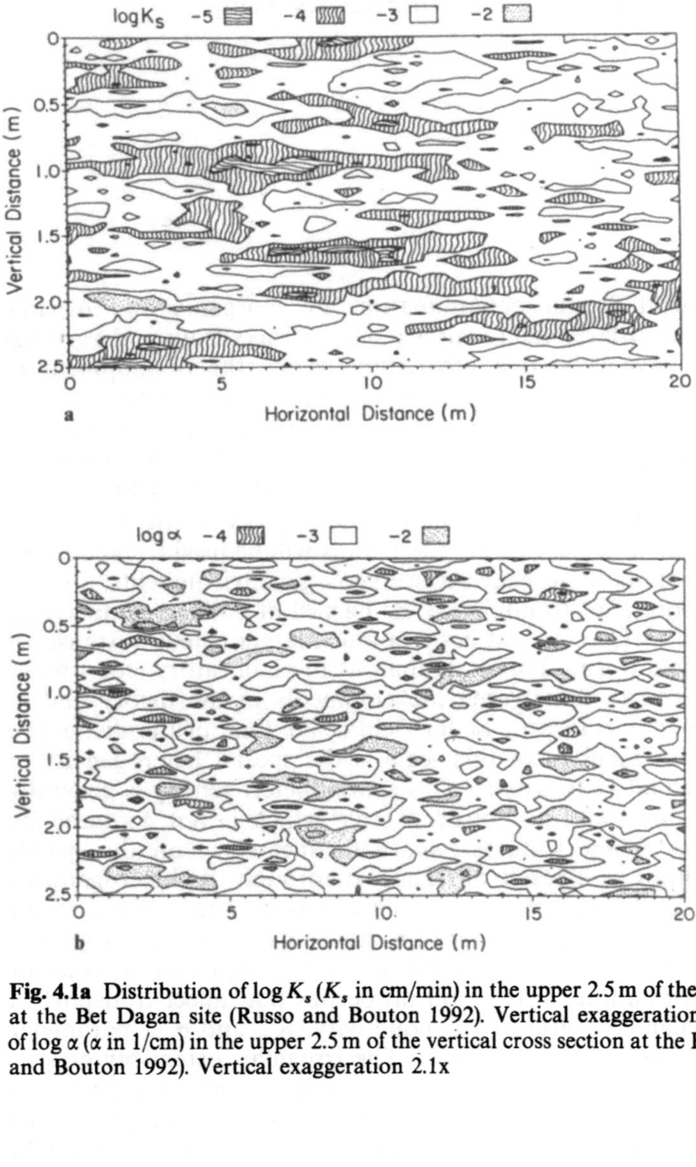

Fig. 4.1a Distribution of log K_s (K_s in cm/min) in the upper 2.5 m of the vertical cross section at the Bet Dagan site (Russo and Bouton 1992). Vertical exaggeration 2.1x. **b** Distribution of log α (α in 1/cm) in the upper 2.5 m of the vertical cross section at the Bet Dagan site (Russo and Bouton 1992). Vertical exaggeration 2.1x

natural soil, vegetative, and climatic conditions at Creux de Chippis, Switzerland (Schulin et al. 1987). Clearly, the chloride concentrations show considerable variability. A fundamental question raised by the experimental evidence displayed in Figs. 4.1 and 4.2 is concerned with how one can incorporate in predictive models the impact of spatial variability in soil properties on flow and transport at the field scale. In the following, a few advances in this area will be presented.

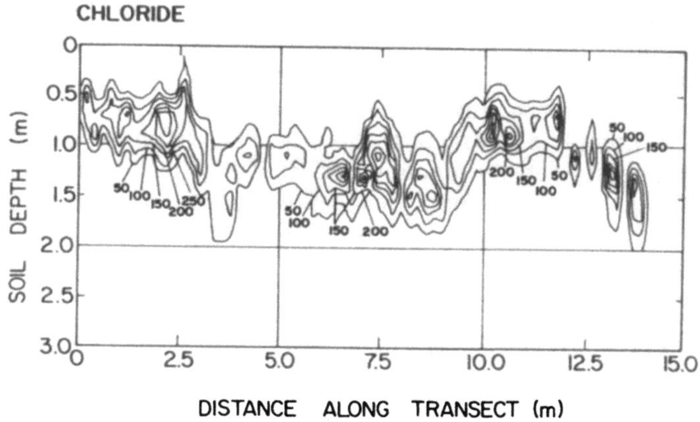

Fig. 4.2. Distribution of observed chloride concentration (in μg/g dry soil) in unsaturated soil at Creux de Chippis site (Schulin et al. 1987). Vertical exaggeration 2.5x

4.2 Basic Concepts and Definitions

The uncertainty of soil properties affecting transport (due to their inherent erratic nature and to a paucity of measurements) generally precludes the use of the traditional, deterministic modeling approach for the prediction of flow and transport at the field scale. In an alternative approach, uncertainty is set in a mathematical framework by regarding the relevant formation properties (e.g., hydraulic conductivity and water retention functions) as random space functions (RSFs). As a consequence, the flow and the transport equations are of a stochastic nature, and the dependent variables (e.g., fluid pressures and saturations, contaminant concentrations) are also RSFs. The aim of the stochastic approach, therefore, is to evaluate the statistical moments of variables of interest, given the statistical moments of the formation properties. In general, this is a formidable task and more often the scope is usually restricted to finding the first two moments: the mean value and the covariance function.

In a stochastic framework, a given heterogeneous soil property, $p(\underline{x})$, is treated as if it were a sample (or a realization) drawn at random from an ensemble of physical plausible RSFs $P(\underline{x})$. The ensemble concept is convenient for defining statistical properties of $P(\underline{x})$. Physically, the ensemble mean can be understood as the arithmetic average of repeated measurements of a property at a given spatial point and under the same external conditions. The ensemble mean is the minimum variance unbiased estimate of the actual property, provided that this property can be described by a probability distribution with finite moments (Jazwinsky 1978). The ensemble variance is a measure of the uncertainty associated with this estimate. In practice, however, only one realization of $P(\underline{x})$ will be available; the ergodic hypothesis must be invoked. According to this hypothesis (Lumley and Panofsky 1964), inferences about the statistical structure of $P(\underline{x})$ may be based on a substitution of the ensemble average with spatial averages obtained from a single realization of $P(\underline{x})$.

A complete characterization of $P(\underline{x})$ requires knowledge of the joint probability density function (PDF) of $P(\underline{x})$ at established coordinate points throughout the domain of the flow system. This formidable task is eased greatly by restricting the quantification of $P(\underline{x})$ to its first and second statistical moments only. Furthermore, it is assumed that $P(\underline{x})$ can be described by the general model:

$$P(\underline{x}) = M(\underline{x}) + W(\underline{x}), \tag{1}$$

where $M(\underline{x})$ is the drift function, and $W(\underline{x})$ is a zero mean, second-order stationary RSF characterized completely by a covariance function, $C_{ww}(\underline{x}_1, \underline{x}_2)$, which, in turn, depends on the separation vector $\underline{s} = \underline{x}_2 - \underline{x}_1$ and not on $\underline{x}_1$ and $\underline{x}_2$ individually. In other words, according to Eq. (1), $P(\underline{x})$ is characterized by variations at two different scales. The larger scale variations are viewed as slowly varying deterministic trends, around which there are more localized variations that are viewed as locally stationary.

The porous formation may be visualized as being composed of a three-dimensional structured arrangement of blocks of different soil matetials that may exhibit specific sizes but are not completely regular. The covariance function, $C_{ww}(\underline{s})$, therefore depends on the amount of overlap between these blocks. When the overlapping blocks are allowed to vary in size at random, the covariance function is described by an exponential model, i.e.,

$$C_{ww}(\underline{s}) = C_n[1 - H(s)] + C_s \exp(-s'), \tag{2a}$$

where $\underline{s}$ is the separation vector with components in the three Cartesian coordinates, s_x, s_y, and s_z; $s = |\underline{s}|$, and s' is a scaled separation distance defined as:

$$s' = [(s_x/I_{wx})^2 + (s_y/I_{wy})^2 + (s_z/I_{wz})^2]^{1/2}. \tag{2b}$$

I_{wx}, I_{wy}, and I_{wz} are characteristic length scales (correlation scales) of $W(\underline{x})$ in the x, y, and z directions, respectively; $H(s) = 1$ if $s = 0$, and $H(s) = 0$ if $s > 0$; and C_n and C_s are "unstructured" (nugget) and "structured" components of the variance, σ_w^2, of $W(\underline{x})$, respectively, i.e., $\sigma_w^2 = C_{ww}(0) = C_n + C_s$.

The "structured" variance represents property fluctuations attributable to spatial separation. On the other hand, the nugget variance may arise from a contribution of measurement errors and variance over distances much shorter than the smallest sampling interval. It should be emphasized that in stochastic treatment of flow and transport, where the pertinent formation property serves as input for integration (Dagan 1989), C_n is irrelevant in that its measure is zero.

4.3 Modeling of Solute Transport in Heterogeneous Porous Media

Stochastic models of solute transport predict the ensemble macrodispersive flux that accounts for differences among the trajectories of random concentration replicates over the ensemble. They should be distinguished from the macroscopic spread of a single plume for particular site-specific applications (e.g., transport experiments). As pointed out by Sposito et al. (1986), a prediction of the spreading of a single plume, based on ensemble mean statistics, may be inappropriate for particular site-specific applications, unless the relevance of ensemble behaviour to the single field behaviour can be establised explicitly. In principle, ergodicity (i.e., the interchange between ensemble and spatial averages) can be achieved if the characteristic length of either the solute input zone or the averaging volume is sufficiently large, or after an exceeding large travel time, for which the dispersive effect ensures spreading over a large volume.

In subsurface hydrology, stochastic models of solute transport at the formation scale have been developed, based on a theoretical framework (e.g., Dagan 1982, 1984; Gelhar and Axness 1983). Unfortunately, these models have been restricted to saturated groundwater flow. Although their underlying physics are similar, groundwater transport and vadose-zone transport do differ with respect to the scale of the flow domain, flow regime, and the direction of principal flow components relative to porous formation strata. Theoretical developments in groundwater transport modeling have been facilitated by the fact that under steady flow (that is, a reasonable approximation for the flow in aquifers) the only relevant flow parameter is saturated conductivity, K_s, which, in turn, is an inherent formation property. For the sake of completeness the results that have been obtained in the past by the theory of transport by groundwater and that are relevant to the present analysis, will be reviewed briefly. For a comprehensive review, see e.g., Dagan (1987, 1989).

4.3.1 Transport in Saturated Porous Media

The theory is based on two distinct, independent stages. In the first stage, which is independent of whether the flow is saturated or unsaturated, the transport theory relates the spatial moments of the solute body to the velocity field. It is assumed that transport through a heterogeneous porous formation can be described as dispersion by continuous motions (Dagan 1989), as was first suggested by Taylor (1921). Implicit in the transport theory are the following assumptions:

1. Lagrangian and Eulerian stationarity and homogenity.
2. Given statistics of the velocity field.
3. Negligible fluctuations of particle displacement about the mean trajectory.

Employing the aforementioned assumptions for ergodic conditions, and neglecting pore-scale dispersion, the first two moments of the PDF of the solute particle displacement can be expressed (Dagan 1982, 1984) as:

$$\frac{d\langle \underline{X}(t)\rangle}{dt} = \underline{U} \tag{3a}$$

and

$$\frac{dX_{ij}(t)}{dt} = 2\int_0^t C_{uu}(\underline{U}st')dt', \tag{3b}$$

where $\langle \underline{X}(t)\rangle$ is the expected value of the particle position (i.e., the average trajectory of the center of gravity of the solute body); $X_{ij}(t)$ $(i,j = 1,2,3)$ is the particle displacement covariance tensor (i.e., the average "moment of inertia" or second spatial moment of the solute body, with respect to its mean center of gravity); $\underline{U}$ is the expected value (assumed to be constant and uniform in time and space) of the Eulerian velocity field, $\underline{v}(\underline{x},t)$, which, in turn, is assumed to be time independent and $C_{uu}(\underline{s}) = \langle \underline{u}(\underline{x}_1)\underline{u}(\underline{x}_2)\rangle$ is the velocity covariance function, where $\underline{s} = \underline{x}_1 - \underline{x}_2$ is the separation vector, and $\underline{u}(\underline{x},t)$ the velocity fluctuation. Once the velocity field has been determined, either by measurement or by computation, so that the first two moments of the joint PDF of the velocity field are estimated, the first two spatial moments of the solute body, $\langle \underline{X}(t)\rangle$ and $X_{ij}(t)$, can be evaluated from Eqs. (3a) and (3b), respectively. The latter is related directly to the macrodispersion tensor, D_{ij} (Smith and Schwartz 1980; Dagan 1982, 1984; Gelhar and Axness 1983).

The second stage, which is dependent on the flow regime, consists of relating the statistical moments of the velocity field to those of the properties of the heterogeneous formation. For saturated groundwater flow this task has been accomplished (Dagan 1982, 1984; Gelhar and Axness 1983; Neuman et al. 1987) using Darcy's law and the continuity equation for water flow. In these studies it was assumed that the porosity, ϕ, of the heterogeneous formation in constant and deterministic, whereas the other relevant formation property, saturated conductivity K_s, is stationary and log-normally distributed, and is characterized completely by the expected value of $\log K_s$, F, and by the covariance function of the fluctuation of $\log K_s$, $C_{ff}(\underline{s})$. For a domain of large extent, compared with the correlation scale of $\log K_s$, I_f, for a uniform average head gradient, $\underline{J}(J_1,0,0)$, and by a first-order analysis in $\sigma_f^2 = C_{ff}(0)$, expressions for the mean velocity, $\underline{U}$, and the velocity covariance, $C_{uu}(\underline{s})$, in terms of F and $C_{ff}(\underline{s})$, were obtained (see e.g.; Dagan 1984):

$$\underline{U} = \underline{J}K_G/\phi \tag{4a}$$

$$C_{uu}(\underline{s}) = \underline{U}^2 C_{ff}(\underline{s}) + B. \tag{4b}$$

Here, $K_G = e^{\langle F\rangle}$ is the geometric mean of K_s, while the additional undefined terms denoted by B in Eq. (4b) are given elsewhere (Dagan 1984).

Combining the two aforementioned stages, Dagan (1984) derived closed-form analytical expressions for the longitudinal (X_{11}) and the transverse

(X_{22}) components of the ensemble time-dependent displacement covariance tensor for a tracer solute moving through a heterogeneous, infinite aquifer with an isotropic structure under two- and three-dimensional, steady-state, saturated flows, in terms of the variance, σ_f^2, and the correlation scale, I_f, of the input formation property, $\log K_s$. Later, Dagan (1988) extended the theory to a more generalized case, considering flow and transport through heterogeneous anisoropic structures. The components of the time-dependent ensemble displacement covariance tensor, derived from Dagan's (1984) model, compared favorably with those obtained from an analysis of the temporal behavior of the spatial moments of the tracer plume at the Borden site (Freyberg 1986; Sudicky 1986).

4.3.2 Transport in Unsaturated Porous Media

4.3.2.1 Simplified Stochastic Approach

Water flow in the vadose zone (that might be transient and nonmonotonic as well) is much more complicated than groundwater saturated flow. In this case, the relevant flow parameters – hydraulic conductivity, K, and water capacity, C – depend both on few formation properties (e.g., saturated conductivity, pore-size distribution, and saturated and residual water contents) and on dependent flow variables – water saturation, S, or capillary pressure head, ψ – in a highly nonlinear fashion. Consequently, under unsaturated flow, the evaluation of variations of the porous medium properties on transport is extremely complex, and so far has not been studied in a theoretical framework, except for a few investigations of water flow. These were based on a small perturbation approximation of a stochastic, linearized flow equation (Yeh et al. 1985a, b; Mantoglou and Gelhar 1987a, b, c), and were restricted to analyses of steady unsaturated water flow and transient unsaturated water flow, respectively, with vertical mean infiltration through an unbounded, infinite, heterogeneous porous medium.

Current, simplified vadose-zone stochastic solute transport models (e.g., Dagan and Bresler 1979; Bresler and Dagan 1981, 1983; Amoozegar-Fard et al. 1982; Jury 1982; Simmons 1982; Destouni and Cvetkovic 1989, 1991) may be classified as either mechanistic or nonmechanistic. Mechanistic [represented by the model of Dagan and Bresler (1979)] is taken here to imply that the model incorporates the most fundamental mechanisms of the process, as understood at present (for example, describing water flow and solute transport at the core scale by the Richards' equation and by the convection-dispersion equation, CDE, respectively). Nonmechanistic [represented by the transfer function (TF) model of Jury (1982)] implies that the model disregards the internal physical mechanisms contributing to the transport process, and concentrates on the relationships between an input function and an output response function. Both groups of models assume that the heterogeneous inputs to the models are stationary RSFs, and are ergodic over the region of interest, so that their ensemble moments can be derived from sample spatial statistics. In addition, steady water flow is assumed and thus, neither water content nor water flux change with time or depth. Most important, both groups of models treat the soil as though it were composed of a series of isolated stream tubes (vertically homogeneous, independent soil columns) with different velocities. In the mechanistic approach, in each vertical soil column the velocity is related to the soil properties by using Darcy's law and assuming a unit heat gradient. It is further assumed that both saturated water content, θ_s, and β, a parameter related to the soil pore-size distribution, are constant and deterministic, whereas K_s, the saturated conductivity, is proportional to the square of a log-normally distributed scaling factor, δ, viewed as a stationary RSF. Because of the assumption of independent, homogeneous vertical soil columns, the correlation scale of the input formation poperty, $Y = \log\delta$, in the vertical direction is infinite and the variance, σ_Y^2, describes soil property variations among the soil columns, regardless of the position of the columns in the horizontal plane. In other words, the spatial variability in properties of the porous medium is expressed in terms of the variance of $Y = \log\delta$ only, rather than by its covariance function [i.e., Eq. (2)].

Assuming the log-normally distributed scaling factor, δ, the model of Dagan and Bresler (1979) can be used to calculate the first two moments of the particle travel time PDF to a horizontal control plane located at depth $z = Z$:

$$\langle T(Z)\rangle = \int_{-\infty}^{\infty} T(Y,Z) f(Y) dY \tag{5a}$$

and

$$\langle T'(Z)^2\rangle = \int_{-\infty}^{\infty} [T(Y,Z) - \langle T(Z)\rangle]^2 f(Y) dY, \tag{5b}$$

where $f(Y)$ is the PDF of the formation property $Y = \log\delta$ and $T(Y,Z)$ is the particle travel time given by:

$$T(Y,Z) = \frac{Z\theta_s}{R_0}\left(\frac{R_0}{K_s^*}\right)^{\beta} \exp(-2\beta Y) \text{ for } R_0 < K_s^* \exp(2Y) \tag{6a}$$

$$T(Y,Z) = \frac{Z\theta_s}{K_s^*} \exp(-2\beta Y) \text{ for } R_0 \geqslant K_s^* \exp(2Y), \tag{6b}$$

where R_0 is the recharge rate at the soil surface, θ_s is the saturated water content, and $K_s^* = [\int (K_s)^{1/2} f(K_s) dK_s]^2$. Similarly, moments of the travel distance can be calculated. Equations (5) and (6) predict that the second moment of the travel time PDF, i.e., the ensemble variance of the travel time, is proportional to the square of the displacement distance (Fig. 4.3a) or, in other words, that ensemble longitudinal macrodispersivity grows

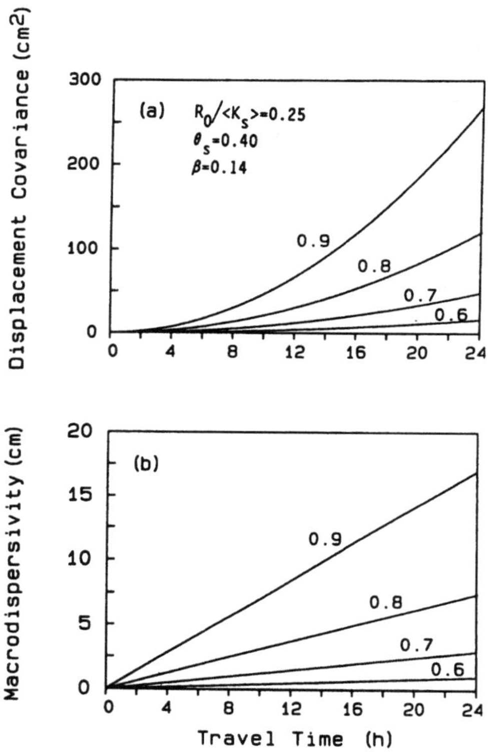

Fig. 4.3a Longitudinal component of the displacement covariance tensor as a function of travel time for different values of the variance of the log-saturated conductivity σ_f^2 (the numbers labeling the curves) calculated by the model of Dagan and Bresler (1979). Note that $\sigma_f^2 = 4\sigma_Y^2$, where Y is the log-scaling factor δ; θ_s is saturated water content; β is a soil parameter related to pore size distribution; $\langle K_s \rangle$ is the expected value of saturated conductivity; and R_0 is a recharge rate. **b** Longitudinal macrodispersivity coefficient as a function of travel time for different value of the variance of the log-saturated conductivity σ_f^2 (the numbers labeling the curves) calculated by the model of Dagan and Bresler (1979)

linearly with solute residence time and displacement distance, as the dispersion process develops (Fig. 4.3b). The same result can be obtained by using a nonmechanistic approach (e.g., the TF model of Jury 1982) if an assumption is made that convective solute movement dominates lateral mixing (Jury et al. 1986).

4.3.2.2 Simulation of Solute Transport

Simulation is a powerful tool that can be regarded as a "numerical experiments" which, in turn, can provide much more detailed information (that is unaffected by uncertainty due to measurement errors or inadequate sampling) about the effects of spatial variability in soil properties on the response of the flow system than is attainable in practice from field investigations. Furthermore, results of transport simulations may be used to test the applicability of current stochastic vadose-zone transport models to more realistic field situations.

The problem of transport of a conservative, nonreactive solute in a vertical cross-section of a hypothetical, partially saturated, heterogeneous soil under transient water flow was simulated recently by Russo (1991). The simulation was based on the assumption that, locally, water flow and solute transport can be described by the Richards' equation and by the CDE equation, respectively. Furthermore, the assumption of scale heterogeneity (Philip 1967) was employed by Russo (1991) so that the logarithm of the scaling factor, $Y = \log \delta$, was considered to be a realization of an RSF characterized by an anisotropic, axisymmetric, two-dimensional exponential covariance function.

The movement of a single chloride pulse in the vertical cross-section of the hypothetical, yet realistic, heterogeneous soil was simulated for 34 days under a transient flow regime generated by daily irrigations. Simulated water content and solute concentration distributions in the vertical cross-section of the heterogeneous soil at different time lapses were quantified in terms of space averages and covariance functions. Time evolution of the solute plume was quantified in terms of its first two spatial moments, and from this the time dependence of the longitudinal and transverse components of the velocity vector and the spatial covariance tensor were estimated. The simulated transport and the accompanying analysis (Russo 1991) were based on single realizations of the formation input property, as well as of the output variables. It was assumed, therefore, that ergodicity requirements were at least partially satisfied. This was justified in view of the size of the solute inlet zone (that extended over 12 correlation scales of the input property formation, $\log \delta$, in the transverse horizontal direction) and the

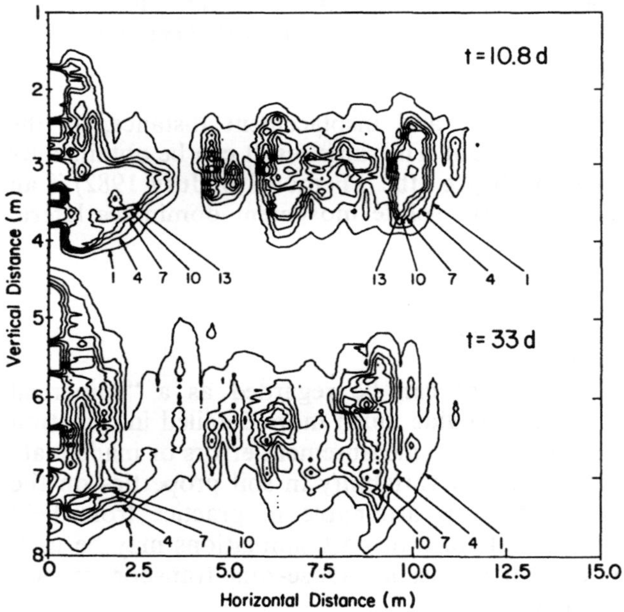

Fig. 4.4. Distribution of simulated chloride concentration (in mEq/1) in a hypothetical scale-heterogeneous unsaturated soil at two different elapsed times (t) (Russo 1991) Vertical exaggeration 2.1

vertical extent of the domain where flow took place (over several tens of the correlation scale of $\log \delta$ in the longitudinal vertical direction). Hence, an interchange between the concentration spatial moments and their ensemble means was assumed, although the estimated value of the latter might be affected by uncertainty due to the finiteness of the solute inlet zone (Dagan 1990).

Results of the simulated transport (Russo 1991) are depicted in Fig. 4.4 and demonstrate the considerable variability in the solute concentration in space and time due to the complex heterogeneity of the soil hydraulic properties in both the vertical and the horizontal direction, in agreement with field data (see Fig. 4.2, data from Schulin et al. 1987). Because of the complex heterogeneity of the soil hydraulic properties, the movement of the simulated solute plume was characterized by a compression–expanison phenomenon. This is illustrated by the time behaviour of the longitudinal component of the spatial covariance tensor (Fig. 4.5a) and the related longitudinal macrodispersion coefficient (Fig. 4.5b) derived from the spatial moment analysis of the simulated solute plume. These time fluctuations are in qualitative agreement with field data (see Fig. 4.6, data from Ellsworth 1989).

The components of the time-dependent spatial covariance tensor, obtained from the analysis of the temporal behaviour of the spatial moments

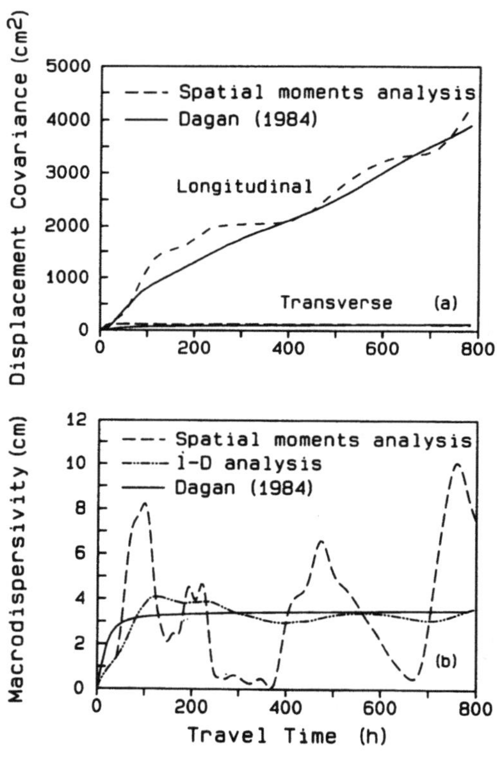

Fig. 4.5a Longitudinal (X_{11}) and transverse (X_{22}) components of the displacement covariance tensor estimated from the respective components of the second spatial moment of the simulated solute plume about its center of gravity in a hypothetical scale-heterogeneous unsaturated soil as a function of travel time, t (Russo 1991). *Solid lines* are $X_{11}(t)$ and $X_{22}(t)$ derived by Dagan (1984) for two-dimensional, steady, saturated flow. **b** Longitudinal vertical component of macrodispersivity tensor λ_{11} as a function of travel time, t. Values of $\lambda_{11}(t)$ evaluated from the first derivative of cubic splines fitted to the $X_{11}(t)$ data (Fig. 4.5a) are given by the *dashed and dotted* line. Values of $\lambda_{11}(t)$ evaluated from the depth-moment integrals of the depth-scaled, one-dimensional vertical CD equation, using the horizontally averaged values of concentration as a function of depth, are given by the *dashed line*. The *solid line* is $\lambda_{11}(t)$, derived by Dagan (1984) for two-dimensional, steady, saturated flow

of the simulated solute plume in unsaturated flow, compared favorably with the components of the ensemble spatial covariance tensor derived from Dagan's (1984) model for transport by saturated groundwater flow (Fig. 4.5a). The momentary discrepancy between the actual spread of the single plume and the one based on the ensemble average of the solute body, was explained by the finiteness of the solute inlet zone (Dagan 1990) and by the unsteadiness and the compressibility of the unsaturated flow (as compared with the constant porosity, incompressible, steady, saturated flow). In addition, it was found by Russo (1991) that the time-dependent longitudinal macrodispersivity, λ_{11}, determined from depth moment integrals of the depth-scaled, one-dimensional CD equation (Jury and Sposito 1985) using horizontally averaged concentrations, was in qualitative agreement with $\lambda_{11}(t)$ determined from the stochastic groundwater transport model of Dagan (1984) (Fig. 4.5b). Furthermore, the asymptotic value found by Russo (1991), $\lambda_{11}(t \to \infty) = 3.47$ cm, was in very good agreement with the asymptotic value of the longitudinal component of the macrodispersivity tensor derived by Dagan (1982, 1984; and also by Gelhar and Axness 1983) for two- and three-dimensional saturated, steady flows, as:

$$\lambda_{11}(t \to \infty) = D_{11}(t \to \infty)/U_1 = \sigma_f^2 I_{fz}, \tag{7}$$

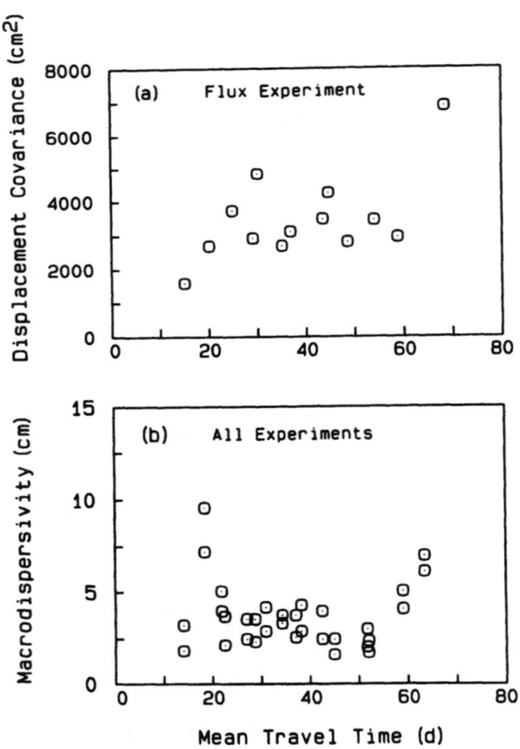

Fig. 4.6a Longitudinal (X_{11}) component of the displacement covariance tensor estimated from the second spatial moment of the measured solute plume about its center of gravity at the Etiwanda site as a function of travel time, *t*. (Ellsworth 1989). **b** Longitudinal vertical component of macrodispersivity tensor λ_{11}, estimated from the second spatial moment of the measured solute plume about its center of gravity at the Etiwanda site as a function of travel time, *t*. (Ellsworth 1989)

where U_1 and D_{11} are the longitudinal components of the mean velocity vector and the dispersion tensor, respectively, in the vertical direction, and $\sigma_f^2 = \langle f'^2 \rangle$ and I_{fz} are the variance and the correlation scale in the longitudinal direction, respectively, of the input formation property, $\log K_s$. For the scale-heterogeneous soil analyzed by Russo (1991), $\sigma_f^2 = 0.29$ and $I_{fz} = 12$ cm, and hence, from Eq. (7), $\lambda_{11}(t \to \infty) = 3.48$ cm.

Based on the analysis of the first two spatial moments of the simulated solute plume, Russo (1991) concluded that the simplified, stochastic vadose-zone transport models (e.g., Dagan and Bresler 1979; Jury 1982), which predict that the variance of the travel distance is proportional to the square of the time of observation (Fig. 4.3a), may be applicable to relatively shallow depths (≈ 1.8 m or about 15 correlation scales of the input formation property in the longitudinal vertical direction). However, these simplified transport models may fail to describe the spread of a solute plume when the transport takes place to relatively great depths, mainly because of the neglect of vertical heterogeneity in the soil hydraulic properties, the neglect of the higher dimensionality of the flow system, and the possible effects of transverse fluxes and/or lateral solute mixing. These conclusions are also supported in part by the field data of Ellsworth (1989) (Fig. 4.6).

The surprising agreement between the effective dispersion in the simulated transport (Russo 1991) in unsaturated flow and previous theoretical developments in saturated flow (e.g., Dagan 1984) (see Fig. 4.5) was explored by Russo and Dagan (1991) in a general Lagrangian formulation by using small-perturbation, first-order approximations. The results of the analysis of the velocity field in the unsaturated flow domain of the hypothetical scale-heterogeneous soil (Russo 1991) suggested good agreement between the results of the Lagrangian kinematic analysis and of the analysis of the spatial moments of the concentration distribution, in agreement with the stochastic theory of transport in saturated flow. This was explained by the finding that the product of the conductivity variance and its correlation scale (and, concurrently, the product of the velocity variance and its correlation scale) was of similar magnitude for both saturated and unsaturated flow regimes. Results of the analysis by Russo and Dagan (1991) suggested that despite the complexity of unsaturated flow, as compared with saturated flow, it might be amenable to a similar analysis. They emphasized, however, that the question of whether this conclusion is a result of general validity or rather a particular one pertaining to flow conditions in the simulation by Russo (1991), should be investigated further.

4.3.2.3 General Stochastic Approach

A general stochastic transport model, proposed recently (Russo 1992), might lead to a simple formulation of transport in unsaturated, heterogeneous soils. The study by Russo (1992) was motivated by the recognition that the Lagrangian kinematic analysis, which renders the dependence of the particle displacement covariance tensor on the velocity covariance function and travel time [Eq. (3)], is of a general nature, and applies to both saturated and unsaturated flows (Russo and Dagan 1991). Hence, the framework and results developed in the past for groundwater flow (see e.g., Dagan 1989) can be adopted to analyze vadose-zone transport. This task was accomplished by combining a general Lagrangian formulation, relating the spatial moments of the solute body to the velocity field [Eq. (3)], with the stochastic theory of Yeh et al. (1985a, b) for steady, unsaturated flow, which relates the statistical moments of the velocity field to those of properties of the heterogeneous formation.

The stochastic model by Russo (1992) considers transport of a conservative, nonreactive solute through a three-dimensional, heterogeneous, porous formation with an anisotropic structure, under unidirectional vertical, steady, unsaturated flow. In addition to the assumptions implicit in the Lagrangian kinematic analysis, outlined in Section 4.3.1, the following assumptions were adopted:

1. At the core scale, the relations between conductivity, K, and the capillary pressure head, ψ, at a given point of the spatial coordinate system, x, are isotropic and can be described by the expression of Gardner (1958):

   $K(\psi, \underline{x}) = K_s(\underline{x}) \exp[-\alpha(\underline{x})\psi]$.

2. The intrinsic formation properties, K_s and α, as well as the dependent flow variable, ψ (considered as a positive quantity), are second-order stationary RSFs and are ergodic over the region of interest.
3. The covariance functions of both $\log K_s$ and α are described by the three-dimensional, statistically anisotropic exponential function [Eq. (2) with $W = \log K_s$ or $W = \alpha$, and $C_n = 0$].
4. At the core scale, Darcy's law for unsaturated flow is valid.
5. For a given mean pressure head, H, the volumetric water content effective for water flow, θ_e, is constant and uniform in space.

By employing the aforementioned assumptions, expressing the velocity in terms of mean $\underline{U}$ and perturbation $\underline{u}$, utilizing Taylor's expansion, and by retaining first-order terms, expressions were obtained by Russo (1992) for the components of the velocity covariance in the direction of the mean gradient, i.e., $\langle u_1(\underline{x}) u_1(\underline{x}') \rangle$:

$$C_{u_1 u_1}(\underline{x}, \underline{x}') = \left(\frac{K_m}{\theta_e} \right)^2 (J_1^2 C_{yy}(\underline{x}, \underline{x}') + 2J_1 C_{yj_1}(\underline{x}, \underline{x}') + C_{j_1 j_1}(\underline{x}, \underline{x}')) \tag{8a}$$

and in the direction perpendicular to the mean gradient, i.e., $\langle u_2(\underline{x}) u_2(\underline{x}') \rangle$:

$$C_{u_2 u_2}(\underline{x}, \underline{x}') = \left(\frac{K_m}{\theta_e} \right)^2 (C_{j_2 j_2}(\underline{x}, \underline{x}')). \tag{8b}$$

Here, $K_m = \exp(Y)$ is the geometric mean of unsaturated conductivity K; $\theta_e = \theta_e(\psi)$ is the volumetric water content effective for water flow; $J_1 = -\partial(H + x_1)/\partial x_1$ is the mean hydraulic gradient in the vertical direction,

x_1; and $C_{yy}(\underline{x}, \underline{x}') = C_{yy}(\underline{s})$ is the covariance function of log-unsaturated conductivity perturbation, given by:

$$C_{yy}(\underline{s}) = C_f(\underline{s}) + H^2 C_{aa}(\underline{s}) + A^2 C_{hh}(\underline{s}) - 2AC_{fh}(\underline{s}) + 2AHC_{ah}(\underline{s}), \tag{9}$$

where A is the mean value of α; $C_{ff}(\underline{s})$ and $C_{aa}(\underline{s})$ are covariance functions of the perturbation of the formation properties $\log K_s$ and α, respectively; $C_{hh}(\underline{s})$ is the covariance function of the capillary pressure head perturbation: $C_{fh}(\underline{s})$ and $C_{ah}(\underline{s})$ are cross-covariance functions between perturbations of $\log K_s$ and ψ, and α and ψ, respectively; and $Cyj_1(\underline{x}, \underline{x}') = Cyj_1(\underline{s})$ is the cross-covariance function between perturbations of log-unsaturated conductivity and the capillary pressure head gradient in the x_1 direction ($j_1 = \partial h/\partial x_1$), given as:

$$Cyj_1(\underline{s}) = Cfj_1(\underline{s}) - HCaj_1(\underline{s}). \tag{10}$$

Here, $Cfj_1(\underline{s})$ and $Caj_1(\underline{s})$ are cross-covariance functions between perturbations of $\log K_s$ and j_1, and α and j_1, respectively; $Cj_1j_1(\underline{x}, \underline{x}') = Cj_1j_1(\underline{s})$ and $Cj_2j_2(\underline{x}, \underline{x}') = Cj_2j_2(\underline{s})$ are covariance functions of the capillary pressure gradient perturbations in the direction of the mean gradient, and perpendicular to the mean gradient, respectively.

Details of the evaluation of the various covariance and cross-covariance functions required to calculate the covariance functions of the components of velocity in the logitudinal vertical, u_1 and in the transverse lateral, u_2 directions from Eq. (8) are given elsewhere (Russo 1992). By using the covariance functions of u_1 and u_2, and Eq. (3b) Russo (1992) estimated the respective components of the particle displacement covariance tensor [$X_{11}(t)$ and $X_{22}(t)$, respectively] and investigated the effects of statistical properties of the relevant formation properties, log K_s and α (i.e., $\sigma_f^2, \sigma_a^2, I_{fz}, I_{az}, \rho_f = I_{fx}/I_{fz}$, and $\rho_a = I_{ax}/I_{az}$), as well as of the flow-dependent variable (i.e., mean capillary pressure head or mean water saturation) on both longitudinal and transverse components of the solute body spread.

The dependence of the dimensionless components of the displacement covariance tensor, $X'_{11} = X_{11}/I_{fz}\sigma_f^2$ and $X'_{22} = X_{22}/I_{fz}\sigma_f^2$, and of the ratios $X'_{11}/2t'$ and $X'_{22}/2t'$ upon dimensionless travel time, $t' = tU_1/I_{fz}$, for a few values of the effective water saturation S_e is illustrated in Figs. 4.7 and 4.8, respectively. For purposes of comparison, the closed-form analytical expressions for $X'_{11}(t')$ and $X'_{22}(t')$ and for ratios $X'_{11}/2t'$ and $X'_{22}/2t'$, derived by Dagan (1984) for steady, three-dimensional flow in isotropic, saturated porous formation, are also depicted in these figures. It should be emphasized that from a physical point of view, the ratios $X'_{11}/2t'$ and $X'_{22}/2t'$ may be regarded as apparent dispersion coefficients that would lead, in the solution of the CD equation, to the same X_{11} and X_{22} as the actual time-dependent ones (Dagan 1988). For $t \to \infty$, these ratios tend to the constant macrodispersion coefficients (Aris 1956) defined as $D_{ii} = (1/2)(dX_{ii}/dt)$ ($i = 1, 2$, no sum on i).

As shown in Fig. 4.7, the dimensionless longitudinal $X'_{11}(t')$ and transverse $X'_{22}(t')$ components of the displacement covariance tensor calculated by Russo (1992) for the unsaturated, heterogeneous porous formation are in

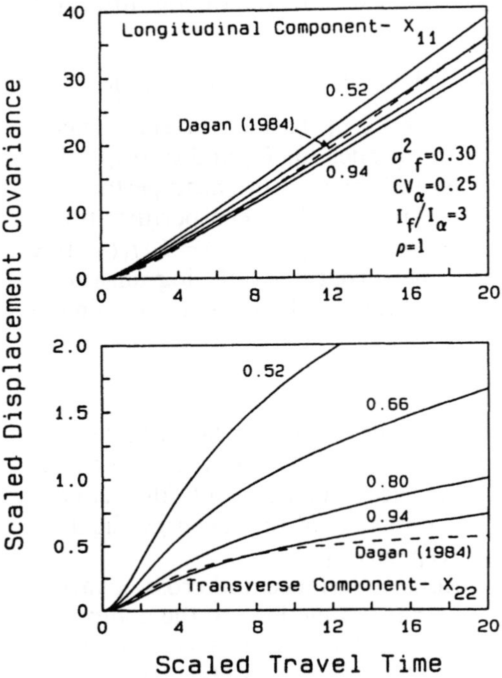

Fig. 4.7. Scaled (*top*) longitudinal ($X'_{11} = X_{11}/I^2_{f1}\sigma^2_f$) and (*bottom*) transverse ($X'_{22} = X_{22}/I^2_{f1}\sigma^2_f$) components of the displacement covariance tensor as a function of scaled travel time, $t' = tU_1/I_{f1}$, for different water saturations, S_e, denoted by the numbers labeling the curves (Russo 1992). *Dashed lines* represent closed-form analytical expressions of $X'_{11}(t')$ and $X'_{22}(t')$ derived by Dagan (1984) for steady saturated flow in a three-dimensional, isotropic structure. U_1 is the mean velocity (vertical), σ^2_f is log-saturated conductivity variance, CV_α is the coefficient of variation of α, I_{f1} and $I_{\alpha 1}$ are the correlation scales in the vertical direction of $\log K_s$ and α, respectively, and $\rho = I_f/I_{f1} = I_\alpha/I_{\alpha 1}$ is the aspect ratio

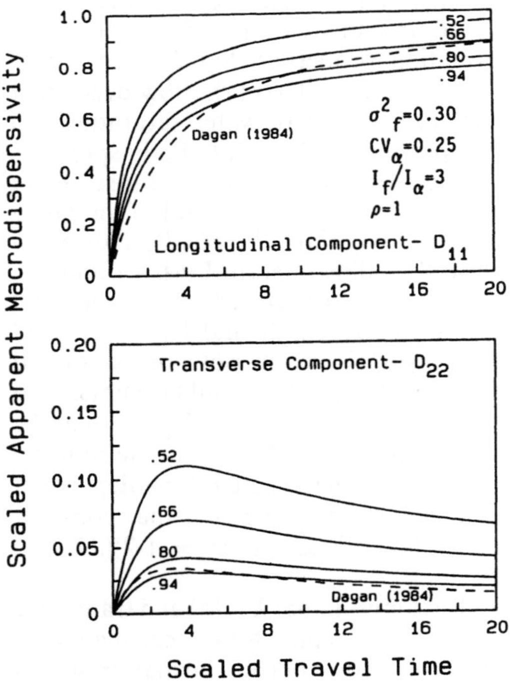

Fig. 4.8. Scaled (*top*) longitudinal ($D'_{11} = X_{11}/2tU_1I_{f1}\sigma^2_f$) and (*bottom*) transverse ($D'_{22} = X_{22}/2tU_1I_{f1}\sigma^2_f$) components of the apparent macrodispersivity coefficient as a function of scaled travel time, t', for different water saturations, S_e (denoted by the *numbers* labeling the curves). *Dashed lines* represent closed-form analytical expressions of $D'_{11}(t')$ and $D'_{22}(t')$ derived by Dagan (1984) for steady saturated flow in three-dimensional, isotropic structure. Values and symbols are as in Fig. 4.7

agreement with the expressions derived by Dagan (1984) for saturated, heterogeneous porous formations. However, water saturation affects both the magnitude and time behavior of $X'_{11}(t')$ and $X'_{22}(t')$. For a given dimensionless travel time, t', and as water saturation decreases, both X'_{11} and X'_{22} increase; the extent depends on the magnitude of the statistical parameters of the formation properties, $\log K_s$ and α. In addition, as water saturation decreases, both $X'_{22}(t')$ and $d(X'_{11})/dt'$ tend asymptotically to constant values at a slower rate. Generally, for a given water saturation, both $X'_{11}(t')$ and $X'_{22}(t')$ increase as the variability in both formation properties, $\log K_s$ and α, increases, and as the ratio between the correlation scales of $\log K_s$ and α decreases. On the other hand, $X'_{11}(t')$ is essentially robust to the statistical anisotropy of the porous formation (expressed in terms of the aspect ratio, $\rho = I_f/I_{f1}$), while $X'_{22}(t')$ increases and tends asymptotically to constant values at a slower rate, as ρ increases, in agreement with the findings of Dagan (1988) for saturated flow.

It was shown by Russo (1992) that in cases where the correlation scale of $\log K_s(I_f)$ is large compared with the correlation scale of $\alpha(I_a)$, the effect of water saturation on $X_{11}(t')$ is considerably smaller than its effect on $X'_{22}(t')$ (Fig. 4.7). The relative robustness of the longitudinal component $X'_{11}(t')$ [and, concurrently, of the asymptotic constant longitudinal macrodispersion coefficient, $D_{11} = d(X'_{11}/dt')_{t\to\infty}$] to water saturation was explained by the fact that in cases where $I_f \gg I_a$, an increase in the variance of $\log K$ (and, concurrently, in the velocity variance) as water saturation decreases in compensated in part by a decrease in the correlation scale of $\log K$ (and, concurrently, in the correlation scale of the longitudinal component of the velocity). In other words, in cases where $I_f \gg I_a$, the product of the variance and the correlation scale of log-conductivity, for both saturated and unsaturated conditions, are of similar magnitude, i.e., $\sigma_y^2 I_{y1} \approx \sigma_f^2 I_{f1}$, so that the constant longitudinal macodispersivity coefficient $D_{11}/U_1 = \sigma_f^2 I_{f1}$ is of similar magnitude for both saturated and unsaturated conditions (Fig. 4.8). In cases where $I_f \leqslant I_a$, however, the product of $\sigma_y^2 I_y$ and, concurrently, the value of D_{11}/U_1 increases considerably as water saturation decreases.

Experimental data on the spatial variability in α are sparse. However, results of a recent field study at Bet Dagan (Russo and Bouton 1992), based on 417 undistributed soil cores taken from a vertical cross section (20 m long and 2.5 m deep; see Fig. 4.1) suggested that the estimated correlation scales of $\log K_s$ in both vertical and horizontal directions were approximately three times larger than the respective correlation scales of α. In view of previous analyses of vadose-zone transport (Russo 1991; Russo and Dagan 1991) and the results of the stochastic analysis by Russo (1992), this experimental information is encouraging in the sense that it further supports the conclusion that the spread of solutes in real, unsaturated, heterogeneous porous formations might be amenable to a relatively simple analysis.

4.4 Summary and Conclusions

This contribution discusses the effect of spatial variability in soil hydraulic properties on transport of solutes in the unsaturated zone through the mechanism of convection by water flow, focusing on some recent studies of, and developments in, the theory and models of vadose-zone transport at the field scale. The starting point is the considerable variability in the point values of the solute concentration in space and time, attributed to the complex heterogeneity of the soil hydraulic properties in both vertical and horizontal directions. This was illustrated by field-scale transport experiments (i.e., Schulin et al. 1987; Butters et al. 1989; Ellsworth 1989) and simulation (Russo 1991). Therefore, for practical purposes (including the assessment of transport model formulation), larger scale, integrated measures of the solute transport, such as the first two spatial moments of the point concentration distribution, are most appropriate in that these moments are subject to a much lesser degree of uncertainty than are the point values. The main conclusions of this chapter can be summarized as:

1. Existing, simplified stochastic vadose-zone transport models that treat the soil as though it were composed of a series of isolated vertical stream tubes with different velocities [represented by the models of Dagan and Bresler (1979) and Jury (1982)] may describe the solute spread at shallow depths near the solute entry zone ("near-field"), but may fail to describe the spread of a solute plume at deeper depths ("far-field"), mainly because of the neglect of vertical heterogeneity in the soil hydraulic properties.
2. The stochastic model proposed by Russo (1992) for the transport of the ensemble-average concentration of a tracer solute in a heterogeneous, partially saturated porous formation, which takes into account the multidimensionality of variations in the hydraulic properties, can describe "far-field" solute spread. This, in turn, approaches an asymptotic Fickian behavior after the solute body has travelled a sufficient distance (i.e., a few tens of conductivity correlation scales). Results obtained by using the stochastic transport model of Russo (1992) are consistent with results obtained by the theory of transport by groundwater flow (e.g., Dagan 1984).
3. Results of a recent transport simulation (Russo 1991) and its accompanying analysis (Russo and Dagan 1991), as well as the results of the stochastic model of Russo (1992), suggest that despite the complexity of unsaturated flow, as compared with saturated flow, under certain circumstances (i.e., gravitational flow) vadose-zone transport might be amenable to a similar analysis as saturated groundwater transport. However, because of the complexity of unsaturated flow, as compared with saturated flow, in unsaturated flow the necessary input data for the analysis should include statistical parameters of both formation properties – $\log K_s$ and α, as well as mean values of the response variables – water saturation or capillary pressure head. Based on the analysis by Russo (1992), it may be concluded that the findings of Russo

and Dagan (1991), regarding the relative robustness of the product of the conductivity variance and its correlation scale to water saturation, may be considered a valid, general result as long as the correlation scale of α is relatively small compared with the correlation scale of $\log K_s$.

4. There is an immediate need for additional field studies to characterize the spatial variability in soil properties relevant to vadose-zone transport, and for field-scale, controlled, transport experiments to further validate existing stochastic transport models. Inasmuch as stochastic models of solute transport predict the ensemble-average spread of a solute body, while field experiments provide information on the spread of a single solute body for particular, site-specific applications (i.e., a single realization from an ensemble of many plausible realizations), care should be taken in the design of transport experiments. For purposes of model validation, the design problem can be formulated with the objective of reducing uncertainties about the ensemble-average solute flux and accumulated mass at a given control plane. Since these uncertainties decrease with the ratio of the size of the solute inlet zone to the size of the correlation scales of the relevant soil properties normal to the mean flow direction, the required size of the solute inlet zone in the transport experiment can be evaluated from the correlation scale of the log-unsaturated conductivity using the Lagrangian framework developed by Dagan et al. (1992).

Finally, the results of the stochastic models described in this chapter together with additional field data of soil properties relevant to vadose-zone solute transport and data from field-scale transport experiments, may improve our understanding of field-scale solute spread in partially saturated soils, and may serve for the subsequent development of field-scale solute transport models for complex, real life scenarios.

Acknowledgments. This is contribution no. 3473-E, 1992 series, from the Agricultural Research Organization, The Volcani Center, Bet Dagan, Israel. Research was supported in part by a grant from the United States–Israel Binational Agricultural Research and Development Fund (BARD). The author is grateful to Mr. Asher Laufer for technical assistance during this study.

References

Amoozegar-Fard A, Nielsen DR, Warrick AW (1982) Soil solute concentration distribution for spatially varying pore water velocities and apparent diffusion coefficients. Soil Sci Soc Am J 46: 3–8

Aris R (1956) On the dispersion of a solute in a fluid through a tube. Proc R Soc A 235: 67–78

Bresler E, Dagan G (1981) Convective and pore scale dispersive solute transport in unsaturated heterogeneous fields. Water Resour Res 17: 1683–1689

Bresler E, Dagan G (1983) Unsaturated flow in spatially variable fields. 3. Solute transport models and their application to two fields. Water Resour Res 19: 429–435

Butters GL, Jury WA, Ernst FF (1989) Field scale transport of bromide in an unsaturated soil. 1. Experimental methodology and results. Water Resour Res 25: 1575–1581

Dagan G (1982) Stochastic modeling of groundwater flow by unconditional and conditional probabilities. 2. The solute transport. Water Resour Res 18: 835–848
Dagan G (1984) Solute transport in heterogeneous porous formations. J Fluid Mech 145: 151–177
Dagan G (1987) Theory of solute transport by groundwater. Annu Rev Fluid Mech 19: 183–215
Dagan G (1988) Time-dependent macrodispersion for solute transport in anisotropic heterogeneous aquifers. Water Resour Res 24: 1491–1500
Dagan G (1989) Flow and transport in porous formations. Springer, Berlin Heidelberg New York
Dagan G (1990) Transport in heterogeneous porous formations: spatial moments, ergodicity and effective dispersion. Water Resour Res 26: 1281–1290
Dagan G, Bresler E (1979) Solute transport in unsaturated heterogeneous soil at field scale. 1. Theory. Soil Sci Soc Am J 43: 461–467
Dagan G, Cvetkovic V, Shapiro AM (1992) A solute flux approach to transport in heterogeneous formation. 1. The general framework. Water Resour Res 28: 1369–1376
Destouni G, Cvetkovic V (1989) The effect of heterogeneity on large scale solute transport in the unsaturated zone. Nordic Hydrol 20: 43–52
Destouni G, Cvetkovic V (1991) Field scale mass arrival of sorptive solute into the groundwater. Water Resour Res 27: 1315–1325
Ellsworth TR (1989) Field-scale spatial and temporal characterization of solute plume transport through unsaturated porous media. PhD. Thesis. University of California, Riverside, CA
Freyberg DL (1986) A natural gradient experiment on solute transport in a sand aquifer. 2. Spatial moments and the advection and dispersion of nonreactive tracers. Water Resour Res 22: 2031–2046
Gardner WR (1958) Some steady state solutions of unsaturated moisture flow equations with application the evaporation from a water table. Soil Sci 85: 228–232
Gelhar LW, Axness C (1983) Three-dimensional stochastic analysis of macrodispersion in aquifers. Water Resour Res 19: 161–190
Jazwinsky AH (1978) Stochastic processes and filtering theory. Academic Press, San Diego, CA
Jury WA (1982) Simulation of solute transport using a transfer function model. Water Resour Res 18: 363–368
Jury WA, Sposito G (1985) Field calibration and validation of solute transport models for the unsaturated zone. Soil Sci Soc Am J 49: 1331–1441
Jury WA, Sposito G, White RE (1986) The transfer function model of solute transport through soil. I. Fundamental concepts. Water Resour Res 22: 243–247
Lumley JL, Panofsky A (1964) The structure of atmospheric turbulence. John Wiley, New York
Mantoglou A, Gelhar LW (1987a) Stochastic modeling of large-scale transient unsaturated flow systems. Water Resour Res 23: 37–46
Mantoglou A, Gelhar LW (1987b) Capillary pressure head variance, mean soil water content, and effective specific soil moisture capacity of transient unsaturated flow in stratified soils. Water Resour Res 23: 47–56
Mantoglou A, Gelhar LW (1987c) Effective hydraulic conductivity of transient unsaturated flow in stratified soils. Water Resour Res 23: 57–67
Neuman SP, Winter CL, Newman CN (1987) Stochastic theory of field-scale Fickian dispersion in anisotropic porous media. Water Resour Res 23: 453–466
Philip JR (1967) Sorption and infiltration in heterogeneous media. Aust J Soil Res 5: 1–10
Russo D (1991) Stochastic analysis of vadose-zone solute transport in a vertical cross section of heterogeneous soil during nonsteady water flow. Water Resour Res 27: 267–283
Russo D (1992) Stochastic modeling of macrodispersion for solute transport in a heterogeneous unsaturated porous formation. Water Resour Res (in press)
Russo D, Bouton M (1992) Statistical analysis of spatial variability in unsaturated flow parameters. Water Resour Res 28 (in press)
Russo D, Dagan G (1991) On solute transport in a heterogeneous porous formation under saturated and unsaturated flows. Water Resour Res 27: 285–293
Schulin R, van Genuchten MTh, Fluhler H, Ferlin P (1987) An experimental study of solute transport in a stony field soil. Water Resour Res 23: 1785–1794

Simmons CS (1982) A stochastic-convective transport representation of dispersion in one-dimensional porous media systems. Water Resour Res 18: 1193–1214
Smith L, Schwartz FW (1980) Mass transport. 1. Stochastic analysis of macrodispersion. Water Resour Res 16: 303–313
Sposito G, Jury WA, Gupta VK (1986) Fundamental problems in the stochastic convection-dispersion model of solute transport in aquifers and field soils. Water Resour Res 22: 77–88
Sudicky EA (1986) A natural gradient experiment on solute transport in a sand aquifer: spatial variability of hydraulic conductivity and its role in the dispersion process. Water Resour Res 22: 2069–2082
Taylor GI (1921) Diffusion by continuous movements. Proc Lond Math Soc A20: 196–211
Yeh T-C, Gelhar LW, Gutjahr AL (1985a) Stochastic analysis of unsaturated flow in heterogeneous soils. 1. Statistically isotropic media. Water Resour Res 21: 477–456
Yeh T-C, Gelhar LW, Gutjahr AL (1985b) Stochastic analysis of unsaturated flow in heterogeneous soils. 2. Statistically anisotropic media with variable α. Water Resour Res 21: 457–464

5 Solute Lifetime Correlations in Chemical Transport Through Field Soils

G. Sposito[1]

5.1 Introduction

One of the many soil physics problems that Eshel Bresler researched so well was the prediction of dissolved solute movement in unsaturated soil profiles. An especially interesting and fruitful approach was pioneered by Eshel and Gedeon Dagan a dozen years ago when they developed a physically based stochastic model of the vertical transport of nonreactive solutes in field soils that exhibit lateral spatial variability in their hydraulic properties. This seminal work, which has been reviewed by Bresler et al. (1982), Addiscott and Wagenet (1985), and Sposito et al. (1986), is the topic of Chapter 1 by Gedeon Dagan.

The principal interest for the present chapter is in the contrast of mechanistic detail between the Bresler-Dagan model (convection-dispersion equation, Buckingham-Darcy flux law, Richards equation, etc.) and the purely stochastic approach to predictive theory taken by WA Jury a decade ago, in which the description of tracer solute transport was encapsulated in a nonlinear transfer function (Jury 1982). The Jury transfer-function model combined a simple physical picture of convective tracer movement with an input-output analysis cast in probabilistic terms. Thus, no detailed concepts of solute transport mechanisms entered. Calibration and predictive use of the model were described by Jury et al. (1982) and later in more generality by Jury and Sposito (1985). Jury et al. (1986), White et al. (1986), and Sposito and Jury (1988) extended the model to remove its dependence on any particular mechanism of solute transport; to include dissolved solutes that can undergo both chemical and biological transformations; and to show how the probability density function (pdf) for the solute lifetime (its residence time in a monitored soil body) can be constructed directly from experimental data without appeal to parametric modeling (e.g., a two-parameter log-normal probability distribution).

The predictive application of the Jury transfer-function model to field soil profiles has been discussed in a series of papers by White (1985, 1987, 1989) and, more on the theoretical side, in a recent monograph by Jury and Roth (1990). They invoke the scaling hypothesis, proposed originally by

[1] Department of Soil Science, University of California, 108 Hilgard Hall, Berkeley, CA 94720, USA

Jury (1982), that "the probability that a solute molecule added at $t=0$ at the inlet end $z=0$ will arrive at a depth z in a time t_0 less than or equal to t is the same as the probability that it will arrive at a depth ℓ in a time t_0 less than or equal to $t\ell/z$". Jury and Roth (1990) show that this basic hypothesis, or the equivalent assumption of perfect statistical correlation between the solute lifetimes in two vertically contiguous soil zones, is sufficient to relate solute lifetime means and variances uniquely in the two zones. Thus, model calibration in one zone results in unique statistical predictions about the solute lifetime in the other zone.

The objective of the present chapter is to establish, in terms of solute lifetime correlations, a general theoretical framework for the predictive use of the Jury transfer-function model. The reader is assumed to be acquainted with the fundamental idea of the solute lifetime pdf and its important special case, the travel-time pdf, as described by Jury et al. (1986) and Sposito and Jury (1988). Statistical properties of solute travel times in layered soils have been discussed by Jury and Roth (1990), who provided special cases of some of the definitions to be given in Section 5.2.

5.2 Fundamental Statistical Concepts

A field-scale body of soil through which dissolved solute moves can be imagined as the union of N vertically stacked spatial domains of arbitrary size, shape, and orientation, but which do not overlap. Each domain is termed a "layer", although it need not correspond to any stratigraphic feature of the soil body. A set of $N+1$ nonnegative random variables $\{T_k, k=0,1,\ldots,N\}$ is associated with the N layers, such that T_0 is the time at which a solute molecule first appears in the transporting portion of the soil body ("solute input time") and $T_k (k>0)$ is the time interval during which the solute molecule exists in layer k ("solute lifetime"). It may be assumed that T_0 is statistically independent of the other T_k, but no additional probabilistic structure is imposed. A fixed value t' of T_0 is selected, so that all probabilisitic quantities relating to the $T_k (k>0)$ will be conditional (cf. Sposito and Jury 1988).

Complete statistical information about the set $\{T_k, k>0\}$ is available once the (normalized) conditional joint pdf, $f(t_1,\ldots,t_N|t')$, is known (Feller 1968). For example, the *solute lifetime* pdf for layer k is defined by the equation (cf. van Kampen 1981):

$$g_k(t|t') \equiv \int_0^\infty \cdots \int_0^\infty \delta(t-t_k) f(t_1,\ldots,t_N|t') dt_1 \cdots dt_N, \tag{1}$$

where $\delta(\cdot)$ is the Dirac delta-"function". The pdf $g_k(t|t')$ is a generalization, to the case of N layers, of the solute lifetime pdf described by Jury et al. (1986). Three important (conditional) moments of this pdf are ($k>0$):

$$\int_0^\infty g_k(t|t')dt = 1 \tag{2a}$$

$$\int_0^\infty t g_k(t|t')dt = E[T_k|T_0 = t'] \equiv M_k(t') \tag{2b}$$

$$\int_0^\infty [t - M_k(t')]^2 g_k(t|t')dt = E[(T_k - M_k)^2|T_0 = t'] \equiv \sigma_k^2(t'), \tag{2c}$$

where $E[\cdot|T_0 = t']$ is a conditional expectation value (cf. Feller 1968).

Correlation structure among the $T_k (k > 0)$ can be epitomized in the *correlation coefficient* (Feller 1968):

$$\rho_{jk}(t') = E[Y_j Y_k|T_0 = t'], \tag{3}$$

where

$$Y_k = (T_k - M_k)/\sigma_k \quad (k = 1, \ldots, N). \tag{4}$$

If $\rho_{jk} = 0$, the random variables T_j and T_k are not correlated, although they still may be statistically dependent on one another (Feller 1968). If $|\rho_{jk}| = 1$, there is perfect (negative or positive) correlation between T_j and T_k. In general, $-1 \leqslant \rho_{jk} \leqslant 1$ for any t', and for $j \neq k$, ρ_{jk} will depend on the indices j, k as determined by the partitioning into N layers and the mechanisms of solute transport and transformation in the soil body.

The fate of a solute molecule as it passes through layers 1 to m in a soil body is described by the *cumulative solute lifetime* pdf, defined by the equation:

$$h_m(t|t') \equiv \int_0^\infty \cdots \int_0^\infty \delta(t - s_m) f(t_1, \ldots, t_n|t')dt_1 \ldots dt_N, \tag{5}$$

where s_m is a realization of the nonnegative random variable

$$S_m = \sum_{k=1}^{m} T_k, \tag{6}$$

which is the sum of solute lifetimes in the sequence of m layers. Evidently $h_1(t|t') = g_1(t|t')$, but otherwise $h_m(t|t')dt$ is the (conditional) probability that a solute molecule exists in the first m layers for a time interval between t and $t + dt$, by analogy with the single-layer interpretation of $g_k(t|t')dt$ (Jury et al. 1986). The formal relationship of the pdf $h_m(t|t')$ to the observed fraction of input solute mass retained in the portion of a soil body comprised in the first m layers can also be established by analogy with the derivation for $g_k(t|t')$ outlined in the Appendix of Jury et al. (1986).

The moments of $h_m(t|t')$ reveal its close connection to the $g_k(t|t')$. For example,

$$\int_0^\infty th_m(t|t')dt = \int_0^\infty \cdots \int_0^\infty s_m f(t_1,\ldots,t_N|t')dt_1\cdots dt_N$$

$$= \sum_{k=1}^m \int_0^\infty \cdots \int_0^\infty t_k f(t_1,\ldots,t_N|t')dt_1\cdots dt_N$$

$$= \sum_{k=1}^m \int_0^\infty tg_k(t|t')dt = \sum_{k=1}^m M_k(t'), \tag{7}$$

which shows that $E[S_m|T_0 = t']$ is simply the sum of m expectation values, $E[T_k|T_0 = t']$. Similarly for the variance of S_m (cf. Feller 1968):

$$\int_0^\infty (t - E[S_m|T_0 = t'])^2 h_m(t|t')dt \equiv \mathrm{var}[S_m]$$

$$= \int_0^\infty \cdots \int_0^\infty \left(s_m - \sum_{k=1}^m M_k\right)^2 f(t_1,\ldots,t_N)dt_1\cdots dt_N$$

$$= \int_0^\infty \cdots \int_0^\infty \left[\sum_{k=1}^m (t_k - M_k)\right]^2 f(t_1,\ldots,t_N)dt_N$$

$$= \sum_{j=1}^m \sum_{k=1}^m \sigma_j\sigma_k E[Y_j Y_k|T_0 = t'] = \sum_{j=1}^m \sum_{k=1}^m \sigma_j\sigma_k\rho_{jk}, \tag{8}$$

where Eq. (4) has been used in the penultimate step and the dependence on t' is suppressed in the last step. Thus, the statistics of solute transport through m layers depend on those of the individual layers, as well as on the correlations between the solute lifetimes in different layers. Once the layer moments and the time-correlations are known, the average behaviour of solute in the soil body can be predicted. In this context, a central theoretical problem is to deduce mathematical relationships of a general nature between the $h_m(t|t')$ and the $g_k(t|t')$, as well as among the $g_k(t|t')$, on the basis of assumptions (or experimental data) about the layer moments and the time-correlations.

5.3 Perfect Correlations Among Solute Lifetimes

If the solute lifetimes in the individual layers are correlated perfectly, $|\rho_{jk}| = 1$ for all j, k. Consider the random variable (again suppressing t'):

$$X_{jk} \equiv \rho_{jk} Y_j - Y_k, \tag{9}$$

where the factors on the right side are defined in Eqs. (3) and (4). From these latter two equations it follows that (cf. Feller 1968):

$$E[X_{jk}^2 | T_0 = t'] = 1 - \rho_{jk}^2, \tag{10}$$

and, therefore, that the mean value of X_{jk}^2 vanishes if the correlation between T_j and T_k is perfect. Since the mean value of X_{jk} is identically zero, perfect correlation implies that the variance of X_{jk} is also zero. Thus, only one value can be taken by X_{jk}, i.e.,

$$Y_k = \rho_{jk} Y_j = \pm Y_j, \tag{11}$$

with the minus sign corresponding to perfect negative correlation. With respect to solute lifetimes, Eq. (11) implies, for perfect positive correlation:

$$T_j = (\sigma_j/\sigma_k) T_k + [M_j - (\sigma_j/\sigma_k) M_k] \tag{12}$$

and

$$S_m = \sum_{k=1}^{m} (\sigma_k/\sigma_1) T_1 + \sum_{k=1}^{m} [M_k - (\sigma_k/\sigma_1) M_1], \tag{13}$$

by repeated application of Eq. (12), setting $k = 1$, to Eq. (6).

The effect of perfect (positive) correlation on the solute lifetime pdf can be seen by substituting Eq. (12) with $k = 1$ into Eq. (1):

$$\begin{aligned} g_k(t|t') &= \int_0^\infty \cdots \int_0^\infty \delta\{t - (\sigma_k/\sigma_1)t_1 - [M_k - (\sigma_k/\sigma_1)M_1]\} f(t_1,\dots,t_N|t')dt_1 \cdots dt_N \\ &= (\sigma_1/\sigma_k) \int_0^\infty \cdots \int_0^\infty \delta\{(\sigma_k/\sigma_1)t - t_1 + [M_1 - (\sigma_1/\sigma_k)M_k]\} \\ &\qquad \times f(t_1,\dots,t_N|t')dt_1 \cdots dt_N \\ &= (\sigma_1/\sigma_k) g_1[(\sigma_k/\sigma_1)t + M_1 - (\sigma_1/\sigma_k)M_k|t'], \end{aligned} \tag{14}$$

where the scaling relationship of the delta-"function" (Butkov 1968),

$$\delta(ax) = \delta(x)/|a| \quad (a \neq 0), \tag{15}$$

has been used in the second step. Equation (14) shows that perfect correlation leads to proportionality between any $g_k(t|t')$ and the solute lifetime pdf for layer 1 ("the entry layer") with an appropriately scaled and translated time argument. In the same way, the combination of Eqs. (13) and (5) yields the result:

$$h_m(t|t') = g_1[(t - \beta_m)/\alpha_m|t']/\alpha_m, \tag{16}$$

where

$$\alpha_m = \sum_{k=1}^{m} (\sigma_k/\sigma_1) \quad \beta_m = \sum_{k=1}^{m} (M_k/\alpha_m) - M_1. \tag{17}$$

Thus, for perfectly correlated solute lifetimes, the individual layer moments and the entry-layer lifetime pdf provide full statistical information about stochastic solute behavior.

The special case in which the N layers have been chosen so as to have *identical first and second moments* leads to a simplification of Eqs. (14) and (16):

$$g_k(t|t') = g_1(t|t') \quad (k = 1, \ldots, N) \tag{18a}$$

$$h_m(t|t') = g_1(t/m|t')/m \quad (m = 1, \ldots, N). \tag{19a}$$

In this special case, perfect correlation implies identical solute lifetime pdfs for all layers, and the important result is found that the cumulative lifetime pdf is merely a scaled version of the entry-layer lifetime pdf. The much weaker assumption of *identical coefficients of variation* in all layers leads to the result:

$$g_k(t|t') = g_1[(\sigma_k/\sigma_1)t|t']/(\sigma_k/\sigma_1) \tag{18b}$$

$$h_m(t|t') = g_1(t/\alpha_m|t')/\alpha_m, \tag{19b}$$

which preserves the time-scaling relationship between $g_1(t|t')$ and all other lifetime pdfs. It follows from Eq. (19) [or Eq. (8)] that the variance of S_m is proportional to the variance of T_1. Thus, the first two moments for $h_m(t|t')$ are scaled values of those for $g_1(t|t')$.

Jury (1982) proposed an expression equivalent to Eq. (19a) in his original formulation of the transfer-function model, based on a simple physical picture of the vertical movement of a tracer through a field soil. His expression has met with success in applications to field-scale experimens on tracer transport (Jury et al. 1982; Jury and Sposito 1985; Butters and Jury 1989). The derivation of Eq. (19a) given here, however, makes clear the point brought out by Jury and Roth (1990), that the time-scaling relationship between the cumulative lifetime pdf and the entry-layer lifetime pdf requires only assumptions of a statistical nature and thus implies no particular mechanisms of solute transport or transformation.

The invariance property of Eq. (19) and the Jury hypothesis, quoted in Section 5.1, can be appreciated by considering the probability that the cumulative lifetime lies in the range $\alpha_m t_1$ to $\alpha_m t_2$, where α_m is given by Eq. (17):

$$\int_{\alpha_m t_1}^{\alpha_m t_2} h_m(t|t')dt = (1/\alpha_m) \int_{\alpha_m t_1}^{\alpha_m t_2} g_1(t/\alpha_m|t')dt$$

$$= \int_{t_1}^{t_2} g_1(\tau|t')d\tau, \qquad (20)$$

where the last step involves the change of variable, $\tau = t/\alpha_m$. Equation (20) states that the probability that a solute lifetime lies between $\alpha_m t_1$ and $\alpha_m t_2$ in an m-layer system is the same as the probability that a solute lifetime lies between t_1 and t_2 in the entry layer of the system. Thus, solute lifetimes in the m-layer system are merely scaled by α_m relative to those in the entry layer, and Eq. (20) reflects this property as a kind of *time-scaling invariance*. The two conditions sufficient to produce this symmetry are perfect correlation between the solute lifetimes in the layers and identical coefficients of variation among the lifetimes. The stronger second condition of identical first and second moments among the lifetimes simplifies the scaling factor to m [see Eq. (19a)]. In this case, solute lifetimes in the m-layer system are magnified m times relative to those in the entry layer, and their statistical features are invariant under this magnification: physical and temporal scaling are congruent operations.

The results in Eqs. (14), (16), (18), and (20) have been obtained without reference to the explicit mathematical form of $g_1(t|t')$. This pdf, therefore, can be selected by criteria that are independent of the correlation structure for the solute lifetimes. For example, $g_1(t|t')$ can be measured directly in narrow-pulse solute input experiments, or it can be modeled parametrically by fitting analytical functions to experimental data (White et al. 1984, 1986; White 1985, 1987, 1989; Jury and Sposito 1985). One candidate parametric model is the inverse Gaussian pdf (Tweedie 1956, 1957a,b; Johnson and Kotz 1970):

$$p(t; \mu, \tau) = (\mu^2/2\pi\tau t^3)^{1/2} \exp[-(t-\mu)^2/4\tau t], \qquad (21)$$

where μ and τ are positive constants and $t > 0$. The mean and variance of the variate T described by $p(t; \mu, \tau)$ are (Johnson and Kotz 1970):

$$E[T] = \mu \quad E[(T-\mu)^2] = 2\mu\tau. \qquad (22)$$

A physical interpretation of the inverse Gaussian pdf is obtained by making the associations:

$$\mu \equiv L/v \quad \tau \equiv D/v^2, \qquad (23)$$

which show $p(t; \mu, \tau)$ to be a solute *travel-time* pdf (for solute input at $t' = 0$ and the travel distance L) based on a one-dimensional convention-dispersion equation for a tracer solute with uniform, constant convective velocity v and dispersion coefficient D (Jury and Sposito 1985). Equations (22) and (23) indicate that the mean solute travel time is L/v and that the variance of the travel time is proportional to D. The dimensionless ratio μ/τ is equal to the Péclet number P:

$$\mu/\tau = vL/D \equiv P, \tag{24}$$

and the coefficient of variation is $(2P)^{1/2}$. The parameter τ in Eq. (23) defines the intrinsic time scale in the convection-dispersion equation (Sposito and Jury 1988). If the time variable in the inverse Gaussian pdf is translated relative to its mean value μ and scaled relative to τ, then it is straightforward to demonstrate that $p(t; \mu, \tau)$ becomes (Johnson and Kotz 1970):

$$p(z; P) = (P^2/4\pi(z+P)^3)^{1/2} \exp[-z^2/4(z+P)], \tag{25}$$

where $z \equiv (t-\mu)/\tau$ and P is the Péclet number, which can now be seen to determine the shape of the pdf.

If solute travel times in an m-layer system meet the conditions underlying Eqs. (18b) and (19b), and data for the entry layer can be fitted well to Eq. (21), then it follows immediately that $h_m(t|0)$ is an inverse Gaussian pdf, with the time variable scaled by α_m, and that the Péclet number is uniform along the m-layer system described by $h_m(t|0)$. Because of scaling property of the inverse Gaussian pdf (Johnson and Kotz 1970):

$$p(\lambda t; \mu, \tau) = p(t; \lambda\mu, \lambda\tau), \tag{26}$$

where λ is any real number except 0, it is evident from Eq. (22) that the m-layer mean travel time will be α_m times larger than that for the entry layer and that the travel-time variance will be α_m^2 times larger. Perfect correlation and a uniform coefficient of variation act to "stretch" the two natural time scales in the pdf by an equal amount depending on the variances (or the means) in the individual layers as prescribed by Eq. (17). On the other hand, the ratio of these time scales (the Péclet number) is uniform, so Eq. (24) implies a linear growth of the dispersion coefficient with travel distance ("scale effect") if the convective solute velocity is the same in all layers. These predictions of the model can be tested experimentally. But the important point here is that *such tests examine the correctness of assuming perfect correlation and a uniform coefficient of variation, and not the correctness of the inverse Gaussian model for the travel-time pdf*. The issue of which correlation properties of solute lifetimes are accurate for a given transport problem is independent of the issue of which parametric lifetime pdf is most suitable for the problem. The former issue is approached by investigating the scaling properties of solute lifetime statistics, whereas the latter issue is approached by investigating solute movement in the entry layer. Once a lifetime pdf is selected, say, on the basis of goodness-of-fit criteria (Jury and Sposito 1985), its parameters can be used to facilitate the study of lifetime correlations. Alternatively, the scaling relations in Eqs. (16), (18), and (19) can be tested directly with only numerical values of the lifetime pdfs as determined from solute transport data (White et al. 1986).

5.4 Less Than Perfect Correlations Among Solute Lifetimes

If there is absolutely no correlation among solute lifetimes in an m-layer system, $\rho_{jk} \equiv 0$ for $j \neq k$ and Eq. (8) simplifies to the expression (Feller 1968):

$$\mathrm{var}[S_m] = \sum_{k=1}^{m} \sigma_k^2. \tag{27}$$

This reslut is also obtained if the solute lifetimes are statistically independent. *Provided that only the first two moments of the lifetime pdfs are considered*, useful results can be derived on the basis of the much stronger assumption of mutual independence. In that case, by definition, the conditional joint pdf in Eq. (1) factorizes into a product of N solute lifetime pdfs:

$$f(t_1,\ldots,t_n|t') = \prod_{k=1}^{m} g_k(t_k|t'), \tag{28}$$

wherein each $g_k(t|t')$ may have a different mathematical form, with corresponding different statistical moments for each layer. Although Eq. (1) is obviously simplified by mutual independence, Eq. (5) becomes a rather complicated convolution of the $g_k(t|t')$ [see, e.g., Cox (1970)] despite the appearance of Eq. (27). Thus, in this case, there is no necessary relationship between the mathematical form of $h_m(t|t')$ and that of the $g_k(t|t')$, such as was found in the case of perfect correlation. If the solute lifetimes are assumed to be identically distributed statistically, however, then at least for certain parametric models of $g_k(t|t')$, theorems about the sums of independent, identically distributed random variables may be invoked to connect the form of $h_m(t|t')$ with the $g_k(t|t')$ [see Johnson and Kotz (1970)].

As an illustrative example of a relation between $h_m(t|t')$ and the $g_k(t|t')$, consider again the inverse Gaussian model in Eq. (21). Suppose that each $g_k(t|t')$ can be described by this model. Under what conditions will $h_m(t|t')$ also be an inverse Gaussian pdf? [Note the much stronger condition, that *all* $g_k(t|t')$ must be inverse Gaussian, than was needed for perfect correlation which required information only about $g_1(t|t')$.] Tweedie (1957a) has shown that the necessary and sufficient requirement is, for mutually independent T_k, that the parameter τ in Eq. (21) be uniform for all $g_k(t|t')$: $\tau_1 = \tau_2 = \ldots = \tau_N$. In physical terms, this means that the intrinsic convection-dispersion time scale is the same in all the layers (although the mean travel times in the layers may be different). Moreover, from Eq. (23) it follows that, if the solute convective velocity is unifom, the dispersion coefficient must also be the same in all layers. This result is strikingly different from the predicted linear growth of the dispersion coefficient with travel distance for the case of perfect correlation. Thus, if the inverse Gaussian model is adequate to describe solute travel-time pdfs, distinct, testable predictions about the "scale effect" of the solute dispersion coefficient result from different choices about the travel-time correlations. Note once again that *these predictions help to resolve the nature of the correlations, and not the nature of the travel-time pdf.*

If correlations among solute lifetimes are imperfect but not vanishingly small, the full complexity of the N-by-N matrix, comprising the elements ρ_{jk}, must be confronted. Models informed by data on the j, k dependence of these elements for $j \neq k$ must be developed in order to describe a relationship between $h_m(t|t')$ and the $g_k(t|t')$. There is, however, a general statement of a formal nature that can be made about imperfect lifetime correlations: *Every set* $\{T_k, k > 0\}$ *of imperfectly correlated lifetimes is in one-to-one correspondence with a set* $\{U_k, k > 0\}$ *of uncorrelated random variables.* Thus, in a formal sense, the factorization in Eq. (28) is always possible to the extent that only second-moment statistical properties are at issue.

To see this point in detail, it is necessary to consider the Gram-Schmidt procedure for creating uncorrelated random variables [see, e.g., Sposito (1970)]. A new random variable U_k is created by operating on the random variable Y_k in Eq. (4) with a set of "projection operators" P_j:

$$U_k \equiv Y_k - \sum_{j=1}^{k-1} \frac{E[U_j Y_k]}{E[U_j^2]} U_j \equiv \left(I - \sum_{j=1}^{k-1} P_j\right) Y_k, \tag{29}$$

where I is the identity operation. Each P_j in Eq. (29) is defined by the operation:

$$P_j Y_k \equiv \frac{E[U_j Y_k]}{E[U_j^2]} U_j \quad (j = 1, \ldots, k-1), \tag{30}$$

which removes from each Y_k the part that is correlated to a U_j having an index j less than k. The portion of Y_k remaining after all P_j for $j < k$ have operated will not be correlated to any U_j for $j < k$.

As an example, consider a three-layer system:

$$U_1 = Y_1 \tag{31a}$$

$$U_2 = Y_2 - \frac{E[U_1 Y_2]}{E[U_1^2]} U_1 = Y_2 - E[Y_1 Y_2]\, Y_1 = Y_2 - \rho_{12} Y_1 \tag{31b}$$

$$\begin{aligned} U_3 &= Y_3 - \frac{E[U_1 Y_3]}{E[U_1^2]} U_1 - \frac{E[U_2 Y_3]}{E[U_2^2]} U_2 \\ &= Y_3 - \rho_{13} Y_1 - \frac{\rho_{23} - (\rho_{12}\rho_{13})}{(1 - \rho_{12}^2)} (Y_2 - \rho_{12} Y_1) \\ &= Y_3 - \frac{(\rho_{23} - \rho_{12}\rho_{13})}{(1 - \rho_{12}^2)} Y_2 - \frac{(\rho_{13} - \rho_{12}\rho_{23})}{(1 - \rho_{12}^2)} Y_1. \end{aligned} \tag{31c}$$

It is straightforward (but increasingly tedious) to verify that the three U_k are not correlated. For example,

$$E[U_1 U_2] = E[Y_1 Y_2] - \rho_{12} E[Y_1^2] \equiv 0, \tag{32}$$

according to Eq. (31) and the normalization of the Y_k. The mean values of the U_k, like those of the Y_k, equal 0. The variances then follow from Eq. (31) and the absence of correlation among the U_k:

$$E[U_1^2] = E[U_2^2] = 1 - \rho_{12}^2$$

$$E[U_3^2] = 1 - \rho_{13}^2 - \frac{(\rho_{23} - \rho_{12}\rho_{13})^2}{(1 - \rho_{12}^2)}$$

$$E[U_k^2] = 1 - \sum_{j=1}^{k-1} \frac{(E[U_j Y_k])^2}{E[U_j^2]} \qquad k = 1, 2, \ldots, N. \tag{33}$$

Notable in Eq. (31) is the "correlation chain" structure of the coefficients of Y_2 and Y_1 in the expressions related to U_3. This latter random variable receives contributions from Y_1 both through "direct correlations" (ρ_{13}) between layer 3 and layer 1 and through "correlation chains" ($\rho_{12}\rho_{23}$) between these two layers via mediation by layer 2. The same applies to the contributions of Y_2 to U_3, with "correlation chains" ($\rho_{12}\rho_{13}$) by mediation through layer 1. In specific models of lifetime correlations, it is possible that correlation chains can be neglected to simplify calculations if they are deemed weaker than direct correlations. The range of k in ρ_{jk} for fixed $j \neq k$ may also be limited by similar considerations.

Less than perfect correlations among solute lifetimes in an N-layer system suggests that the relationship among the solute lifetime pdfs will be something less than the strong scaling expression in Eq. (14), but certainly more than the complete diversity of mathematical form implied by Eq. (28). The explicit relationship can be developed with reasonable simplicity after introducing the conditional pdf, $p_{1/k-1}(t_k | t_1, \ldots, t_{k-1}; t')$, whose product with dt_k is interpreted as the probability that the solute lifetime in layer k lies in the range t_k to $t_k + dt_k$ when the solute lifetimes in the layers 1 through $k-1$ are equal to $t_1, \ldots, t_{k-1}$ and the solute input time is t' [cf. van Kampen (1981)]. To appreciate the significance of this conditional pdf, consider a two-layer system for which the joint pdf is $f_2(t_1, t_2 | t')$:

$$f_2(t_1, t_2 | t') \equiv p_{1/1}(t_2 | t_1; t') g_1(t_1 | t'), \tag{34}$$

where the conventional definition of a conditional pdf based on Eq. (1) and $f_2(t_1, t_2 | t')$ has been invoked. Given the normalization condition:

$$\int_0^\infty p_{1/k-1}(t_k | t_1, \ldots, t_{k-1}; t') dt_k = 1, \tag{35}$$

it follows from Eqs. (1), (34), and (35) that:

$$g_2(t | t') = \int_0^\infty p_{1/1}(t | t_1; t') g_1(t_1 | t') dt_1, \tag{36}$$

is the relationship between the $g_k(t|t')$ for the two layers. Thus $p_{1/1}(t_2|t_1;t')$ acts as a "transition" pdf for solute lifetimes in the two layers, containing in its mathematical dependence on t_1 and t_2 whatever degree of correlation may exist between the solute lifetimes T_1 and T_2.

Now consider a three-layer system, for which the generalization of Eq. (34) is [cf. van Kampen (1981)]:

$$f_3(t_1,t_2,t_3|t')=p_{1/2}(t_3|t_1,t_2;t')f_2(t_1,t_2|t'). \tag{37}$$

Multiplication of Eq. (37) on both sides by $\delta(t-t_3)$, followed by integration over all the $t_j (j=1,2,3)$, will not produce a concise expression like that in Eq. (34) because of the dependence of $p_{1/2}$ on both t_1 and t_2. On the other hand, if lifetime correlations are assumed to extend only over strictly *adjacent* layers (termed "the Markov property"), then:

$$p_{1/2}(t_3|t_1,t_2;t')\equiv p_{1/1}(t_3|t_2;t'), \tag{38}$$

and Eq. (37) can be developed into the expression:

$$\begin{aligned} g_3(t|t') &= \int_0^\infty p_{1/1}(t|t_2;t')g_2(t_2|t')dt_2 \\ &= \int_0^\infty\int_0^\infty p_{1/1}(t|t_2;t')p_{1/1}(t_2|t_1;t')g_1(t_1|t')dt_1dt_2, \end{aligned} \tag{39}$$

by making use of Eq. (36) in the last step. Equation (39) shows how each $g_k(t|t')$ can be constructed from the entry-layer lifetime pdf and the Markovian transition pdfs. This possibility is epitomized in the model definition for the (Markovian) joint lifetime pdf:

$$f_M(t_1,\ldots,t_N|t')\equiv\prod_{k=2}^{N} p_{1/1}(t_k|t_{k-1};t')g_1(t_1|t') \quad (N>1). \tag{40a}$$

Complete statistical information about $(T_k, k>0)$ is then contained in the entry-layer pdf and the set of $N-1$ Markovian transition pdfs. Knighton and Wagenet (1987) have developed a model of reactive solute transport in an N-layer soil based on the transition probabilities $p_{kj}(\tau,t)\equiv \int_0^t p_{1/1}(t_j|t_k=\tau;0)dt_j$ as central elements. Their approach to simulating the time dependence of the $p_{1/1}$ is one way of realizing Eq. (40a) for specific applications.

5.5 Concluding Remarks

The intent of this chapter has been to show the importance of the joint lifetime pdf to the predictive features of the transfer-function model of solute transport in a field-scale soil. The problem of calculating single-layer lifetime pdfs – the $g_k(t|t')$ – in terms of the joint pdf via Eq. (1) can be reduced to modeling the pdfs on the right side of Eq. (40a), if lifetime correlations exist only between adjacent layers. Even Eq. (14) can be included as a special case if the formal constraint expression ($k > 1$):

$$\int_0^\infty \cdots \int_0^\infty \delta(t - t_k) \prod_{j=2}^{k} p_{1/1}(t_j | t_{j-1}; t') dt_2 \cdots dt_k$$

$$= \delta\{t - (\sigma_k/\sigma_1)t_1 - [M_k - (\sigma_k/\sigma_1)M_1]\}, \qquad (40b)$$

is added to Eq. (40a). On the other hand, if the lifetime correlations vanish uniformly (or are made to do so formally), Eq. (28) can be applied to calculate second moments, but the $g_k(t|t')$ are unrelated mathematically. In any event, it is always necessary to find a model form for the entry-layer pdf, $g_1(t|t')$. This model form [Eq. (21) offers an example] can be developed from solute transport data, as discussed by White et al. (1986) and Jury and Sposito (1985). It is important to remember, however, that *the functional form of* $g_1(t|t')$ *has no necessary relationship to the calculation of the cumulative lifetime pdf* in Eq. (5), whose essential behavior derives from lifetime correlations as represented in Eq. (8) for example. Parametric models of $g_1(t|t')$ serve to facilitate interpretation of the effects of different correlation structures, but their accuracy as physical representations of the solute transport process is not tested by examining how well a model form of $h_m(t|t')$ predicts solute movement in a soil profile.

Acknowledgements. The research on which this chapter is based was largely completed during the tenure of a visit to the University of Oxford. Thereafter, frequent discussions with Professor R.E. White (University of Melbourne) and Professor W.A. Jury (University of California at Riverside) helped to refine the physical interpretation of the equations derived. Preparation of this chapter was supported in part by USDA-CSRS Regional Research Project W-82, "Pesticides and Other Organics in Soil and Their Potential for Groundwater Contamination." Gratitude is also expressed to Mr. Daniel Heap for his excellent typing of the manuscript.

References

Addiscott TM, Wagenet RJ (1985) Concepts of solute leaching in soils: a review of modelling approaches. J Soil Sci 36: 411–424

Bresler E, McNeal BB, Carter DL (1982) Saline and sodic soils. Springer, Berlin Heidelberg New York, pp 147–165

Butkov E (1968) Mathematical physics. Addison-Wesley, Reading MA, chap 6

Butters GL, Jury WA (1989) Field scale transport of bromide in an unsaturated soil. 2. Dispersion modeling. Water Resour Res 25: 1583–1589

Cox DR (1970) Renewal theory. Methuen, London, chap 1

Feller W (1968) An introduction to probability theory and its applications. John Wiley, New York, chap 9

Johnson NL, Kotz S (1970) Continuous univariate distributions. 1. John Wiley, New York, chap 51

Jury WA (1982) Simulation of solute transport using a transfer function model. Water Resour Res 18: 363–368

Jury WA, Roth K (1990) Transfer functions and Solute movement through soil. Birkhäuser, Basel, chaps 2 and 5

Jury WA, Sposito G (1985) Field calibration and validation of solute transport models for the unsaturated zone. Soil Sci Soc Am J 49: 1331–1341

Jury WA, Stolzy LH, Shouse P (1982) A field test of the transfer function model for predicting solute transport. Water Resour Res 18: 369–375

Jury WA, Sposito G, White RE (1986) A transfer function model of solute transport through soil. 1. Fundamental concepts. Water Resour Res 22: 243–247

Knighton RE, Wagenet RJ (1987) Simulation of solute transport using a continuous time Markov process. 1. Theory and steady state application. Water Resour Res 23: 1911–1916

Sposito G (1970) An introduction to quantum physics. John Wiley, New York, chapter 3

Sposito G, Jury WA (1988) The lifetime probability density function for solute movement in the subsurface zone. J Hydrol 102: 503–518

Sposito G, Jury WA, Gupta VK (1986) Fundamental problems in the stochastic convection-dispersion model of solute transport in aquifers and field soils. Water Resour Res 22: 77–88

Tweedie MCK (1956) Some statistical properties of inverse Gaussian distributions. Virginia J Sci 7: 160–165

Tweedie MCK (1957a) Statistical properties of inverse Gaussian distributions. I. Ann Math Statist 28: 362–377

Tweedie MCK (1957b) Statistical properties of inverse Gaussian distributions. II. Ann Math Statist 28: 696–705

van Kampen NG (1981) Stochastic processes in physics and chemistry. North-Holland, Amsterdam, chap 1

White RE (1985) The transport of chloride and non-diffusible solutes through soil. Irrig Sci 6: 3–10

White RE (1987) A transfer function model for the prediction of nitrate leaching under field conditions. J Hydrol 92: 207–222

White RE (1989) Prediction of nitrate leaching from a structured clay soil using transfer functions derived from externally applied or indigenous solute fluxes. J Hydrol 107: 31–42

White RE, Thomas GW, Smith MS (1984) Modelling water flow through undisturbed soil cores using a transfer function model derived from ^{3}HOH and Cl transport. J Soil Sci 35: 159–168

White RE, Dyson JS, Haigh RA, Jury WA, Sposito G (1986) A transfer function model of solute transport through soil. 2. Illustrative applications. Water Resour Res 22: 248–254

Part II
Solutions of Flow and Transport in Unsaturated Media by Deterministic Models

6 HYSWASOR – Simulation Model of Hysteretic Water and Solute Transport in the Root Zone

C. Dirksen[1], *J.B. Kool*[2], *P. Koorevaar*[1], *and M.Th. van Genuchten*[3]

6.1 Introduction

In the context of a book in memory of Eshel Bresler, the merit of numerical computer models to simulate simultaneous water and solute transport in the unsaturated and saturated zones does not need to be documented. We mention just two areas of application. Sound irrigation management aims at optimizing field water and salt balances such that a healthy environment for crop growth is created and maintained. To accomplish this requires not only a quantitative understanding of the basic processes involved, but also of the interactions between them. This complex problem can hardly be tackled without the help of computers. Regulating the movement of harmful solutes, such as pesticides, herbicides, and hydrocarbons, through the environment is an even more complex problem. Not only routes, but also travel times and concentration changes due to physical (e.g., adsorption) and chemical processes need to be known. From the beginning Bresler contributed significantly to the development of numerical models of water and solute transport (Bresler and Hanks 1969; Bresler 1973) and later of root water uptake (Bresler et al. 1982).

Bresler also contributed to early attempts to include the hysteretic nature of soil water retention in descriptions and calculations of soil water transport (Bresler et al. 1969; Hanks et al. 1969). These authors made use of work by Rubin (1967) which showed that hysteresis retards redistribution of water after infiltration, because pressure heads can vary considerably with relatively small water content changes while traversing so-called scanning curves between the main wetting and main drying soil water characteristics. As a result, at a given pressure head the water content is larger during drying than during wetting, and vice versa. Nonhysteretic redistribution according to either the main drying or wetting curve is much larger. Early hysteretic water transport models (e.g., Bresler et al. 1969; Staple 1969; Dane and Wierenga 1975) involved rather simple forms of scanning curves. In the course of time, various theoretical studies have provided insight into the process of hysteresis (e.g., Poulovassilis 1962; Topp 1971a; Mualem 1974,

[1] Department of Water Resources, Wageningen Agricultural University, Wageningen, The Netherlands
[2] Hydro Geologic Inc, Herndon, Virginia, USA
[3] USDA-ARS, US Salinity Laboratory, Riverside, California, USA

1984). Results, as the so-called dependent and independent domain theories, have provided a basis for deriving scanning curves from the main wetting and drying curves. This is not just convenient, but almost mandatory. Measured sets of scanning curves are, and will remain, scarce (e.g., Topp 1969, 1971b; Vachaud and Thony 1971) owing to the sheer amount of work and time required, and the rather poor accuracy which can be expected. It is not surprising, then, that also empirical hysteresis models have been proposed. The model of Scott et al. (1983) is, in fact, a rather simple algebraic algorithm which produces scanning curves from the main wetting and drying curves exhibiting the general features of the few experimental data available. Kool and Parker (1987) found that results obtained with this empirical model and with the theoretical model of Mualem (1984) did not differ significantly, while the empirical algorithm presented significant reductions in computation time.

Under the now widely used high-frequency irrigation, the root zone is alternately exposed to wetting and drying. Under these conditions, accurate descriptions and calculations of water and salt transport require that hysteresis also be taken into account. Detailed experimental data on root water uptake under nonuniform matric and osmotic soil water potential distributions in the root zone could not be explained without hysteresis (Dirksen 1985, 1987). The apparent nonexistence of a numerical computer model which could deal with such a complex problem prompted the development of the simulation model for *hys*teretic *wa*ter and *so*lute transport in the *r*oot zone, HYSWASOR, presented in this chapter.

The basis of HYSWASOR is formed by WORM, an efficient mass-lumped, fully implicit Galerkin finite element code for one-dimensional, isothermal transport of water and solutes in rigid porous media (Van Genuchten 1987). The water transport module of this code was at first implemented with the empirical hysteresis model of Scott et al. (1983), as described by Kool and Parker (1987). However, this model generates open scanning loops, facilitating the artificial phenomenon of "water pumping" (e.g., Jaynes 1984) under cyclical reversals between wetting and drying such as occur under high-frequency irrigation. To prevent this, the Scott model was later modified such that after two reversals the scanning curve joins the original scanning or main curve at the point of the first reversal (Sect. 6.2.2). The only major change in the solute transport module, in comparison with WORM, is the inclusion of nonlinear adsorption. The base code WORM and its descendant, HYDRUS, have been documented elsewhere (Van Genuchten 1987; Kool and Van Genuchten 1991). These reports contain detailed descriptions of the governing equations and the numerical implementations and solutions of the water and solute transport equations. These have not been changed significantly for HYSWASOR, except for the hysteresis algorithm, and need not be repeated here.

Both WORM and HYDRUS are written in FORTRAN77. An early version of HYDRUS was converted from FORTRAN77 to TURBO-PASCAL, redesigned and expanded (Dirksen and Augustijn 1988; Koorevaar et al. 1988) and, in the course of time, developed into the present model HYSWASOR. Profiting from very short compilation times and outstanding debugging facilities, the TURBO-PASCAL code was simplified and computation times reduced. Also, additional input options were introduced in user-friendly form, in particular for the boundary conditions. Different computation of water capacity improved water balances, especially during ponded infiltration. In addition to a more accurate hysteresis model, HYSWASOR incorporates a root water uptake function that allows for better discrimination between the effects of water and salinity stress on root water uptake. It can also simulate single-species solute advection and dispersion according to Fick's law, allows for solute adsorption to the solid phase according to a Freundlich equilibrium isotherm, and can accommodate first-order decay with separate rate constants for the dissolved and adsorbed phase. Finally, HYSWASOR provides runtime graphical displays of simulation results which make it a valuable teaching tool.

In this chapter we summarize the governing equations, describe only the new closed scanning loop hysteresis algorithm and the root water uptake function in some detail, and present HYSWASOR from the user's point of view, with emphasis on aspects of input and output. For further details the reader is referred to the program code which can be obtained from the authors. It is provided with liberal comments to enable the user to grasp the structure and operation of the program, and to make changes according to individual needs and wishes. Finally, simulation results are compared with experimental data for a soil column grown with alfalfa and irrigated daily and at 6-day intervals with saline water.

6.2 Governing Equations

6.2.1 Water Transport

Water transport is assumed to be one-dimensional, vertical through a rigid porous media, as described by:

$$C[\theta]\frac{\partial h}{\partial t} = \frac{\partial}{\partial z}\left(k[\theta]\frac{\partial h}{\partial z} - k[\theta]\right) - S[z,t]. \tag{1}$$

Here, h is pressure head (L); $C[\theta] = d\theta/dh$ is differential soil water capacity (L^{-1}), with θ representing volumetric water content ($L^3\,L^{-3}$); k is hydraulic conductivity (LT^{-1}); $S(z,t)$ is volumetric root water uptake rate (T^{-1}); t is time (T); and z is depth (L), increasing downward. Then, hydraulic head $H = h - z$ (L), and downward fluxes are positive.

The main wetting and drying soil water retention functions $\theta[h]$ are described by the parametric function (Van Genuchten 1980):

$$S_e = [\theta - \theta_r]/[\theta_s - \theta_r] = (1 + \alpha|h|^n)^{-m}, \tag{2}$$

where θ_s and θ_r are saturated and residual water contents respectively; and α (L^{-1}, reciprocal of head unit), n, and m ($m = 1 - 1/n$) are shape factors.

Wetting and drying curves are described by different values for the parameters θ_s, θ_r, α, n, and m. Similar equations are used in the algebraic algorithm to generate closed hysteretic scanning curves (Section 6.2.2). To prevent problems, the main wetting and drying curves are slightly modified (see Fig. 6.1). At the dry end, the residual water content for the wetting curve is adjusted until this curve joins the main drying curve tangentially. If this does not occur above $h = -10\,000$ cm, the main wetting curve is terminated at that pressure head. For smaller pressure heads, the program assumes nonhysteretic water transport according to the drying curve. At the wet end, the wetting "satiated" water content (wcsw) does not approach the drying saturated water content (wcsd) closer than 0.001.

The hydraulic conductivity function can be described by (Mualem 1976; Van Genuchten 1980):

$$k = k_s S_e^{1/2}[1-(1-S_e^{1/m})^m]^2, \tag{3}$$

where k_s is the saturated hydraulic conductivity. The parameter $m = 1 - 1/n$ can be fitted with Eqs (2) and (3) to experimental water retention and/or hydraulic conductivity data. HYSWASOR also allows independent selection of n, or the use of a $k[\theta]$ table to describe this function by maximally ten linear segments of log k vs. θ (Sect. 6.3.2).

6.2.2 Closed Scanning Loop Hysteresis Algorithm

A typical excursion of pressure head and water content in a soil exhibiting hysteresis is shown schematically in Fig. 6.1. The main soil water retention curves and scanning curves are designated by their order and type (e.g., 2w is a secondary wetting curve). The excursion involves soil originally at wcrd, wetted along 0w (for wc < wcmin, 0w = 0d) to reversal point rev0, dried along 1d to rev1, wetted along 2w to rev2, dried along 3d to clos3, dried along 1d to clos1, and dried along 0d to wcrd. All the curves are described by Eq. (2) with (αd, nd) for drying and (αw, nw) for wetting. The difference between the curves is effected in the translation from S_e to θ (in this section designated by Se and wc, respectively, corresponding to the program code), and vice versa, by substituting in Eq. (2) for θ_r and θ_s the appropriate "residual" water contents (wcrd, wcrs1, etc.) and "satiated" water contents (wcsd, wcsw, wcss1, etc.), respectively. These water contents can be calculated when two points of the new scanning curve are known. A new scanning curve must be generated at each reversal point and, during initialization, at each point (wci, hi) within the main loop.

For reversals at scanning curves of order 1 and higher, two pairs of water content and pressure head are available in the current reversal point (wc, h) and the previous reversal point stored as the closure point (wcclos, hclos) for a future scanning curve. Indicating the parameters for the two points by 1 and 2, Se1 and Se2 are first calculated directly by substituting h1 and h2, respectively, in Eq. (2). Substitution of (Se1, wc1) and (Se2, wc2) in the

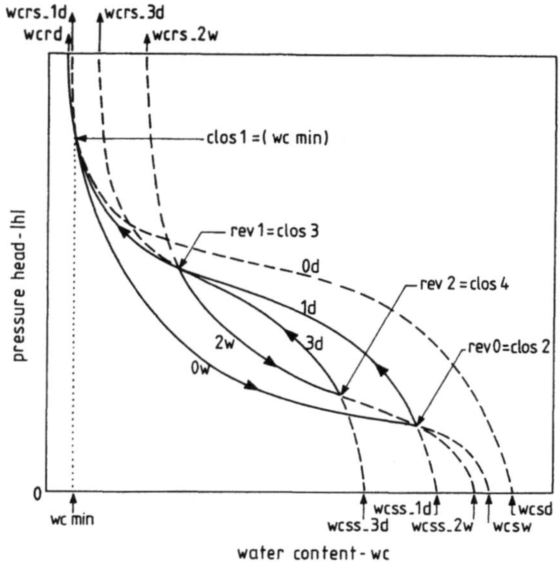

Fig. 6.1. Typical excursion of pressure head and water content at a nodal point, schematically indicated by *solid lines* and *arrows*. The generated scanning curves are shown by *dashed lines* with the associated parameters. (For further explanation, see text)

same equation then yields two equations with the two unknowns:

$$\text{wcrs} = [\text{Se2} * \text{wc1} - \text{Se1} * \text{wc2}]/[\text{Se2} - \text{Se1}] \tag{4}$$

$$\text{wcss} = \text{wcrs} + [\text{wc1} - \text{wcrs}]/\text{Se1}. \tag{5}$$

Substitution of wcrs and wcss in Eq. (2) for θ_r and θ_s, respectively, then yields the definition of the desired scanning curve.

For reversals from the main curves, and for generating first-order scanning curves during initialization through points (wci, hi) within the main envelope, closure (previous reversal) points do not exist. For first-order drying (1d) curves, the point (wcmin, hmin) at which 0w is joined with 0d (Sect. 6.2.1) is taken as the closing point; Eqs. (4) and (5) are then used to calculate wcrs(= wcrs1d) and wcss(= wcss1d). For 1w curves, the second point is chosen at the x-axis ($h = 0$) between wcsd and wcsw (strictly, not a closing point). The corresponding water content wcss1w is calculated according to (Aziz and Settari 1979; Kool and Parker 1987):

$$\text{wcss} = \text{wcsd} - [\text{wcsd} - \text{wc}]/[1 + \text{A}(\text{wcsd} - \text{wc})] \tag{6}$$

$$\text{A} = 1/[\text{wcsd} - \text{wcsw}] - 1/[\text{wcsd} - \text{wcrd}], \tag{7}$$

with wci instead of wc in the case of initialization. In the case of initialization, the point (if it exists) at which the newly generated 1d curve intersects with 0w, or the 1w curve with 0d, is also calculated as the closure point for a future 2w or 2d curve.

The loop between 2w and 3d is closed immediately after one reversal, while that between 0w and 1d is closed only after the excursion reaches rev1 for the second time and continues without further reversals. This shows that actual closures reduce the order of scanning curves by two until a main curve is reached. Consequently, all reversal/closure points must be retained until such time. The loop between 1d and 2w was interrupted at rev2 and did not reach the potential closure point clos2. Successive reversals form loops of increasing order nested inside each other, thereby reducing the effect of (potential) closures as the order increases. In view of this, and the order reduction at closures, the maximum order of scanning curve allowed is presently set at 5. Whenever the maximum would be exceeded, the order is reduced instead of increased by 1, causing oscillations between 4 and 5 until a closure is attained. This appears to be adequate for most situations.

6.2.3 Solute Transport

Solute transport is described by the advection-dispersion equation:

$$\frac{\partial}{\partial z}\left(\theta D\frac{\partial c}{\partial z}\right)-\frac{\partial qc}{\partial z}-\lambda\theta c=\frac{\partial\theta Rc}{\partial t}, \tag{8}$$

where c is solute concentration (ML^{-3}); $D=L_{\mathrm{dis}}q/\theta+D_p$ is the dispersion coefficient (L^2T^{-1}); L_{dis} is the dispersion length (L) of the medium; q is water flux density (LT^{-1}); and D_p is the effective molecular diffusion coefficient of ions in the soil solution (L^2T^{-1}). The adsorbed concentration, $s(MM^{-1})$, is assumed to be in local equilibrium with the solute concentration according to the Freundlich isotherm $s=\mathrm{k}_1c^\eta$. The overall decay coefficient $\lambda(T^{-1})$ and retardation factor R are then defined as:

$$\lambda=\lambda_1+(\rho_b\mathrm{k}_1\theta^{-1}c^{\eta-1})\lambda_2 \tag{9}$$

$$R=1+\rho_b\mathrm{k}_1\theta^{-1}c^{\eta-1}. \tag{10}$$

Further, ρ_b is bulk density (ML^{-3}), and λ_1 and λ_2 are first-order decay coefficients (T^{-1}) for the dissolved and adsorbed phases respectively.

6.2.4 Root Water Uptake

The volumetric root water uptake rate $S[z,t]$ can be specified in various ways (e.g., Molz 1981), depending on the particular problem and available data. HYSWASOR was designed to study specifically the influence of soil water pressure and osmotic potential on root water uptake. Osmotic potential can be expressed as osmotic head, h_0. For mixtures of NaCl and $CaCl_2$ – used in the experiments simulated as described in Section 6.5 – h_0 can be assumed to be a linear function of solute concentration, c, and electrical conductivity of the soil solution, EC (US Salinity Laboratory Staff 1954):

$$h_0(\mathrm{cm})=-40c(\mu\mathrm{mol\,cm}^{-3})=-400EC(\mathrm{dS\,m}^{-1}). \tag{11}$$

Analogous to crop salt tolerance data in terms of relative yields (Maas and Hoffman 1977), Van Genuchten (1987) proposed a root water uptake function with a dependence on the (weighted) sum of pressure and osmotic head relative to the (total) head at which the uptake is reduced to 50%. Dirksen and Augustijn (1988) extended the analogy with Maas and Hoffman by replacing the weighting factors with separate threshold values for pressure head and osmotic head, h^* and h_0^*, respectively, above which root water uptake is not reduced. The effect of lower heads is also characterized by separate values, $h^{1/2}$ and $h_0^{1/2}$, at which root water uptake is reduced to half the maximum value when all other factors are optimal. The total reduction is set proportional to the product of the separate terms, rather than to a sum. The root water uptake function presently used in HYSWASOR is then:

$$S[z,t] = E_p[t] \cdot R[z,t] \cdot \mathrm{RedTerm} \tag{12}$$

$$\mathrm{RedTerm} = \left(\frac{\phi_{\mathrm{air}}}{\mathrm{MinAir}}, \text{ if } < 1\right) \cdot \frac{1}{1+\left(\dfrac{h^* - h[z,t]}{h^* - h^{1/2}}\right)^p} \cdot \frac{1}{1+\left(\dfrac{h_0^* - h_0[z,t]}{h_0^* - h_0^{1/2}}\right)^p}, \tag{13}$$

where E_p is potential transpiration rate (LT^{-1}), and $R[z,t]$ is relative root density or activity (L^{-1}) defined such that the integral of $R[z,t]$ over the root zone depth equals 1. The reduction in root water uptake because of a lack of aeration under wet conditions (Feddes et al. 1978) is assumed to be linear with the air volume fraction, ϕ_{air}, between the threshold MinAir and zero. For potentials above the threshold values, the integral of S over the root zone equals E_p; under stress, it is smaller. The crop salt tolerance data collected by Maas and Hoffman (1977) for alfalfa indicate a threshold value for the soil water solution of 4 dS/m, or $h_0^* = -1600$ cm, and a 50% value of about 17.5 dS/m, or $h_0^{1/2} = -7000$ cm. The exponent p determines the rate at which the water uptake diminishes when the threshold values are exceeded. Van Genuchten (1987) expected p to be approximately 3, based on crop yield data. The osmotic reduction term for these values, and $h > h^*$, is plotted against h_0 in Fig. 6.2.

Leaf water potentials closely reflect the effect of soil water potentials on water uptake integrated over the root zone and may be expected to yield better relationships for the present purpose than crop salt tolerance data. The dependence of total leaf water heads on soil water heads found for alfalfa (Dirksen 1985, Figs. 8 and 9) exhibits the same form as the osmotic reduction term in Eq. (13). Using relative total leaf water heads between −7000 and −21 000 cm (instead of relative yields), an "eyeball" average curve through the leaf total head and soil water osmotic head data in Fig. 8 of Dirksen (1985) is described well by $h_0^* = -2800$ cm, $h_0^{1/2} = -3900$ cm, and $p = 1.5$ (Fig. 6.2). Similarly, the leaf water potential data for the nonsaline column A in Fig. 9 of Dirksen (1985) were fitted more arbitrarily to $h^* = -400$ cm, $h^{1/2} = -1200$ cm, and $p = 1.5$. The additional effect of h^* and $h^{1/2}$ on RedTerm is a simple linear reduction. For instance, for $h = h^{1/2}$, all values are multiplied by 0.5 (Fig. 6.2). Both sets of parameters in Fig. 6.2 are used with the simulations (Table 6.1). The

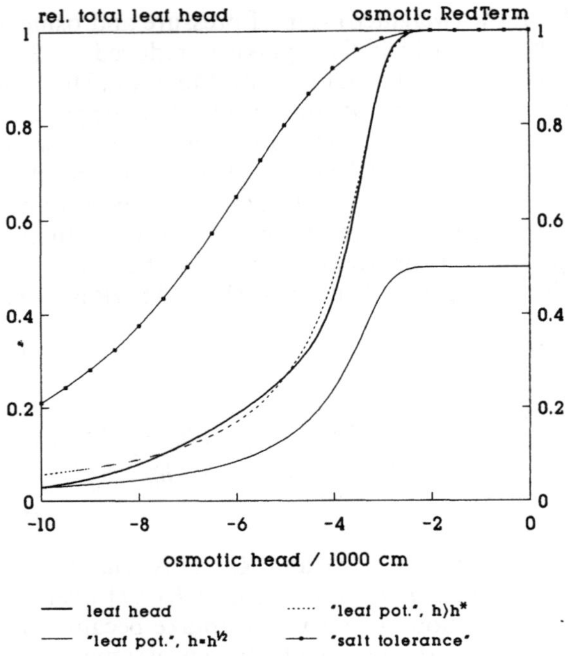

Fig. 6.2. Comparison between the dependence on soil water osmotic head of (average) relative total leaf water heads of alfalfa (Dirksen 1985), and that of osmotic water uptake reduction according to Eq. (13) for the "leaf potential" set of parameters ($h_0^* = -2800$ cm, $h_0^{1/2} = -3900$ cm, and $p = 1.5$), when $h > h^*$. The latter osmotic water uptake reduction, when $h = h^{1/2}$, is also shown, as well as that for the "salt tolerance" set of parameters ($h_0^* = -1600$ cm, $h_0^{1/2} = -7000$ cm, and $p = 3.0$), when $h > h^*$

total leaf water potential versus soil water potential data for alfalfa (Fig. 9 in Dirksen 1985) support the selection of individual threshold values for pressure head and osmotic head. Unfortunately, such data are almost nonexistent, in contrast to crop salt tolerance data. Analogous to these data, p can be expected to vary between crops and possibly differ for h and h_0.

6.3 Input

Input file names may be without extension or must have the format *.DAT. Table 6.1 shows a typical input file. The top three lines are available to identify the particular experiment or problem, etc. This is echoed to the *.OUT file (* stands for the name of the input file), but is not used in the program. The rest of the file is divided into four sections separated by = = = lines, with indicative text. These, and the text lines within the sections, are for the convenience of the user and may be altered. Only the numerical values below the labels are read into the memory.

6.3.1 General Data

The first line of numerical values contains miscellaneous parameters. The element size (MaxDepth/NrOfElem) is uniform throughout the soil column.

Table 6.1. Input file for simulation of 21-day growth period of alfalfa irrigated daily with saline water. The boundary and initial conditions are not complete (see text and Figs. 6.3 and 6.4)

```
Daily irrigation @ 0.35 cm/h. J5251 @ 12.00 - J5272 @ 14.00   April 10, 1992
Alfalfa cut: J5252 @ 900 h and J5272 @ 900 h. Lights on: 500 - 2100 h.
Fluxbot = 0.   Max Ep = 0.110.   Slightly altered rdd.  "Leaf potential" set
------------------ General Data ------------------------- must be line 4 -------
MaxDepth   NrOfElem   ConcGrWat   MariotteHead
 105          84         0.0           1
ExCode TranspCode GraphCode PltCode GraphInt PltInt ScreenInt  Debug MaxSteps
 11        0          0        1        2       4       4        0    5000
Sunrise  Sunset  DelTIn  DelTMax   TMax   OneDay
 -7        9     0.005     0.25     506     24
Times for output (increasing order, max. nr.: 50 incl. Tmax; 2 lines)
  9 14 96 100 166 171 251 269 336 394 489 506

Depths (z) in hydraulic head table, H(t,z), (max. 8):
  0  10  20  30  45  60  75   90
Water uptake parameters (all heads in cm):
  Matric: Threshold   50% ;   Osmotic: Threshold   50%   UptakeExponent
            -700    -1200                -2800    -3900        1.5
Root density/activity (incr. depth, incl. 0 and MaxDepth; max. nr.: 20):
  depth: 0.0   8.0  11.0  14.0  18.0  25.0  50.0   80.0   105
  dens.: 4.0  10.0  10.0   8.0   5.0   4.0   2.0    0.0    0
--------------------------- Soil parameters ----------- must be line 22 -------
Soil 1:     ad        nd     wcsd     wcrd      aw       nw    wcsw   SatCond   nk
         0.00837 1.931  0.360   0.059   0.0767  1.596  0.350     2.0     1.7
        CondCode  Rhob  Difp  Ldis  lamb1  lamb2  FreundK  FreundExp  MinAir
          20       1.3  0.02   0.2    0      0       0        1        0.04
        k-theta table  (increasing order of theta, max. nr. 10)
          Theta:    0.1      0.3      0.35
            k:    6.0e-5     1.8      4.0
----------- Boundary conditions -------- must be line (22+8*NrOfSoils) --------
BCTime TopCode TopVal TopLin TopCon BotCode BotVal BotLin PotTrans PTLin Evapor
 0.0      3     0.0     0     0.0      1      0      0     0.110    0      0
 11.0     1     0.35    0    30.0      1      0      0     0.000    0      0
 14.88    3     0.0     0     0.0      1      0      0     0.110    0      0
 21.0     3     0.0     0     0.0      1      0      0     0.030    0      0
 35.0     1     0.35    0    30.0      1      0      0     0.000    0      0
 38.88    3     0.0     0     0.0      1      0      0     0.035    0      0
179.0     1     0.35    0    30.0      1      0      0     0.000    0      0
183.08    3     0.0     0     0.0      1      0      0     0.065    0      0
419.0     1     0.35    0    30.0      1      0      0     0.000    0      0
423.49    3     0.0     0     0.0      1      0      0     0.101    0      0
491.0     1     0.35    0    30.0      1      0      0     0.000    0      0
495.49    3     0.0     0     0.0      1      0      0     0.110    0      0
501.0     3     0.0     0     0.0      1      0      0     0.030    0      0
------ Initial conditions ------ must be line (24+8*NrOfSoils+NrOfBC) ---------
  depth    soil   kap           hi       wci       ci
   0.0      1     -1           0.0      0.192     140
  11.0      1     -1           0.0      0.203     136
  15.0      1     -1           0.0      0.194     170
  20.0      1     -1           0.0      0.155     119
  25.0      1     -1           0.0      0.120     154
  30.0      1     -1           0.0      0.090     130
  40.0      1     -1           0.0      0.075     144
  50.0      1     -1           0.0      0.060     114
  60.0      1     -1           0.0      0.060     140
  75.0      1     -1           0.0      0.072      93
 105.0      1     -1           0.0      0.115     141
```

The maximum number of nodes is 250. ConcGrWat is the solute concentration of the groundwater. MariotteHead, needed for TopCode = 2, is the height (positive or negative) of the Mariotte control point with respect to the soil surface.

The second line of numerical values contains various control parameters:

ExCode = 0/1	: solve for water flow only, without/with hysteresis.
Excode = 10/11	: idem, for water transport and solute transport.
TranspCode = 0/1	: transpiration constant/sinusoidal during light.
GraphCode = 0/1	: do not/do create ASCII graphs in *.OUT file.
PltCode = 0/1	: do not/do create *.PLT output file.
GraphInt	: node interval of ASCII graphs in *.OUT file.
PltInt	: node interval of columns in *.PLT output file.
ScreenInt	: number of timesteps between refreshed screen graphs.
Debug = 0/1	: do not/do create diagnostic output in *.OUT file.
MaxSteps	: maximum number of timesteps.

The third line of numerical values contains time parameters. Sunrise and Sunset must be specified relative to the start of simulation. One must be < = zero; for instance, if simulation starts during light, Sunrise < = zero. Simulations start at BCTime [1] with initial timestep DelTIn and are terminated at TMax or after MaxSteps is attained. Depending on the number of iterations per timestep (#) needed to solve the water flux equation, the next timestep DelT is adjusted: when # = 1–2, DelT is increased by 25% up to DelTMax; when # = 3–4, DelT is not changed; when # = 5–10, DelT is reduced by 25%. When # > 10, DelT is reduced to 33% and the same time step is repeated. If convergence is still not attained for # > 10 and DelT < 0.001*DelTIn, simulation is terminated. With high infiltration rates (> 0.5 k_s), DelT is changed by only 10% at a time. One Day (1 or 24, number of time units per day) defines the time unit. Changing the length unit (cm) requires more changes within the code and is not recommended.

Maximally 50 output times, including TMax, can be specified in two lines. The (corrected) initial conditions and results at TMax are sent automatically to the output file(s). Output is also produced when all the water specified for TopCode = 2 has infiltrated into the soil column. Maximally eight depths can be specified for which hydraulic heads at the output times are collected in a table at the end of the *.PLT file to facilitate plotting H vs. t for various depths (see Fig. 6.9). The matric and osmotic pressure heads for the root water uptake model [Eq. (13)] must be given in the length unit (cm) used throughout the program. Absolute or relative root densities, or activities, must be given for at least 2 and maximally 20 increasing depths. The program assumes a linear relationship between the given depths, extrapolates if necessary, and normalizes the distribution.

6.3.2 Soil Parameters

The program allows maximally ten different soil materials, which must be specified in seven lines each, separated by a dashed line. A soil column may be composed of soil material layers in any order and of any number of elements. The boundary depths and soil material numbers must be entered in the table of initial conditions. All nodes within a layer, including the bottom boundary, are assigned the soil material code of that layer. The main wetting and drying functions $\theta[h]$ must be described by Eq. (2) because the scanning curve algorithm is based on the same equation. The residual water content for the main wetting curve (wcrw) is calculated (Sect. 6.2.1).

When CondCode = 0/1, the $k[\theta]$ function is described by Eq. (3) based on the main drying/wetting curve, respectively, with the saturated hydraulic conductivity SatCond. CondCode = 10/11 is similar to 0/1, except that nk is used instead of nd/nw. When CondCode = 20, $k[\theta]$ is calculated from the $[\theta, k]$ table, assuming a linear log k vs θ relation between the given data pairs (maximum 10). This option not only often allows a better fit to measured values, but also presents a significant reduction in computation time compared with the use of Eq. (3). Hydraulic conductivities needed beyond the given θ range are calculated by extrapolation from the nearest two given data pairs. A value for SatCond must be given, but is not used in this option. For CondCode < 20, the lines for the $[\theta, k]$ table may be left blank.

The features described above can be combined to simulate a saturated porous plate, for instance, in combination with a negative MariotteHead. A separate soil material number with ad = 0.001 (and slightly larger aw) and nd = nw = 1.002, yields a "porous plate" still nearly saturated ($S_e = 0.999$) at $h = -900$ cm. Moreover, a constant (saturated) k can be imposed with the $k[\theta]$ table (CondCode = 20).

6.3.3 Boundary Conditions

Maximally 400 sets of boundary conditions are allowed, enough for daily values for more than 1 year. Simulations start at BCTime (1), even when this is not zero. Boundary conditions become effective at BCTime (i) and remain constant or vary linearly until BCTime (i + 1), when TopLin/BotLin = 0 or 1, respectively.

The following top and bottom boundary conditions can be specified:

TopCode = 0: head (L).
= 1: flux density (evaporation included, LT^{-1}).
= 2: water layer applied, controlled by MariotteHead (L).
= 3: free evaporation during light (Evapor, LT^{-1}).
BotCode = 0: head (e.g., for groundwater table, $h = 0$).
= 1: flux density (LT^{-1}).
= 2: gravitational flow ($dh/dz = 0$).
= 3: suction candle. BotVal is imposed only when $h >$ BotVal; upward flow is not allowed ($q = 0$, if $dH/dz > 0$).

TopVal and BotVal are the numerical values of head, flux density, or water layer associated with the respective boundary conditions. TopCon is the solute concentration of the infiltration water. For TopCode = 3 and BotCode = 2, these values are zero. If for TopCode = 1, the maximum infiltration rate drops below Top Val, an additional output is created at the "time of ponding", and the program switches to TopCode = 0, with TopVal = 0. For TopCode = 2, TopVal specifies the water layer (pulse) applied under control of MariotteHead, which can be negative and positive. When MariotteHead is larger than TopVal or the water layer left on the soil surface, the Mariotte control is inactive and the water infiltrates under the diminishing positive pressure head of its own height. A new pulse of water is added to any water remaining from old pulses. When the (cumulative) water applied to the soil surface has infiltrated, the program produces an additional output (maximally 5) and switches to TopCode = 3. The evaporative flux density Evapor is cut off between $pF = 5$ and $pF = 6$ according to $(-10^5/h)^2$. PotTrans is the potential transpiration rate effective during light only. It changes linearly to the value at the next BCTime, if PTLin = 1. If TranspCode = 1, the transpiration is varied semi-sinusoidally in time and the maximum rate is adjusted to $(\pi/2)$*PotTrans to make the integral under the half-sinusoid equal to PotTrans*LightPeriod.

6.3.4 Initial Conditions

Soil material number, hysteresis index kap(wetting kap = 1, drying kap = − 1), initial values of pressure head (hi), water content (wci), and solute concentration (ci) can be specified for as many depths as desired, but must be given at least for the top and bottom (or deeper) of the soil column. Values at nonspecified depths are obtained by linear interpolation. Soil layer boundaries can be defined as described under soil parameters. When wci = 0 is given, wci will be assigned the value for the given hi at the main wetting or drying curve, depending on kap. The same can be done, instead, with hi by setting hi = 0. If, for hysteretic simulations (ExCode = 1 or 11), a specified (hi, wci) pair lies outside the main envelope, wci is recalculated according to the nearest main curve. For nonhysteretic simulations (ExCode = 0 or 10), only the main drying curve is used (all kap set at − 1). If hi = 0 is specified, hi is calculated for the given wci; otherwise wci is (re) calculated for the given hi.

In the case of no adsorption, linear adsorption or unfavorable Freundlich adsorption (FreundExp > = 1), ci = 0 is allowed. However, if one layer exhibits favorable adsorption (FreundExp < 1), all depths (!) must be assigned a very small ci to avoid stability problems. Favorable adsorption in the upper layer requires a somewhat higher ci to avoid substantial solute balance errors (due to the large value of the derivative of the adsorption isotherm near $c = 0$). These corrections of ci are made automatically and a warning is sent to the *.OUT file.

6.4 Output

Simulations are started by the command HYSWASOR, followed by the name of the input file. If more input file names are given, successive simulations are carried out for each of these. The program always creates a *.OUT file containing:

1. Contents of the input file, with time and date of simulation added.
2. Initial total amounts of water and solute in the soil column.
3. When Debug = 1, for each reversal and closure of a scanning curve the parameters kap, order, wc, h, he (end of timestep), wcrs, wcss, hclos, and wcclos.
4. Rates, cumulative amounts, and errors of water and solute balance components.
5. When GraphCode = 1, ASCII graphs (depth interval = GraphInt) of water content and solute concentration. This allows for quick scanning of unattended simulations.
6. Special messages, such as the end of infiltration, time of ponding, program interrupted, etc.
7. Total number of steps, and runtime.

When PltCode = 1, a *.PLT file is created with kap, order, h, H, θ, c, and S for each output time at depth interval = PltInt. At the end a $H[t,z]$ table is added. PLT files contain only numerical results to facilitate plotting and other data handling.

6.4.1 Screen Output

Simulations start in screen *text* mode, showing the date, time, and input file name. After initialization, the simulation time (absolute and percentage), DelT, step number, number of iterations for water and solute transport, cumulative infiltration (if TopCode = 2), and rates of infiltration/evaporation, drainage, and transpiration are updated after each time step. At each output time, the water and solute balance data sent to the *.OUT file are also displayed on the screen. Other messages sent to the screen include the time of ponding, end of infiltration time, reduced time step, and failure to converge. Using the screen output menu, which is displayed continuously, a selection can be made from the various options of screen *graphics* mode: initial and current water contents, initial and current solute concentrations, water uptake distribution and transpiration rate, or hydraulic head and pressure head. These are updated at time-step intervals set by ScreenInt. For the first time, the maximum scales are set automatically. Thereafter, the maximum scales of solute concentration, water uptake rate, and heads can be set by the user, with the option to retain the last value. Simulations can be interrupted to study the screen output and/or obtain a screen print.

6.5 Simulations

As stated in the introduction, the development of HYSWASOR was prompted by the analysis of complex experimental data obtained during a rather long-term laboratory experimental program. Alfalfa was grown and irrigated at various frequencies with different amounts and salinities of irrigation water (Dirksen 1985, 1987; Dirksen and Raats 1985). Soil water contents, hydraulic heads, and soil water salinities were measured (semi-)automatically with a two-dimensional gamma ray scanner, tensiometer-pressure transducer systems, and salinity sensors, respectively. Incidentally, leaf water potentials were also measured. We will illustrate, and to a certain extent validate, the capabilities of HYSWASOR with the results obtained in the same soil column for one growth period under daily irrigation and one growth period under six-day-interval irrigation. In both cases, the soil column was initially very dry and saline due to prior severe underirrigation.

6.5.1 Daily Irrigation

The input file for the simulation of daily irrigation is given in Table 6.1. Experimental details and results are described in the references above (in particular Dirksen 1987, Table 1 and Fig. 4). The simulation starts at 1200 h on Julian day 251 of 1975 (J5251), 21 h before the alfalfa was cut, and continues until 5 h after the next cut, totalling 506 h. The lights were on from 500 to 2100 h. Daily irrigations at 0.35 cm/h with water of $EC = 3.2\,\mathrm{dS\,m^{-1}}$ ($c = 32\,\mu\mathrm{mol\,cm^{-3}}$, $h_0 = -1280$ cm) started at 2300 h; the initial amount (duration) of 1.358 cm (3.88 h) was increased to 1.428 cm (4.08 h) after seven irrigations, and again to 1.572 cm (4.49 h) after ten irrigations. This growth period was in the beginning of the experimental program when the plants were still developing. Still, transpiration rates were high compared with typical field conditions, because the cross-sectional area of the isolated canopies was considerably larger than that of the soil column. Moreover, the laboratory was well ventilated and air-conditioned at about 25°C and 50% relative humidity. As a result, the potential transpiration E_p could not be estimated and, of course, also RedTerm [Eq. (13)] was not known ahead of time. Actual transpiration rates increased in time with the growing canopy. The relative values of E_p (PotTrans) in Table 6.1 were derived from Table 1 of Dirksen (1987), which gives total mm per day of 16 h light. The maximum value of PotTrans was based on matching the experimental total transpiration estimated by water balance calculations. The root water uptake parameter values are discussed in connection with Fig. 6.2. At the end of the daily irrigation phase of the experimental program (107 days later), the root density distribution of one column was determined by destructive sampling and line intercept counts. Since, at the time of this simulated growth period, the root system was not yet fully developed, this measured (relative) root distribution was adjusted accordingly. The fits of the hydraulic properties

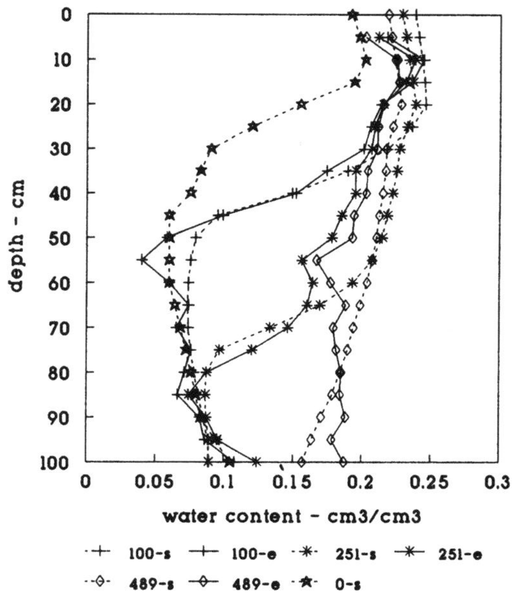

Fig. 6.3. Simulated (*s*) and experimental (*e*) water content profiles at indicated times during daily irrigation of alfalfa, for the "salt tolerance" set of parameter values. For PotTrans and other parameter values, see text and Table 6.1

with the van Genuchten–Mualem equations, without the slight adjustments (Sect. 6.2.1), are presented elsewhere (Dirksen 1991). For the $k[\theta]$ function, we selected two linear log k vs θ segments (CondCode = 20) as a best fit of the experimental data. $L_{dis} = 0.2$ cm is a rather low value; the experimental salinity profiles suggest an even lower value. Most other parameter values are based on experimental information (see References).

Simulations with the "leaf potential" water uptake parameters, PotTrans (max = 0.11 cm/h), and other parameter values in Table 6.1, yielded 18.7 cm total transpiration vs. 18.5 cm determined experimentally. Changing only the water uptake parameters to the set of "salt tolerance" values, the total transpiration was about 17% too high. This is due to the lower value of $h_0^{1/2}$ which causes more water uptake at intermediate salinities. By reducing, in addition, all PotTrans values by 100/117, the total transpiration became 18.9 cm. The simulated water content profiles for these values agree very well (better than for the values in Table 6.1) with experimental data, as shown is Fig. 6.3 (see also Dirksen 1987, Fig. 4). The concentration profiles (Fig. 6.4) do not agree so well. This may be partly due to the limited number of salinity sensors and the large spatial variability. While the top of the soil column was subjected to daily wetting and drying cycles, the column was gradually wetted down to the bottom, displacing the initial high salt concentrations. At $t = 269$ h, the simulated concentrations are much lower above 45 cm, but the maximum $EC = 19.5$ dS/m ($c = 195\ \mu$mol/cm^3) measured at 60 cm depth is approached closely. Later, the simulated maximum salt concentration trails the experimental values. The differences in water contents cannot, of course, be large. Amounts of infiltration and

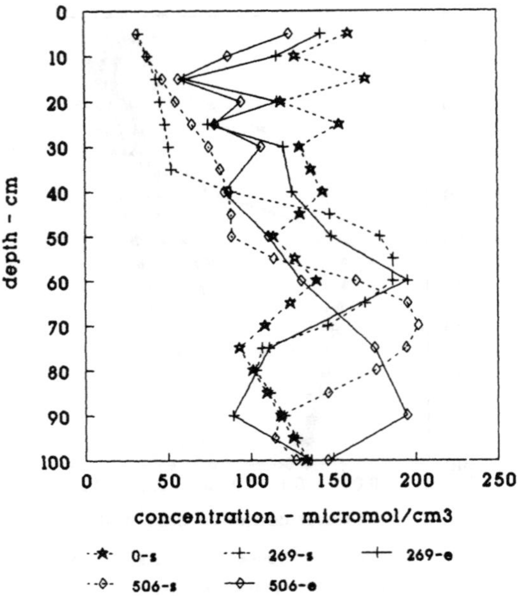

Fig. 6.4. Simulated (*s*) and experimental (*e*) salt concentration profiles at indicated times during daily irrigation of alfalfa, for the "salt tolerance" set of parameter values. For PotTrans and other parameter values, see text and Table 6.1

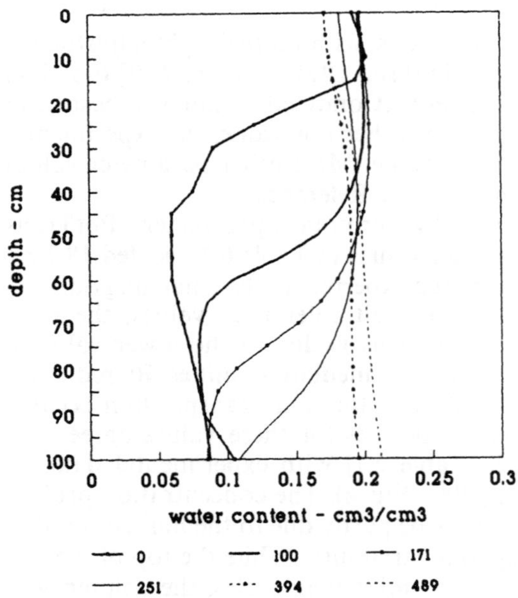

Fig. 6.5. Simulated water content profiles for the same conditions as in Fig. 6.3, however, without hysteresis (ExCode = 10)

drainage were exact and the relative and absolute values of PotTrans were largely based on experimental data. On the other hand, without hysteresis, the simulated water contents are quite different (Fig. 6.5); they were actually nearly uniform after $t = 336$ h. The difference in the salinity profiles with and without hysteresis is much less.

6.5.2 Six-Day-Interval Irrigation

The second set of simulations concerns 8.57-cm irrigations at 0.714 cm/h with water of $EC = 3.0\,\mathrm{dS\,m^{-1}}$ at 6-day intervals during a 23-day growth period. This period was 180 days after the root density distribution used for Table 6.1 was measured. In view of this, the zone of maximum root density was extended to 10–40 cm. Also, light intensities were much higher than during daily irrigation due to additional and replaced lights. As a result, the canopy grew from 12 cm after cutting to about 60 cm in height; its largest cross section was about 2.5 times that of the soil. Doubling the PotTrans values in Table 6.1 (max.0.22 cm/h) the "leaf potential" set of water uptake values yielded 26.0 cm total transpiration vs. 25.7 cm determined experimentally. For the "salt tolerance" set of values, the used reduction of PotTrans was 0.88. Other pertinent parameter values were left as in table 6.1.

Figures 6.6a and 6.7a show comparisons between simulated and experimental water content profiles for the two water uptake parameter sets. Figures 6.6b and 6.7b show similar comparisons for the concentration distributions. The salt originally present in the soil column was displaced downward by the first three irrigations. The fourth irrigation was insufficient for the increased root water uptake; at the end the soil column was again quite dry and saline. The total drainage was 1.30, 1.64, and 1.43 cm for Fig. 6.6, Fig. 6.7, and the experiment, respectively. The total transpiration for Fig. 6.7a, 27.4 cm, is too high but this is solely due to the higher transpiration rate at $t = 413$ h and later. Before that time, the rates were reversed, as can be deduced from Fig. 6.6a and 6.7a. As a result, the simulated water content profile at $t = 544$ h of Fig. 6.7a corresponds better with the experimental data, but at the expense of agreement between the concentration profiles. The influence of the parameter sets on the water uptake are illustrated by the initial, maximum, and final distributions (Fig. 6.8). The integrated water uptake rates at these times, for the "leaf potential" and "salt tolerance" sets, are 0.030 and 0.030, 0.145 and 0.160, and 0.060 and 0.96 cm/h, respectively, while the PotTrans values for the "leaf potential" set were 0.065, 0.200, and 0.220 cm/h respectively. Finally, simulated $H[t,z]$ plots for the irrigation immediately after the simulation period (day 23) can be compared with experimental data of an equivalent irrigation somewhat later in the experimental program (Fig. 6.9). The computation time with a personal computer (20 MHz, 80386, 80387) for these simulations was about 30 minutes with the screen in graphics mode.

6.6 Concluding Remarks

These simulations over 21 and 23 days, starting under extreme dry and saline conditions, would appear to be a good test for the simulation model HYSWASOR. The changes in water content and salinity are large and generally show good agreement with the experimental data. Under daily

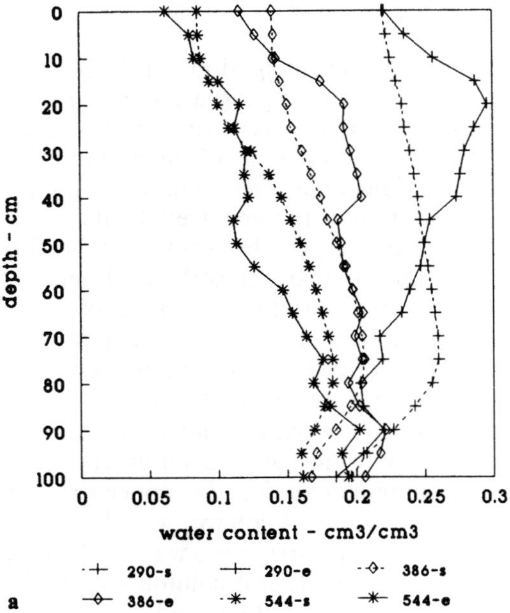

a

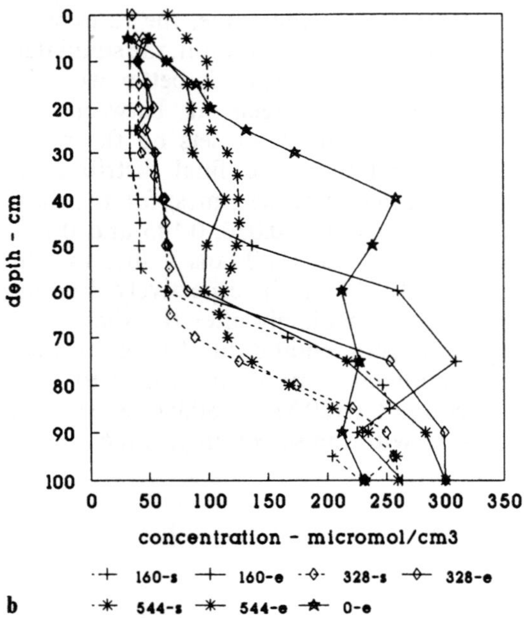

b

Fig. 6.6 a, b. Simulated (*s*) and experimental (*e*) results at indicated times during 6-day irrigation of alfalfa, for "leaf potential" set of parameter values. For other parameter values, see text and Table 6.1. **a** Water content profiles (for initial water contents, see Fig. 6.7a). **b** Salt concentration profiles

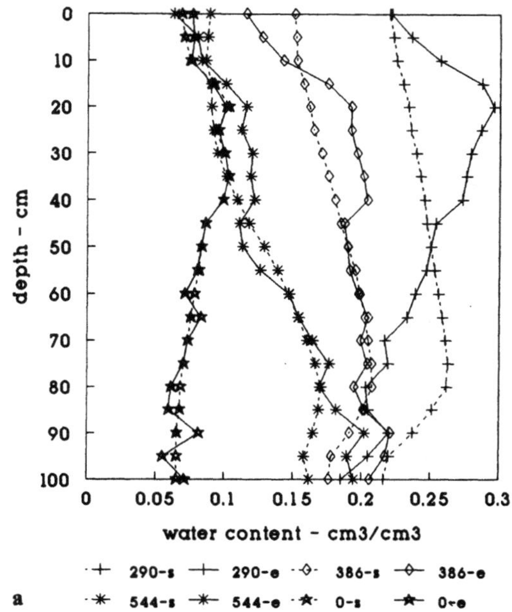

a

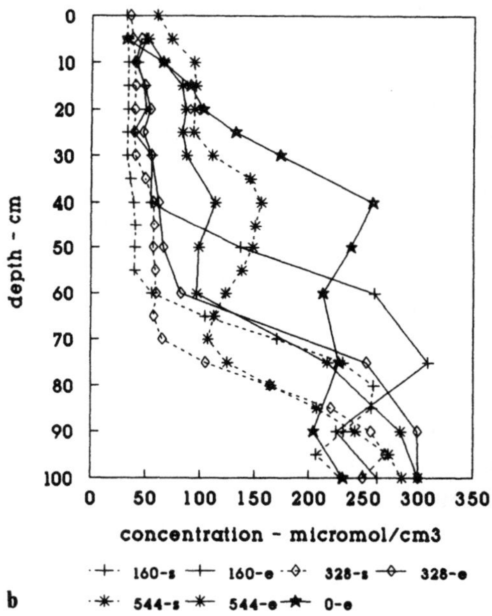

b

Fig. 6.7a, b. Simulated (*s*) and experimental (*e*) results at indicated times during 6-day irrigation of alfalfa, for "salt tolerance" set of parameter values. For other parameter values, see text and Table 6.1. **a** Water content profiles. **b** Salt concentration profiles

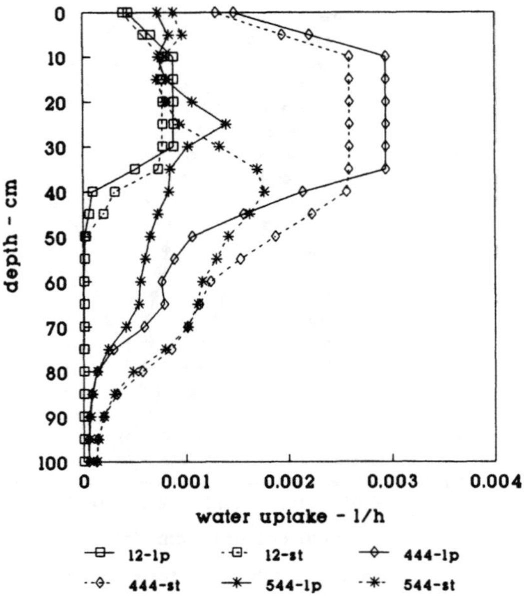

Fig. 6.8. Initial ($t = 12$ h), maximum ($t = 444$ h), and final ($t = 544$ h) simulated distributions of root water uptake rate for "leaf potential" (*lp*) and "salt tolerance" (*st*) sets of parameters

irrigation the simulated salt "front" advances more slowly than in the experiments, but under 6-day irrigation the order is reversed. This dilemma could not be "solved" by small adjustments in the parameter values. The experimental water content and salinity profiles at the bottom of the column under 6-day irrigation appear to be in conflict with each other. Besides this, the major difference between simulation and experiment seems to be in the slope of the water content profiles under wetter conditions (Figs. 6.6a and 6.7a, $t = 290$ h). As discussed by Marshall and Holmes (1979) and confirmed by our simulations, fine-textured soils tend to exhibit the experimental slopes. Small adjustments of the hydraulic property parameters, however, have so far not given better results.

There could be some differences in experimental and simulated results due to the generated scanning curves. These are empirical and their accuracy cannot be judged on the basis of experimental data. We expect this possible effect to be minor. The major effect of hysteresis, retarding water redistribution after irrigations, is not much influenced by the shapes of the scanning curves. Under trickle or small high-frequency irrigations, hysteresis allows steep and even reversed water content gradients to persist. Large irrigations result in rather uniform water content profiles down to the depth of wetting. Hysteretic simulations of redistribution without root water uptake after 10-cm irrigations in the dry soil (results not shown) yielded no significant water content increases below 80 cm, even after 10 days. Without hysteresis, water reached the bottom within 48 h, and after 10 days the largest difference in water content was less than 0.01. Figure 6.5 illustrates the effect of hysteresis with root water uptake.

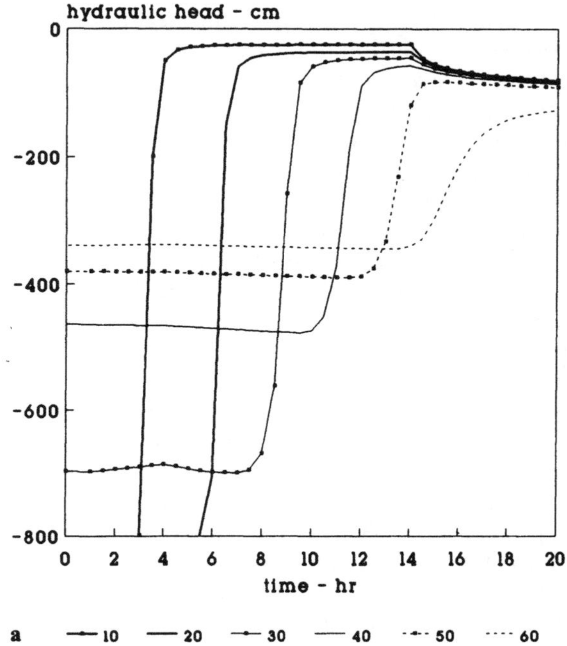

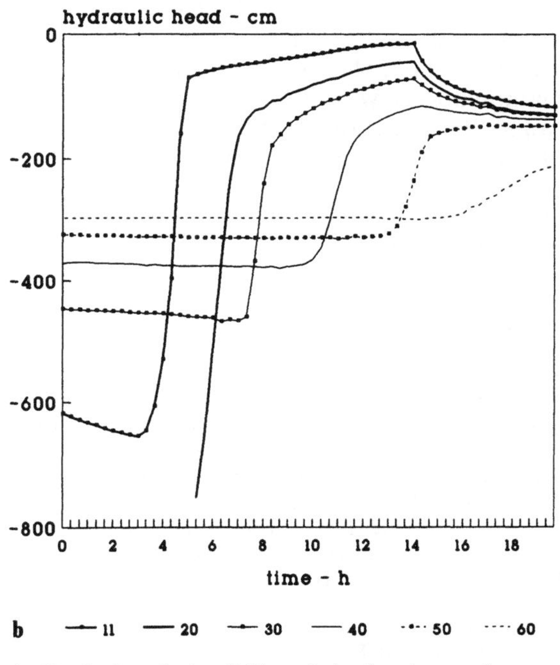

Fig. 6.9a. Simulated and **b** measured $H[t,z]$ plots during 8.57-cm irrigation (see text)

Further manipulation of the input variables to improve the agreement between simulation and experiment does not seem warranted in view of the estimated accuracy of the basic experimental data. The salinity sensors sampled only a very small volume of the soil solution and the rather small number of sensors per column was insufficient in view of the spatial variability found by destructive sampling at the end of the experiments. The water content data were strongly dependent on the collimation of the moving, heavily lead-shielded source and detector of the gamma scanner. This was checked regularly and the results were adjusted on the basis of destructive sampling data at the end, but over the almost 2-year experimental period absolute water content values may not be accurate. Developing root crowns (Dirksen and Raats 1985) and extreme drying may have introduced additional sources of error. The time-domain reflectometry technique now available offers a much better prospect for achieving accurate, automated, simultaneous water content and salinity measurements (Baker and Allmaras 1990; Heimovaara and Bouten 1990; Wraith and Baker 1991).

Further development of the root water uptake model is in order. With Eqs. (12) and (13) S is calculated independently for each depth, whereas plants tend to favor shallow water uptake (Gardner 1983). Even under equal conditions throughout the root zone, significant axial root resistances will favour shallow uptake. This can be compensated ad hoc by correcting the relative root density for activity, but a more rigorous treatment is desirable. The same applies to corrections for contact resistances between roots and soil water by means of relative saturation (Herkelrath et al. 1977).

The osmotic water uptake parameters derived from salt tolerance data, unexpectedly gave better results than those derived from the leaf water potential data. To what extent this result must be attributed to the manipulations with PotTrans, based on the known total transpiration, cannot be ascertained. Whether this result holds for other crops, and/or under less stressed conditions, remains an interesting question. An affirmative answer would be attractive, because salt tolerance data are available for many crops, while leaf water potential data, such as were used here, are scarce.

References

Aziz K, Settari A (1979) Petroleum reservoir simulation. Applied Science, Barking, UK

Baker JM, Allmaras RR (1990) System for automating and multiplexing soil moisture measurement by time-domain reflectometry. Soil Sci Soc Am J 54: 1–6

Bresler E (1973) Simultaneous transport of solute and water under transient unsaturated flow conditions. Water Resour Res 9: 975–986

Bresler E, Hanks RJ (1969) Numerical method for estimating simultaneous flow of water and salt in unsaturated soils. Soil Sci Soc Am Proc 33: 827–832

Bresler E, Kemper WD, Hanks RJ (1969) Infiltration, redistribution and subsequent evaporation of water from soil as affected by wetting rate and hysteresis. Soil Sci Soc Am Proc 33: 832–840

Bresler E, McNeal BL, Carter DL (1982) Saline and sodic soils. Principles – dynamics – modeling. Springer, Berlin Heidelberg New York

Dane JH, Wierenga PJ (1975) Effect of hysteresis on the prediction of infiltration, redistribution and drainage of water in a layered soil. J Hydrol 25: 229–242
Dirksen C (1985) Relationship between root uptake-weighted mean soil water salinity and total leaf water potentials of alfalfa. Irrig Sci 6: 39–50
Dirksen C (1987) Water and salt transport in daily irrigated root zone. Neth J Agric Sci 35: 395–406
Dirksen C (1991) Unsaturated hydraulic conductivity. In: Smith KA, Mullins CE (eds) Soil analysis. Physical methods. Marcel Dekker, New York, pp 209–269
Dirksen C, Augustijn DCM (1988) Root water uptake function for nonuniform pressure and osmotic potentials. Soil Sci Soc Am Agron Abstr: 182
Dirksen C, Raats PAC (1985) Water uptake and release by alfalfa roots. Agron J 77: 621–626
Feddes RA, Kowalik PJ, Zaradny H (1978) Simulation of field water use and crop yield. Halsted, J Wiley, New York, 188 pp
Gardner WR (1983) Soil properties and efficient water use. In: Taylor HM, Jordan WR, Sinclair TR (eds) Limitations to efficient water use in crop production. Am Soc Agron, Madison, WI, pp 56–62
Hanks RJ, Klute A, Bresler E (1969) A numerical method for estimating infiltration, redistribution, drainage and evaporation of water from soil. Water Resour Res 5: 1064–1069
Heimovaara TJ, Bouten W (1990) A computer-controlled 36-channel time-domain reflectometry system for monitoring soil water contents. Water Resour Res 26: 2311–2316
Herkelrath WN, Miller EE, Gardner WR (1977) Water uptake by plants. II. The root contact model. Soil Sci Soc Am J 41: 1039–1043
Jaynes DB (1984) Comparison of soil-water hysteresis models. J Hydrol 75: 287–299
Kool JB, Parker JC (1987) Development and evaluation of closed-form expressions for hysteretic soil hydraulic properties. Water Resour Res 23: 105–114
Kool JB, Van Genuchten MTh (1991) HYDRUS: one-dimensional variably saturated flow and transport model, including hysteresis and root water uptake. Version 3.31. US Salinity Laboratory, Riverside, CA
Koorevaar P, Kool JB, Dirksen C, Augustijn DCM (1988) HYSWASAR: model for one-dimensional hysteretic water and salt transport with root uptake. Soil Sci Soc Am Agron Abstr: 185
Maas EV, Hoffman GJ (1977) Crop salt tolerance – current assessment. J Irr Drain Div Proc Am Soc Civil Eng 103: 115–134
Marshall TJ, Holmes JW (1979) Soil physics. Cambridge University Press, Cambridge, pp 121–126
Molz FJ (1981) Models of water transport in the soil-plant system. A review. Water Resour Res 17: 1245–1260
Mualem Y (1974) A conceptual model of hysteresis. Water Resour Res 10: 514–520
Mualem Y (1976) A new model for predicting the hydraulic conductivity of unsaturated porous media. Water Resour Res 12: 513–522
Mualem Y (1984) A modified dependent domain theory of hysteresis. Soil Sci 137: 283–291
Poulovassilis A (1962) Hysteresis of pore water, an application of the concept of independent domains. Soil Sci 93: 405–412
Rubin J (1967) Numerical method for analyzing hysteresis-affected post-infiltration redistribution of soil moisture. Soil Sci Soc Am Proc 31: 13–20
Scott PS, Farguhar GJ, Kouwen N (1983) Hysteretic effects on net infiltration. Advances in infiltration, ASAE 11-83: 163–170
Staple WJ (1969) Comparison of computed and measured moisture redistribution following infiltration. Soil Sci Soc Am Proc 33: 840–847
Topp GC (1969) Soil water hysteresis measured in a sandy loam compared with the hysteretic domain model. Soil Sci Soc Am Proc 33: 645–651
Topp GC (1971a) Soil water hysteresis in silt loam and clay loam soils. Water Resour Res 7: 914–920
Topp GC (1971b) Soil-water hysteresis: the domain model theory extended to pore interaction conditions. Soil Sci Soc Am Proc 35: 219–225
US Salinity Laboratory Staff (1954) Diagnosis and improvement of saline and alkali soils. USDA Agric Handb 60

Vachaud G, Thony JL (1971) Hysteresis during infiltration and redistribution in a soil column at different initial water contents. Water Resour Res 7: 111–127
Van Genuchten MTh (1980) A closed-form equation for predicting the hydraulic conductivity of unsaturated soils. Soil Sci Soc Am J 44: 892–898
Van Genuchten MTh (1987) A numerical model for water and solute movement in and below the root zone. Res Report, US Salinity Laboratory, Riverside, CA
Wraith JM, Baker JM (1991) High-resolution measurement of root water uptake using automated time-domain reflectometry. Soil Sci Soc Am J 55: 928–932

7 Unstable Flow: A Potentially Significant Mechanism of Water and Solute Transport to Groundwater

D. Hillel[1]

7.1 Nature of the Problem

Our industrial society generates increasing quantities of waste products, some of which can be highly toxic and persistent. Among these are domestic, municipal, industrial, and agricultural wastes which may include pathogenic agents, heavy metals, fertilizer residues, radionuclides, and synthetic compounds of low degradability – all of which threaten over time to poison significant segments of our environment. In recent decades, we have witnessed a growing tendency towards land disposal of wastes, in the expectation that the soil's legendary capacity to filter, retain, buffer, and degrade many components will immobilize pollutants or otherwise render them harmless. In many cases the soil can indeed serve as a recycling and purifying medium (a "living filter"). In too many other cases, however, its powers are exceeded and the soil becomes merely a way station from which pollutants can continue to migrate and invade other realms of the environment via plants, surface waters, or groundwater.

Particularly insidious is the danger of groundwater pollution, a process which is largely invisible, difficult to monitor, and, once begun, extremely difficult to remedy. Water-borne contaminants released from the soil by internal drainage are transmitted to the groundwater via the vadose zone, and may pollute drinking water supplies. Therefore, it is no exaggeration to say that the vadose zone is the major battleground in the war against widespread groundwater pollution. For a long time the notion had prevailed that any and all fluctuations which occur periodically in the soil's surface zone (e.g., as the result of wetting–drying cycles) are quickly damped out in the subsoil, so that the movement of water and solutes in the vadose zone takes place as a slow, steady-state process. Thus, the buffering capacity of the subsoil, and of the vadose zone as a whole, can come into play and gradually help neutralize many potential pollutants before they can reach the water-table.

Recent findings, however, have cast doubt upon the convenient assumption of slow and steady transport through the vadose zone. The role of preferred pathways has come to light. Instead of percolating uniformly and continuously through the matrix of the vadose zone, water infiltrating the soil may concentrate episodically in fissures or passages leading directly and rapidly

[1] Universiy of Massachusetts, Amherst, Massachusetts 10003, USA

to the water-table, thus bypassing the greater volume of the vadose zone (Hillel 1980a, b, 1987). Observations have shown that in some cases transient pulses of contaminant-bearing water may spurt ahead of the mean velocity of the water permeating the vadose zone, and charge the aquifer with repeated plumes of pollution. Models of groundwater recharge that ignore the possibility of such spurts may be misleading, since these spurts may affect the residence time of contaminants in the vadose zone, as well as their spatial distribution and expected arrival time at the water-table. To be realistic a model must include the possibility of preferred pathway flows, define the conditions that might lead to the onset of such flows, and predict their potential contribution to groundwater pollution on a quantitative basis.

Preferred pathway flows, or concentrated flows, may be of various types. Most readily evident are those that follow pre-existing features in the soil that are either formed physically by shrinking and cracking of clay, or formed biologically by burrowing animals or decaying roots. However, quite a different sort of concentrated flow evidently exists that cannot be so readily discerned, as it can take place in soil layers having no such pre-existing features. Such flows may occur spontaneously and sporadically at seemingly random sites within the soil profile. They are associated with a still imperfectly understood phenomenon termed "unstable flow", the potential importance of which has only recently been realized.

A stable flow regime is a distributed flow field in which the entire matrix transmits the fluid, though often at a time-variable and space-variable rate. In the case of infiltration into a uniform medium, a stable flow regime forms a planar wetting front, and any bulges or perturbations that may occur in it from time to time tend to spread out and dissipate gradually as the process continues. Conversely, an unstable wetting front is one that breaks into discrete, spatially discontinuous lobes (commonly called fingers) that race downward while bypassing the greater volume of the soil that lies between the fingers. It appears that the phenomenon of fingering occurs most prevalently (though not exclusively) in distinctly layered profiles. The realistic prediction and tracking of unstable flows constitutes one of the most challenging problems in contemporary environmental research.

7.2 Review of Unstable Flow

Most models of subsurface flow in current use still ignore the existence of preferred pathways in general, and of the unstable flow phenomenon in particular. The convenient assumption of an equivalent homogeneous porous medium (i.e., that transport through the profile can be represented as a one-dimension flow field) has allowed workers to use constant parameter values to characterize porous media for purposes of field sampling and computer simulation of water and contaminant transport. A more recent approach has tended to account for heterogeneous flow patterns in stochastic terms, by assuming that the distribution of the heterogeneities in the subsurface is random. However, as stated by Sposito et al. (1986), "the stochastic convection-dispersion model does not yet warrant unqualified use as a tool for the physically based, quantitative applications of solute transport theory to the management of solute movement at field scales". Williams (1988) recently pointed out that "the identification of the so-called fastest path is critical. It is unreasonable to burry the identification of the fastest path in statistical parameters", in view of the possibility that just a few preferred paths of rapid flow may completely alter a velocity field.

The phenomenon of unstable flow, also termed "wetting-front instability" or "channeling", is known to occur in a variety of situations and to be due to a variety of mechanisms. It occurs most notably in the transition of infiltrating water from a fine-textured to a coarse-textured layer. Instead of advancing as a smooth front, the percolating water may concentrate at specific locations to break into the coarse sublayer in the form of finger-like or tongue-like protrusions which eventually grow to become "pipes". Flow through the coarse-textured layer thereafter takes place through these "pipes" rather than through the entire layer uniformly. As such, wetting-front instability, if indeed prevalent, threatens to invalidate our best-laid models of vertical infiltration, internal drainage, and groundwater recharge. Moreover, it can have inportant practical consequences. Some of the current methods of waste disposal on land, based as they are on establishing clay liners intended to contain or detain and neutralize toxic substances, may have the very opposite effect of inducing fingering and rapid transport of water and contaminants to groundwater. At the very least, the potential occurrence of fingering, unaccounted for in current models, implies the loss of our predictive power, and hence of our ability to manage waste disposal systems with the necessary degree of security.

7.3 Instability Between Fluids Differing in Density or Viscosity

The existence of instability at the moving interface between two immiscible fluids of different densities or viscosities has long been recognized. Most of the early research on the appearance of instabilities in porous systems has concerned flow in fully saturated media. A problem much studied in petroleum engineering is the displacement of oil by water, the latter being the less viscous but denser fluid. It has been shown that if the velocity of the displacing fluid exceeds some critical value, fingers of water will intrude into the oil and "contaminate" the fluid extracted from a well.

This type of fingering was studied by Taylor (1950) and by Saffman and Taylor (1958). Their analysis, applied to the downward propagation of an initially planar interface separating the displacing water phase from the displaced oil, yielded the following stability criterion:

$$(f/K)[(U_w/K_w)+(U_o/K_o)]=(d_o-d_w)g+[(U_w/K_w)-(U_o/K_o)]V, \quad (1)$$

where d, U, and K are the density, viscosity, and effective permeability for the respective fluids. The subscripts w and o refer to water and oil; V is the fluid velocity, g is the gravitational acceleration and f is the frequency of a particular disturbance. If f is negative then the disturbance is unstable and will continue to grow or form a finger. In the absence of gravitational effects this depends upon M, the mobility ratio: $M = (K_w/U_w)/(K_o/U_o)$. A value of less than unity implies no amplification of the disturbance and thus no finger formation, whereas the opposite is indicated for $M > 1$. In the case of vertical flow, a more dense fluid overlying a less dense one will increase the likelihood of finger formation, with a dependence on the displacement velocity V. Then the magnitude of M is not sufficient to define the conditions conducing to fingering.

Experimental investigations of this instability problem have most often used the Hele–Shaw flow analogue, in which a porous system is simulated physically by means of closely spaced glass plates, between which the fluids are caused to flow. The main advantage of the Hele–Shaw cell is that it allows visual inspection of the fingering patterns, and it also avoids the practically inevitable heterogeneities introduced into experimental porous systems in the process of their preparation. However, a Hele–Shaw cell is not a true porous medium in that it precludes hydrodynamic dispersion and diffusion effects, incomplete displacement due to adsorption and to dead-end pores, and differences between the fluids due to their differential wettabilities and capillary pressure functions. Moreover, the Hele–Shaw representation is necessarily limited to two-dimensional flow phenomena.

The original analysis by Saffman and Taylor assumed a non-diffuse interface between the two (immiscible) fluids. Later work recognized that instability can also occur between two miscible fluids, as in the case of oil extraction with a solvent (Perkins et al. 1965) or of the penetration of a saline solution into a medium filled with fresh water.

Wooding (1969) observed that the amplitude of fingers in a Hele–Shaw cell increases at first in proportion to the time squared, but later becomes proportional to time. He also noted the tendency of adjacent fingers to coalesce by mutual entrainment, with a consequent decrease in the number of fingers. His analysis took account of the possibility of a diffusive interface. Bachmat and Elrick (1970) reported the observation of fingers in a saturated sand column from a pool of saline water at the upper surface.

7.4 Instability During Infiltration into Unsaturated Soils

The occurrence of instability during infiltration into an initially unsaturated profile has come to the attention of soil physicists and hydrologists only belatedly. Such instability is most likely to take place in stratified profiles, typically at the transition plane between two dissimilar materials. The earliest reference to the phenomenon was apparently made by Tabuchi (1961) and by Miller and Gardner (1962). Bridge and Collis-George (1972) reported a similar occurrence in their experiments of infiltration into a two-layer profile of fine sand overlying coarse sand. Peck (1965) observed instability patterns during infiltration into an unsaturated sand where the air phase was confined so that its pressure increased as the wetting front moved downwards. Peck noted that irregularities at the wetting front tended to be amplified as the velocity of the front increased with distance down the column.

The intriguing question is why any perturbations which might appear in the wetting front are in some cases flattened out by lateral diffusion so that the front remains stable, while in other cases they bceome self-generating fingers which destabilize the setting front. Hill and Parlange (1972), who studied the instability generated at the interface between layers of different textures, based their analysis on the assumptions of a steep (nondiffuse) wetting front and an initially air-dry soil. They postulated that the formation of fingers is due to an instability at an air–water interface which is gravity driven and is more pronounced as the pore size of the coarse sublayer increases, and suggested that the condition for instability is $V < Ks$, where V is the velocity of the wetting front and Ks is the saturated hydraulic conductivity of the lower (coarse-textured) layer. They further postulated that the phenomenon is governed by a characteristic length scale L of the coarse-textured medium, which they based on the ratio of the conductivity and diffusivity values, and that fingers can appear only if the width and breadth dimensions of the system are sufficiently larger than L. They thus attempted to explain why earlier studies of infiltration through crusts, carried out in narrow columns, failed to detect the phenomenon.

In the experiments of Hill and Parlange, the fingers moved with a constant velocity, corresponding closely to the ratio of the saturated hydraulic conductivity of the coarse sublayer to its antecedent wetness. Moreover, the cross-sectional areas of the fingers were nearly uniform (about $6\,cm^2$ for their particular material). Within these fingers, the infiltrating water moved in saturated cores ("pipes") surrounded by an unsaturated cylinder (sheath) of unsaturated soil. The overall flow rate through the system was governed by the number of fingers per unit area of interface. This study showed that fingering is caused spontaneously by a gravity-driven instability rather than by any preexisting irregularities (nicks or wormholes) at the interface.

A general explanation of fingering was attempted by Raats (1973), who pointed out that profile layering is not a prerequisite for wetting-front instability, which may also be due to increased air pressure ahead of the wetting front (as demonstrated by Peck 1965) or to water repellency of soil layers. Raats based his analysis on the Green and Ampt model (elucidated by Hillel 1971) and offered the hypothesis that stability depends on the change of velocity of the wetting front with distance: small perturbations in an initially planar wetting front will tend to grow if velocity increases with distance and will tend to disappear in the opposite case. Raats explored the consequences of this hypothesis for various sets of conditions likely to cause instability, including infiltration with compression of air ahead of the wetting front and infiltration into a soil in which antecedent wetness increases with depth. Here, however, his prediction that fingering should be enhanced at higher values of antacendent soil moisture has been contradicted by subsequent data and theoretical reasoning (Edelstein 1987; Hillel and Baker 1988).

In the case of a soil covered by a thin surface crust of low conductivity, Raats' hypothesis leads to the following criterion for the onset of instability: $R > (Z - h)/K$. Herein, R is the hydraulic resistance of the crust (i.e., its thickness divided by its hydraulic conductivity), Z is the depth to the wetting front, h is the pressure head at the wetting front, and K is the hydraulic conductivity of the subcrust soil. In the general cases of layered profiles, the wetting front can become unstable if:

$$K_{st} < K_{sb}[(d + H_b - H_c - H_a)/L + 1], \tag{2}$$

where d is depth of ponded water, H_b is barometric pressure, H_c is the capillary pressure head at the wetting front, H_a is the soil air pressure head, L is the soil depth and subscripts t and b refer to the top and bottom layer, respectively. This is roughly equivalent to the Parlange and Hill (1976) criterion, with an expression for the infiltration rate based on the sharp front (Green and Ampt) model, including the possibility of compression of air trapped below the infiltration front. Raats stressed that the unifying property of all unstable flow cases was the increase in velocity of the wetting front with depth, and hence this should be the major criterion to be used for predicting unstable flow.

The problem was subsequently reassessed by Philip (1975a, b), who applied the methods of hydrodynamic stability analysis (Lin 1955). To simplify the analysis, he used the same Green and Ampt ("delta-function") model, but claimed that Raats' criterion for instability (that the wetting-front velocity increases with depth) is correct only for the special case where the moisture content, saturated conductivity, and pressure head of the more permeable sublayer are independent of depth. Where these parameters vary with depth, Philip proposed a different criterion of stability, stating that the physical attribute that is fundamental to the question of stability is the water pressure gradient behind the wetting front. When this gradient assists the flow, the flow is stable; when this gradient opposes it (e.g., if the pressure at the front is greater than at the soil surface), the flow becomes unstable.

Philip characterized perturbations of the wetting front in terms of a wavelength number, and proposed as a second criterion of instability that this characteristic of a perturbation be greater than a critical value, c:

$$c = 2n(2s/dgG), \tag{3}$$

where s is the air–water surface tension, d the water density, g the gravitational acceleration, and G the pressure gradient at the wetting front. Instability may thus be inhibited in laboratory columns narrower than this value.

In examining the implications and summarizing his criteria for particular cases, Philip suggested that flow can become unstable if: (1) the weting front reaches a water-repellent stratum and the pressure head needed to penetrate it is positively, (b) air entrapment and compression ahead of the wetting front cause a sufficient rise in the air pressure there, (3) the air-entry pressure is more negative than the wetting-front moisture potential during redistribution, and (4) hydraulic conductivity increases with depth at a sufficient rate. In all the various instances of instability which Philip identified, it is essential that the flow be gravity assisted. Where gravity plays no part, as in horizontal absorption, or in gravity-opposed flow such as capillary rise, there is no mechanism whereby the pressure gradient can oppose the direction of the flow and, hence, instability cannot arise.

Finally, Philip admitted the limitations of an analysis based on a delta-function model, which may not apply to soils with gradual (rather than sharp) wetting fronts. This is an important qualification. The Green and Ampt model used as the basis for Raats' and Philip's theories is applicable mostly to soils with uniform, large pores. In soils with a wider distribution of pore diameters, wetness varies more gradually with suction, as does the hydraulic conductivity. Hence the wetting fronts are not so sharp and any perturbations in the wetting front are more likely to be suppressed by lateral movement of water due to horizontal suction gradients (and bulges are more likely to spread sideways than to propagate downwards as fingers). In view of the possibility that an analysis based on the Green and Ampt model may not apply to soils with gradual wetting fronts, it appears that a theory of fingering is needed that is not so restricted. An effort in this direction has been made by Hillel and Baker (1988).

Parlange and Hill (1976) expanded their earlier work with a stability analysis which considers that the velocity of the wetting front is related to its curvature. This also considers the influence of soil-water diffusivity on the propagation of a curved wetting front. This analysis repeated their earlier conclusion that fingers tend to develop if the frontal velocity is less than the saturated conductivity of the medium. Parlange and Hill conjectured that an increase in the initial wetness of the sublayer should induce an increase in the width of the developed fingers, but they did not verify this experimentally.

Diment et al. (1982) and Diment and Watson (1983) presented a hydrodynamic stability analysis for flow into unsaturated media, using a linearized perturbation analysis to predict the onset of instability. They presented the results of applying their analysis to specific types of infiltration problems that have been shown to be potentially unstable.

A comprehensive series of laboratory experiments was described recently by Glass et al. (1987, 1989). The theoretical approach was based on an extension of the work described earlier by Parlange and Hill (1976).

The recent analysis by Hillel and Baker (1988), later tested by Baker and Hillel (1990), has emphasized the importance of the "suction of water entry" (SWE) into an initially dry sublayer. This parameter had been ignored in previous treatments of unstable flow. As SWE determines the effective hydraulic conductivity in the sublayer, it also determines whether or not flow is accelerated with distance down the profile. When this occurs, the sublayer cannot conduct throughout its entire volume, since it would then be conducting more water than it is receiving (an obvious impossibility). Hence the spatially distributed flow field with its parallel streamlines must begin to constrict at the critical depth (typically at or just below the transition from a less premeable toplayer to a more permeable sublayer), thus causing the streamlines to converge. In the case of a wide flow field, this forms spatially separated, partial-volume, streams (fingers).

In addition to explaining the phenomenon, the theory by Hillel and Baker also predicts the fractional cross-sectional area of the fingers that form under specified sets of conditons. In the general case, that fraction, F, is:

$$F = [K_t/K_u(S_c)][(H_0 + Z_i + S_e)/Z_i] \tag{4}$$

Here K_t is the saturated hydraulic conductivity of the lass permeable toplayer, K_u is the unsaturated hydraulic conductivity of the sublayer (being a function of the matric suction of water entry into the sublayer, S_e), H_0 is the hydraulic head imposed on the surface by the ponded water, and Z_i is the vertical thickness of the toplayer (depth of the interface below the soil surface).

In the special case where the sublayer is of very coarse texture whose water entry value is zero suction, and if the ponding depth is negligible,

$$F = K_{st}/K_{sb}. \tag{5}$$

Thus, for example, if the conductivity of the sublayer is an order of magnitude ($\times$ 10) greater than that of the toplayer, then we may expect the fingers to occupy some 10% of the cross-sectional area of the sublayer; and if it is two orders of magnitude greater, then the fingers will only occupy some 1% of the soil volume. Under such conditions it would be extremely difficult to detect vadose-zone fingering in the field by random sampling from above.

This estimation of the partial volume occupied by fingers ignores the lateral (radial) spread of water from the saturated or near-saturated core of each finger into the surrounding dry soil. This process, controlled by the soil's sorptivity, tends to form an unsaturated cylindrical sheath of moist soil around each finger. It is, however, a relatively slow process which, moreover, becomes progressively slower with time. If continued for long enough, it may eventually moisten the entire sublayer but, as infiltration often occurs in brief episodes the lateral sorption of water from the fingers is likely to be of limited importance during short-term infiltration.

In principle, a soil layer can vary its fractional conducting volume by varying the diameters of the fingers or by varying the spatial separation between them, or by doing both. The author of this proposal is now working on an optimization approach for the calculation of finger diameters and distances, seeking patterns of least hydraulic resistance. Observations suggest that in practice there exists a minimal diameter of a finger for any particular medium; hence, the principal mechanism for varying the fractional volume of conduction may be to vary the spatial separation between adjacent fingers, i.e., the number of fingers per unit cross-sectional area (say, 1 m^2). One reason why finger diameter cannot decrease below a certain minimum (characteristic of the particular soil) is that, as mentioned above, the fingers are subject not only to gravitational forces, causing them to propagate downwards, but also to lateral absorptive effects resulting from the surrounding soil's sorptivity which, in turn, is affected by texture and antecedent mositure.

Of particular importance is the effect of a sublayer's antecedent moisture on the pattern of infiltration and fingering. In principle, increasing the initial wetness of the sublayer can have a dual effect: (1) it reduces the suction gradient affecting the wetting front, hence retarding infiltration during its transient phase; and (2) it may affect the water-entry suction, allowing sorption of water at a higher suction than if the soil were initially dry. The latter effect can be associated with the phenomenon of hysteresis, whereby the rewetting of a partially drained soil along a scanning curve can occur at a higher suction than the entry of water into a completely dry soil. A higher value of suction at the entry to the sublayer increases the hydraulic gradient and, hence, the transmission rate through the toplayer at the same time that it decreases the flux into the sublayer (by reducing the effective hydraulic conductivity there). Therefore we can deduce that the overall effect of a higher antecedent sublayer wetness will be to reduce the likelihood of a flux discrepancy between the layers, and hence to counter the tendency toward fingering and partial volume flow. This reasonning contradicts the hypothesis of Raats (1973) but accords with the findings by Glass et al. (1987) and those of Edelstein (1987). The same reasoning also suggests that unstable flow is likely to be more prevalent in arid than in humid regions.

It is interesting to note that fingering and partial volume flow can occur not only during infiltration, but also during redistribution and even in the absence of profile layering. As demonstrated by Tamai et al. (1987), an apparently stable planar wetting front infiltrating into a uniform, initially dry medium may break into fingers spontaneously when the supply of water at the surface ceases and the process of redistribution (internal drainage) takes place. We can offer an explanation for this that is consistent with the hypothesis that fingering is caused by an acceleration in the direction of flow. During infiltration, the downward flux is maximal at the soil surface and diminishes continuously in depth as the partially wetted portion of the soil continues to sorb water. When the infiltration stops, the downward flux in the upper part of the profile diminishes rapidly, and a condition is soon established wherein the flux at the wetting front is, for a time, greater than the supply rate from the draining upper layers of the profile. This restriction of supply is analogous to the effect of a less conductive toplayer and may result similarly in the spontaneous separation of the planar wetting front into spatially discrete fingers. However, as this process of redistribution is a self-arresting one (tending to slow down so as to become negligible after some hours or days), fingering during redistribution is likely to be less important than fingering during infiltration.

7.5 Preliminary Evidence from Field Experiments

A few recent field experiments suggest that unstable flow is not merely a curiosity, of interest only to theoreticians and laboratory experimenters.

Indeed, those experiments illustrate the potential importance of this phenomenon, heretofore largely ignored, in actual practice. In one such experiment, conducted in Connecticut, Starr et al. (1978) studied the pattern of leaching in a layered soil consisting of a fine sandy loam overlying a coarse sand, with a water-table at a depth of 1.8 m. Removal of the soil layers following infiltration of a dye solution showed that most of the infiltrating water moved through fingers ranging from 5 to 20 cm in diameter. In a second experiment, they applied a pulse of calcium chloride solution to the soil surface and found that the chloride movement for the 20- to 60-cm soil depths averaged about 50% slower than that determined from the average pore-water velocity at the soil surface. In contrast with this relatively slow rate of chloride movement, several salt pulses reached the water-table soon after or before the mean front had reached the 60-cm depth. This greatly increased flow rate corresponds to the observations in the first experiment where flow was unstable and the solute moved in rapid fingers below the 60-cm depth.

Hendrickx et al. (1988), who conducted a bromide tracer trial in the Netherlands, found that after 5 weeks with 120-mm precipitation the bromide concentration in the groundwater was 13 times higher when an unstable wetting front had formed than could be expected with a stable wetting front. In Wisconsin, Kung (1990) found that water and solutes flowed through less than 10% of the soil volume at the 3.0- to 3.6-m depth layer. Similar occurrences were observed by Hillel and his coworkers, in a series of preliminary experiments that have not yet been published. In a field with a shallow water-table, trials with tracer solutions infiltrated under various boundary conditions revealed fingering where a toplayer of restricted conductivity overlay a coarse-textured sublayer. The fingers occupied a fractional area roughly corresponding to the predictions made by Hillel and Baker (1988).

In view of the potential importance of the unstable flow phenomenon and the danger it poses for groundwater pollution via rapid "channeled" flow, it is disconcerting how little attention has thus far been devoted to it by environmental engineers and hydrologists. Prevalent practices not only ignore the possibility of fingering and concentrated flow phenomena, but may even induce it, albeit inadvertently. The common procedure of establishing a clay liner to "contain" effluent from waste disposal sites creates the sort of hydraulic barrier over a more permeable medium that is most conducive to the onset of fingering, and might thereby exacerbate the very problem it is designed to prevent. Better understanding is needed regarding the vulnerability of aquifers to groundwater contamination by rapid "channeled" flows due to fingering. So far, research on fingering has been done mostly in the laboratory. The time is ripe to move research on unstable flow to the field, to ascertain just how probable and prevalent is the phenomenon of fingering, how realistic are existing theories purporting to predict it, and whether current models of transport through the vadose zone can be improved by inclusion of fingering where the physical conditions conduce to its occurrence.

7.6 Final Comment

Quite recently (December, 1991), the American Society of Agricultural Engineers held a symposium on preferential flow. The proceedings of that symposium contains 47 papers, of which no fewer than ten deal with unstable flow – evidence of a rising interest in this potentially important topic.

References

Bachmat Y, Elrick DE (1970) Hydrodynamic instability of miscible fluids in a vertical porous column. Water Resour Res 6: 156–171

Baker RS, Hillel D (1990) Laboratory tests of a theory of fingering during infiltration into layered soils. Soil Sci Soc Am J 54: 20–30

Bridge BJ, Collis-George N (1973) An experimental study of vertical infiltration into a structurally unstable swelling soil with particular reference to the infiltration throttle. Aust J Soil Res 11: 121–132

Dagan G (1986) Statistical theory of groundwater flow and transport: pore to laboratory, laboratory to formation and formation to regional scale. Water Resour Res 22: 120–134

Diment GA, Watson KK (1983) Stability analysis of water movement in unsaturated porous materials. 2. Numerical studies. Water Resour Res 19: 1002–1010

Diment GA, Watson KK (1985) Stability analysis of water movement in unsaturated porous materials. 3. Experimental studies. Water Resour Res 21: 979–984

Diment GA, Watson KK, Blennerhassett PJ (1982) Stability analysis of water movement in unsaturated porous materials. 1. Theoretical considerations. Water Resour Res 18: 1248–1254

Edelstein DM (1987) The effect of antecedent wetness on flow instability during infiltration into layered soils. MS Thesis, Department of Plant & Soil Sciences, University of Massachusetts (D Hillel, Supervisor)

El-Kadi AI, Brutsaert W (1985) Applicability of effective parameters to unsteady flow in nonuniform aquifers. Water Resour Res 21: 183–198

Fayer M, Hillel D (1982) Field testing of a soil moisture model simulating water-table fluctuations. Soil Sci Soc Am J 46: 396–404

Fayer M, Hillel D (1986a) Air encapsulation. 1. Measurement in a field soil. Soil Sci Soc Am J 50: 568–572

Fayer M, Hillel D (1986b) Air encapsulation. 2. Profile water storage and shallow water-table fluctuations. Soil Sci Soc Am J 50: 572–577

Gee G, Hillel D (1988) Groundwater recharge in arid regions. Hydrol Proc 2: 255–266

Glass RJ, Parlange J-Y, Steenhuis TS (1987) Water infiltration into layered soils where a fine textured layer overlays a coarse sand. Proc. Int. Conf. on Infilt. Development and Applications, Hawaii

Glass RJ, Steenhuis TS, Parlange J-Y (1988) Wetting front instability as a rapid and far-reaching hydrological process in the vadose zone. J Contam Hydrol 3: 207–226

Glass RJ, Steenhuis TS, Parlange J-Y (1989) Wetting front instability. Water Resour Res 25: 1187–1194; 1195–1207

Green WH, Ampt GA (1911) Studies on soil physics: the flow of water and air through soils. J Agr Sci 4: 1–24

Gupta SP, Varnon JE, Greenkorn RA (1973) Viscous finger wavelength degeneration in Hele-Shaw models. Water Resour Res 9: 1039–1046

Hendrickx JMH, Decker LW, Raats PAC (1988) Formation of sand columns caused by unstable wetting fronts. Grondboor en Hamer 6: 173–175 (in Dutch)

Hill ED, Parlange J-Y (1972) Wetting front instability in layered soils. Soil Sci Soc Am Proc 36: 697–702

Hillel D (1971) Soil and water: physical principles and processes. Academic Press, New York
Hillel D (1977) Computer simulation of soil water dynamics. Int. Dev. Res. Centre, Ottawa, Canada
Hillel D (1980a) Fundamentals of soil physics. Academic Press, New York
Hillel D (1980b) Applications of soil physics. Academic Press, New York
Hillel D (1985) Field observations of saturated-unsaturated flow phenomena. Unpublished research results, University of Massachusetts, Dept. Plant & Soil Sciences, Amherst, Mass
Hillel D (1987) Unstable flow in layered soils: a review. Hydrol Proc 1: 143–147
Hillel D, Baker RS (1988) A descriptive theory of fingering during infiltration into layered soils. Soil Sci 146: 51–56
Hillel D, Gardner WR (1969) Steady infiltration into crust-topped profiles. Soil Sci 108: 137–142
Hillel D, Hornberger GM (1979) Physical model of the hydrology of sloping heterogeneous fields. Soil Sci Soc Am J 43: 434–439
Kung K-JS (1990) Preferential solute transport in a sandy vadose zone. Geoderma 46: 12–19
Leland DF, Hillel D (1982) Scale effects on measurements of dispersivity in a shallow unconfined aquifer. Soil Sci Soc Am J 46: 905–912
Lin CC (1955) The theory of hydrodynamic stability. Cambridge University Press, London
Miller DE, Gardner WH (1962) Water infiltration into stratified soil. Soil Sci Soc Am Proc 26: 115–119
Milly PCD (1985) Stability of the Green-Ampt profile in a delta function soil. Water Resour Res 21: 399–402
Parlange J-Y, Hill DE (1976) Theoretical analysis of wetting front instability in soils. Soil Sci 122: 236–239
Peck AJ (1965) Moisture profile development and air compression during water uptake by bounded porous bodies. 3. Vertical columns. Soil Sci 100: 44–51
Perkins TK, Johnston OC, Hoffman RN (1965) Mechanics of viscous fingering in miscible systems. Soc Petrol Eng J 234: 301–317
Perrine RL (1961) Development of stability theory for miscible liquid–liquid displacements. Soc Petrol Eng J 1: 17–25
Philip JR (1975a) Stability analysis of infiltration. Soil Sci Soc Am Proc 39: 1042–1049
Philip JR (1975b) The growth of disturbances in unstable infiltration flows. Soil Sci Soc Am Proc 39: 1049–1053
Pitty E (1980) Penetration of fluid into a Hele-Shaw cell: the Saffman-Taylor experiment. J Fluid Mech 97: 53–64
Raats PAC (1973) Unstable wetting fronts in uniform and nonuniform soils. Soil Sci Soc Am Proc 37: 681–685
Rachford HH (1964) Instability in water flooding oil from water-wet porous media containing connate water. Soc Petro Eng J 231: 133–148
Saffman PG, Taylor GI (1958) The penetration of a fluid into a porous medium or Hele-Shaw cell containing a more viscous liquid. Proc R Soc Lond Ser A245: 312–331
Silliman SE, Wright AL (1988) Stochastic analysis of paths of high hydraulic conductivity in porous media. Water Resour Res 24: 1901–1910
Sposito G, Jury WA, Gupta VK (1986) Fundamental problems in the stochastic convection-dispersion model of solute transport in aquifers and field soils. Water Resour Res 22: 77–88
Starr JL, DeRoo HC, Frink CR, Parlange J-Y (1978) Leaching characteristics of a layered field soil. Soil Sci Soc Am J 42: 386–391
Starr JL, Parlange J-Y, Frink CR (1986) Water and chloride movement through a layered field soil. Soil Sci Soc Am J 50: 1384–1390
Tabuchi T (1961) Infiltration and ensuing percolation in columns of layered glass particles in laboratory (Japanese, with English summary). Nogyo Doboku Kenkyu Besssatsu (Trans Agric Eng Soc Jpn) 1: 13–19
Tamai N, Asaeda T, Jeevaraj CJ (1987) Fingering in two-dimensional, homogeneous, unsaturated porous media. Soil Sci 144: 107–112
Taylor GI (1950) The instability of liquid surfaces when accelerated in a direction perpendicular to their planes. Proc R Soc Lond Ser A201: 192–196

Van Dam JC, Hendrickx JMH, van Ommen HC, Bannink MH, van Genuchten MTh, Dekker LW (1990) Simulation of water and solute transport through a water repellent sand soil. J Hydrol 120: 139–159

Van Ommen HC, Dijksma R, Hendrickx JMH, Dekker LW, Hulshof J, van den Heuvel M (1989) Experimental assessment of preferential flow paths in a field soil. J Hydrol 105: 253–262

White I, Colombera PM, Philip JR (1976) Experimental study of wetting front instability induced by sudden change of pressure gradient. Soil Sci Soc Am J 40: 824–829

White I, Colombera PM, Philip JR (1977) Experimental studies of wetting front instability induced by gradual change of pressure gradient and by heterogeneous porous media. Soil Sci Soc Am J 41: 483–489

Williams RE (1988) Comment on Statistical theory of groundwater flow and transport: pore to laboratory, laboratory to formation, and formation to regional scale. Water Resour Res 24: 1197–1200

Wooding RA (1969) Growth of fingers in an unstable diffusing interface in a porous medium or Hele-Shaw cell. J Fluid Mech 39: 477–495

8 Capillary Barrier at the Interface of Two Layers

H.J. Morel-Seytoux[1]

8.1 The Particular Physical Problem Considered

The system consists of two layers. The soil in the upper layer is fine, which means that its "*effective capillary drive*" (Morel-Seytoux 1987) is high but its saturated hydraulic conductivity is low. The soil of the lower layer, of infinite extent, is coarse, which means that its attributes are, in relative terms, the opposite of those of the fine soil.

A supply rate, r, is applied steadily at the ground surface, of magnitude smaller than the saturated hydraulic conductivity of the fine soil, K_F. The depth of the upper layer, D, is sufficiently large that one can assume that by the time the descending unsaturated wetting front reaches the interface, the flow is taking place strictly by gravity and the water content behind the wetting front is essentially uniform of value θ_0. Figure 8.1 illustrates the geometry of the problem, showing the position of the descending front at relatively early times and at the moment when it first hits the interface.

The question is: what happens at that moment, taken as the origin of time, and at later times t? Only a fraction of the incoming flux, r, will be transmitted. The remaining fraction is reflected and, as a result, a reflected front proceeds upward as shown in Fig. 8.1. Of great interest is the prediction of the evolution with time of the transmitted fraction.

8.2 Conditions at the Interface

Both capillary pressure (expressed as an equivalent height of water) h_c and the water velocity, in the Darcy sense, v_w, must be continuous at the interface. Figure 8.2 shows the capillary pressure curves for the two soils. For convenience, one often deals with normalized water content, defined as:

$$\theta^* = \frac{\theta - \theta}{\tilde{\theta} - \theta_r}, \tag{1}$$

where θ is (volumetric) water content, θ_r is residual water content and $\tilde{\theta}$ is water content at natural saturation. (Subscripts F and C will be used to refer to the fine or coarse soil.)

[1] 57 Selby Lane, Atherton, CA 94027–3926, USA

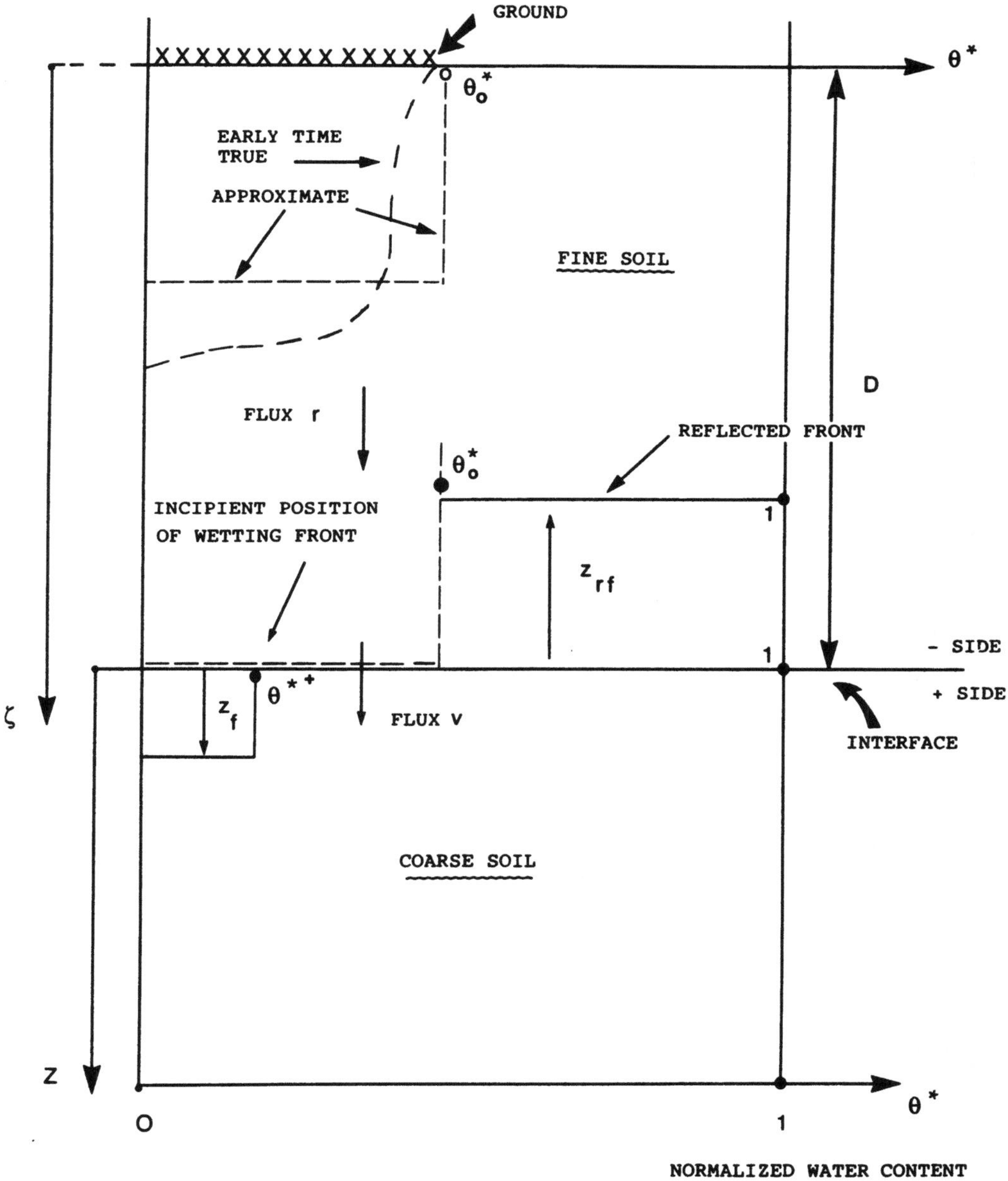

Fig. 8.1. Geometry of problem and positions of reflected and penetrating fronts

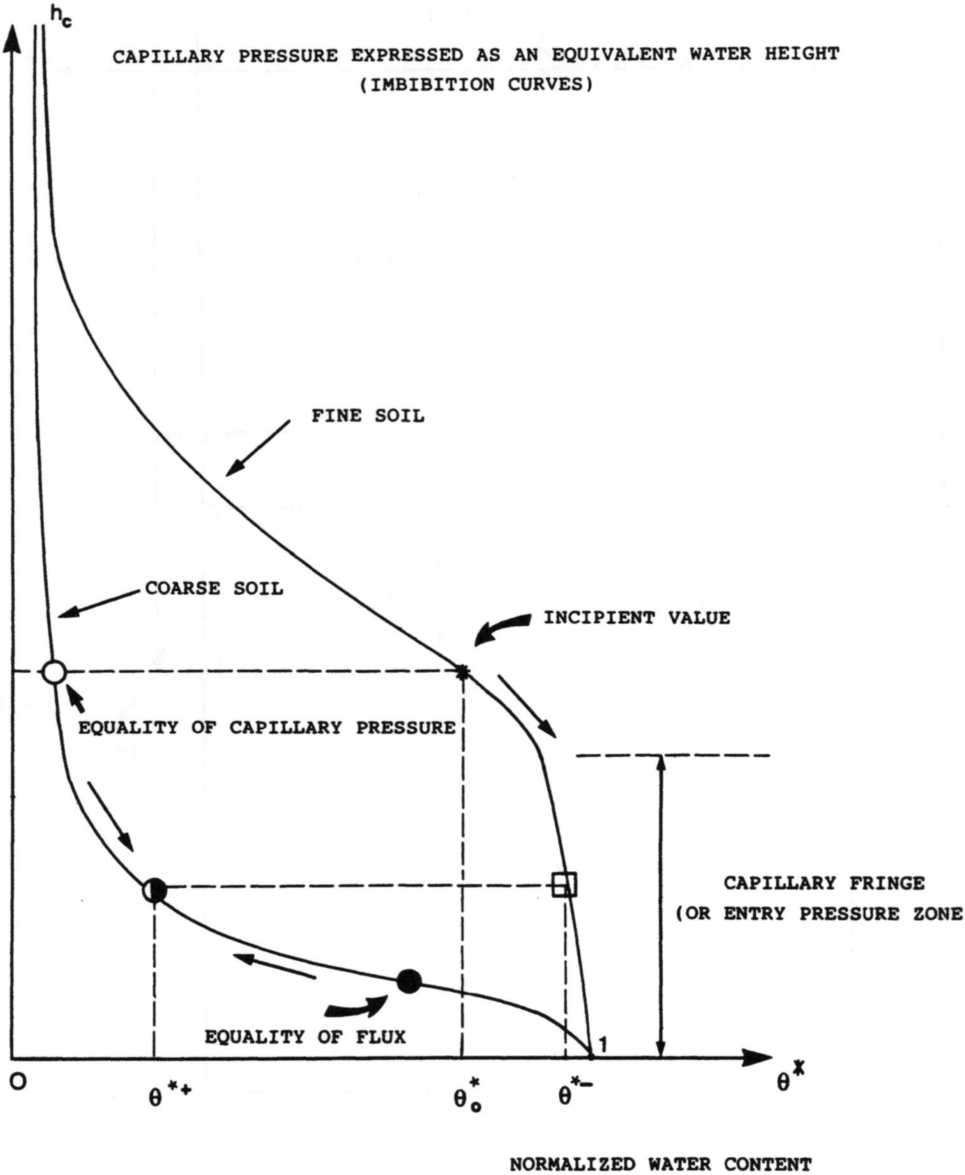

Fig. 8.2. Capillary pressure curves for the fine and coarse soils. Position of system state image points indicating satisfaction of continuity of pressures and fluxes

The point (*) in Fig. 8.2 is the image in the capillary pressure domain of the state of the system at incipient time on the fine soil side (the $^-$side). To satisfy continuity of capillary pressure, the image of the state of the system on the coarse soil side (the $^+$side) is point (○). However, to satisfy continuity of flux, the image should be at point (●), solution of the equation:

$$K_C k_{rwC}(\theta^{*+}) = v_w^- = r, \tag{2}$$

where k_{rw} is relative permeability to water. Clearly, points (○) and (●) do not coincide. This means that the two interface conditions cannot be satisfied if the water content on the fine (soil) side retains its incipient value. It must instantaneously evolve to a greater value (□ in Fig. 8.2) at which the two separate image points (○ and ●) for continuity of capillary pressure and water velocity coalesce into one (◑). At point (□), the water velocity on the $^-$side will have decreased from the incipient value r because now the capillary drive opposes gravity, whereas prior to incipient time capillarity and gravity combine to drive the flux r downward. Due to the relative shape of the two capillary pressure curves, the point (□) will be in the capillary fringe for the fine soil. In other words, θ^{*-} is equal to 1, i.e., $\theta^- = \tilde{\theta}_F$. The common value of the water velocity on the two sides of the interface, v_w^- and v_w^+, will be denoted v.

8.3 Basic Equations

8.3.1 Two-Phase Flow Background and Notations

Total velocity v is defined as the algebraic sum of water and air velocities. Because on the fine soil side of the interface saturation occurs, there water velocity and total velocity coincide. The total velocity can be calculated by application of the usual formula (e.g., Morel-Seytoux and Billica 1984; Morel-Seytoux 1987):

$$v = K_F \frac{\left\{ \int_0^D f_w d\zeta - (H_{cF} - h_{cC}^+) \right\}}{\int_0^D \mu_r d\zeta}, \tag{3}$$

for the fine soil side between levels $\zeta = O$ (ground surface) and $\zeta = D$ (interface location), where ζ is vertical coordinate oriented positive downward with origin at ground surface and H_{cF} is the effectie capillary drive *assuming* zero capillary pressure at the interface. Because the actual capillary pressure at the interface, h_{cC}^+, is not zero, the effective capillary drive is not H_{cF} but

$H_{cF} - h_{cC}^{+}$. The functions f_w and μ_r depend on water content and are defined as:

$$f_w = \frac{k_{rw}}{k_{rw} + \frac{\mu_w}{\mu_a} k_{ra}} \tag{4a}$$

$$\mu_r = \frac{f_w}{k_{rw}}, \tag{4b}$$

where k_{rw} and k_{ra}, μ_w and μ_a are respectively relative permeability and viscosity for water and air.

The effective capillary drive is defined, generally, as:

$$H_c = \int_{h_{c1}}^{h_{c2}} f_w dh_c, \tag{5}$$

where h_{c1} and h_{c2} are prevailing capillary pressure (head) values at two arbitrary elevation levels indicated by subscripts 1 and 2.

8.3.2 Soil Characteristics Representation

8.3.2.1 Capillary Pressure

The analytical representation is the classical power law:

$$h_c = h_{ce}(\theta^*)^{-M} \text{ for } h_c \geqslant h_{ce} \tag{5a}$$

$$\theta^* = 1 \text{ for } h_c \leqslant h_{ce}, \tag{5b}$$

where h_{ce} is the air entry pressure (Corey 1977).

8.3.2.2 Relative Permeability

The chosen analytical representation is:

$$k_{rw} = (\theta^*)^p \text{ for } h_c \geqslant h_{ce} \tag{6a}$$

$$k_{rw} = 1 \text{ for } h_c \leqslant h_{ce}. \tag{6b}$$

8.3.2.3 Effective Capillary Drive

It has been shown (Morel-Seytoux and Khanji 1974) that a reasonable approximation, for a water–air system, is to replace f_w by k_{rw} in Eq. (5).

With this approximation, the H_{cF} appearing in Eq. (3) has the form:

$$H_{cF} = h_{ceF}\left\{1 + \frac{M}{p-M}[1-(\theta_0^*)^{p-M}]\right\}. \tag{7}$$

8.3.3 Propagation Velocities of Fronts

For the front propagating in the coarse soil, that velocity using the general equation (e.g., Morel-Seytoux 1987) is:

$$\frac{dz_f}{dt} = \frac{v}{\theta^+ - \theta_{rC}}. \tag{8}$$

For the particular problem considered, it is assumed that the coarse soil is initially uniformly dry at its residual water content. This assumption is not essential to the solution of the problem. The velocity of propagation for the reflected front is:

$$\frac{dz_{rf}}{dt} = \frac{r - v}{\tilde{\theta}_F - \theta_0}. \tag{9}$$

8.3.4 Total Velocity Expression for Particular Problem

For the particular problem considered here, given the assumed profile shapes, the expression for v is:

$$v = K_F\left\{\frac{f_w(\theta_0)(D - z_{rf}) + z_{rf} - H_{cF} + h_{cC}^+}{\mu_r(\theta_0)(D - z_{rf}) + z_{rf}}\right\}. \tag{10}$$

8.4 System of Equations to Be Solved

To solve the problem, i.e., to find the transmitted flux $v(t)$, one must solve the system of Eqs. (8)–(10) plus the equation:

$$v = k_C k_{rwC}(\theta^{*+}), \tag{11}$$

which expresses the continuity of flux at the interface, for the four unknowns: z_f, z_{rf}, θ^{*+} and v. In Eq. (11) the capillary component which is negligible in a coarse soil, except at very early times, has been neglected.

When the capillary barrier effect is strong, i.e. when the contrast in soil properties is very sharp, at time zero, v will be zero. This implies from Eq. (1)

that the water content on the coarse side must instantaneously assume an initial value such that:

$$h_{cC}(\theta_i^+) = h_{cCi}^+ = H_{cF} - Df_w(\theta_0). \tag{12}$$

Using this result, Eq. (10) can be written in the slightly simpler form:

$$v = K_F \left\{ \frac{[1 - f_w(\theta_0)] z_{rf} + h_{cC}(\theta^+) - h_{cC}(\theta_i^+)}{D\mu_r(\theta_0) + [1 - \mu_r(\theta_0)] z_{rf}} \right\}. \tag{13}$$

However, Eq. (11), which is approximate at very early times, does not yield $v = 0$ for $\theta^+ = \theta_i^+$. For this reason, it is corrected to the form:

$$v = K_C r^* \rho \left\{ \frac{k_{rw}(\theta^+) - k_{rw}(\theta_i^+)}{r^* \rho - k_{rw}(\theta_i^+)} \right\}, \tag{14}$$

where r^* is the ratio r/K_F (normalized supply rate to the upper soil) and $\rho = K_F/K_C$, the permeability contrast. Now Eq. (14) has the proper behaviour for both short and long times. For very long times, v tends to r as it should, having reached its steady-state limit. It is assumed throughout this chapter that the reflected flux is not evacuated laterally. Otherwise, Eq. (9) would require an additional flux term in the numerator of the right-hand side.

8.5 Solution

8.5.1 Procedure

The mathematical problem has been reduced to the solution of Eqs. (9), (13) and (14) for the three unknowns: v, z_{rf} and θ^+. Once v and θ^+ are determined, one can obtain z_f from Eq. (8).

The procedure consists of eliminating v between Eqs. (13) and (14) by equating the right hand-sides. One then expresses z_{rf} from the resulting equation, which is linear in z_{rf}, in terms of θ^+. Next, this expression is differentiated with respect to time, and the new expression for dz_{rf}/dt is equated to the right-hand side of Eq. (9) in which v is expressed in terms of θ^+ using Eq. (14). The result is an ordinary differential equation for the single remaining unknown function of time: $\theta^+(t)$. Because fluxes are of greater interest than water contents, it is more meaningful to write the differential equation for the dependent variables:

$$X = k_{rwC}(\theta^+) \tag{15a}$$

$$Z = X - X_i. \tag{15b}$$

Carrying these procedures, one obtains the differential equation:

$$\frac{dt}{\tilde{\theta}_F - \theta_0} = \frac{(Ah_{ceC} + LZ)dZ}{(Z + X_i)(r - aK_C Z)(b - cZ)} + \frac{c(LZ + Bh_{ceC})dZ}{(r - aK_C Z)(b - cZ)^2}, \tag{16}$$

where, for simplicity in writing, one has defined:

$$a = \frac{r^* \rho}{r^* \rho - X_i} \tag{17a}$$

$$b = [1 - f_w(\theta_0)]\rho \tag{17b}$$

$$L = a\mu_r(\theta_0)D \tag{17c}$$

$$c = a[1 - \mu_r(\theta_0)]. \tag{17d}$$

Equation (16) only applies in a piecewise fashion because the parameters A and B depend upon an intermediate (reference) value X_R between the value of X at the old time t^0, X^0 and its value at the new time t^v, X^v, defined by the relation:

$$X_R = (1 - \lambda)X^0 + \lambda X^v, \tag{18}$$

where λ is a numerical parameter. For $\lambda = 0$, the procedure is explicit; for $\lambda = 1$, it is fully implicit. The parameters A and B are:

$$A = \frac{M_C}{p_C}\rho(X_R)^{-M/p} + \frac{LX_i}{h_{ceC}} \tag{19a}$$

$$B = \rho[X_i^{-M/p} - X_R^{-M_C/p}]. \tag{19b}$$

In Eq. (16), the factors that influence most the behaviour of $Z(t)$ are $(Z + X_i)$ for the short term and $(r - aK_C Z)$ for the long term. The term $(b - cZ)$, on the other hand, remains almost constant and is never close to zero. For this reason, it is convenient to approximate $(b - cZ)$ by $(b - cZ_R)$ in the time interval from t^0 to t^v. With this further approximation, Eq. (16) is easily integrated.

8.5.2 Result of Integration

The final form of the solution is:

$$\frac{t - t^0}{\tilde{\theta}_F - \theta_0} = \frac{(Ah_{ceC} - LX_i)}{(r + aK_C X_i)(b - cZ_R)} \ln\left(\frac{Z + X_i}{Z^0 + X_i}\right) - \frac{cL(Z - Z^0)}{(b - cZ_R)^2 aK_C} + \frac{1}{aK_C(b - cZ_R)}\left\{\frac{rL + aK_C Ah_{ceC}}{r + aK_C X_i} + \frac{c[rL + aK_C Bh_{ceC}]}{aK_C(b - cZ_R)}\right\} \ln\left(\frac{r - aK_C Z^0}{r - aK_C Z}\right). \tag{20}$$

8.5.3 Numerical Procedures

8.5.3.1 Initialization

Given the values of the various input parameters, such as D, r and the parameters of the two soils, one calculates $\theta_0^* = (r^*)^{1/p_F}$, then H_{cF} given by Eq. (7). The values of $f_w(\theta_0)$ and $\mu_r(\theta_0)$ are obtained from their definitions. One then calculates X_i as:

$$X_i = \left[\frac{h_{ceC}}{H_{cF} - Df_w(\theta_0)}\right]^{pc/Mc} . \qquad (21)$$

The parameters ρ, a, b, c and Z are obtained from their definitions.

8.5.3.2 Repetitive Steps

Having proceeded to obtain Z^0 at time t^0, one wishes to determine Z^v at time t^v. Actually, it is easier to select Z^v as $Z^0 + \Delta Z$, and then calculate t^v. Having selected the proper growth increment ΔZ and thus Z^v, Z_R is obtained from its definition and t^v obtained from Eq. (20) by replacing Z by the numerical value Z^v. One then proceeds to the next time interval and the curve $Z(t)$ is thus defined by a set of pairs (Z, t) obtained at a discrete set of values for Z.

8.6 Applications

8.6.1 Values of Parameters for Reference Run

The supply rate r has a value of 0.2 mm/yr. The depth of the top layer is $D = 2$ m. The soil parameters for the fine soil are: $\theta_r = 0.2$, $\tilde{\theta} = 0.5$, $h_{ce} = 400$ mm, $M = 4$, $K = 2$ mm/yr and $p = 11$. For the coarse soil, the values are: $\theta_r = 0.1$, $\tilde{\theta} = 0.4$, $h_{ce} = 20$ mm, $M = 1$, $K = 20$ mm/yr and $p = 5$. The value of the numerical scheme parameter λ is 0.5 (Crank-Nicholson scheme).

8.6.2 Reference Run Results

For this set of parameters (reference run), Fig. 8.3 displays the evolution wih time of the transmitted fraction across the interface. It is interesting to note that for about 130 years the transmission rate is essentially zero. Once sees here a manifestation of the "capillary barrier" effect. Even though the coarse material is quite capable of transmitting the flux r, which in this case is 1% of its saturated hydraulic conductivity, for 130 years essentially none of the flux is transmitted. The water is held back in the fine soil. Indeed,

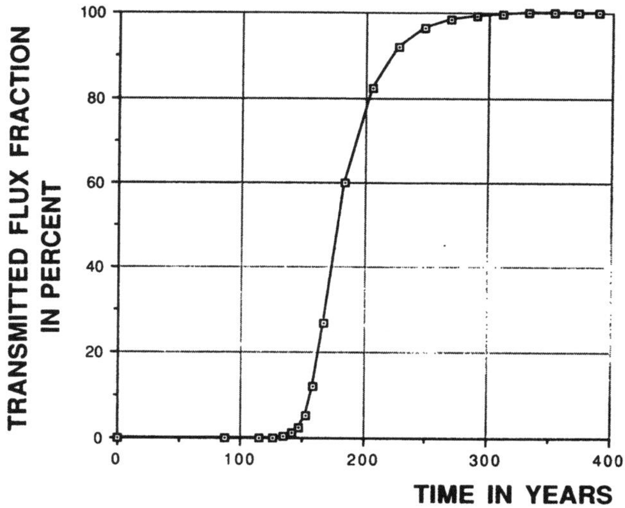

Fig. 8.3. Evolution of transmitted fraction with time for the reference run

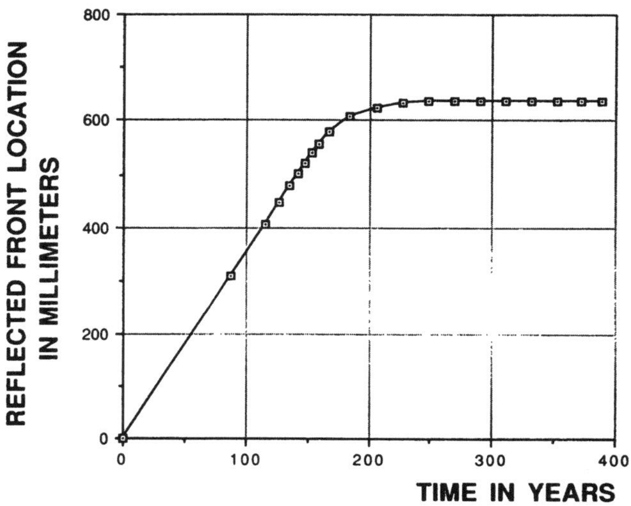

Fig. 8.4. Evolution of water mound above interface with time for the reference run

Fig. 8.4 shows the rise in water depth over the interface. For 130 years the curve is straight. All the incoming water piles up on top of the interface. Figure 8.5 shows the depth of penetration of the wetting front in the coarse soil. For 150 years there is no noticeable penetration. Figure 8.6 shows the water suction of the interface. The suction decreases with time, and at around 130 years its value is about 100 mm. Eventually, for long times (steady state),

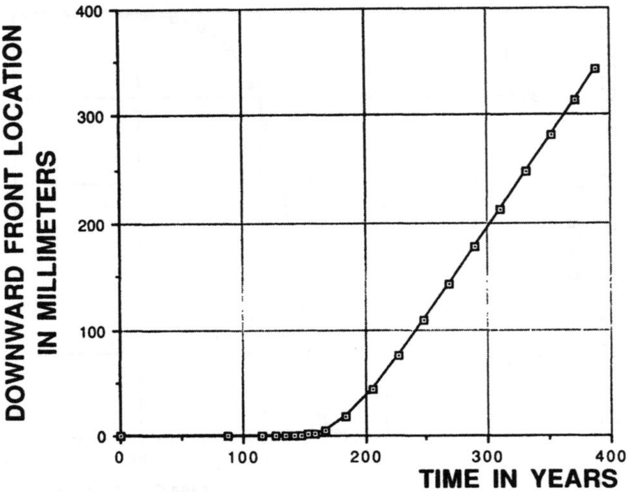

Fig. 8.5. Depth of penetration of wetting front in coarse soil for the reference run

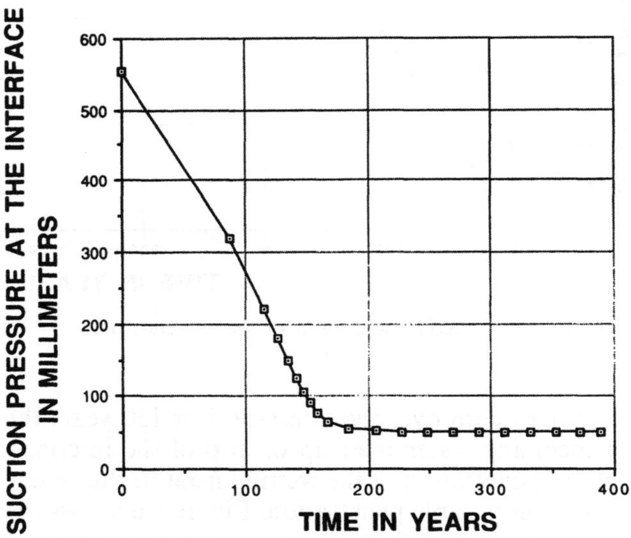

Fig. 8.6. Water suction at the interface for the reference run

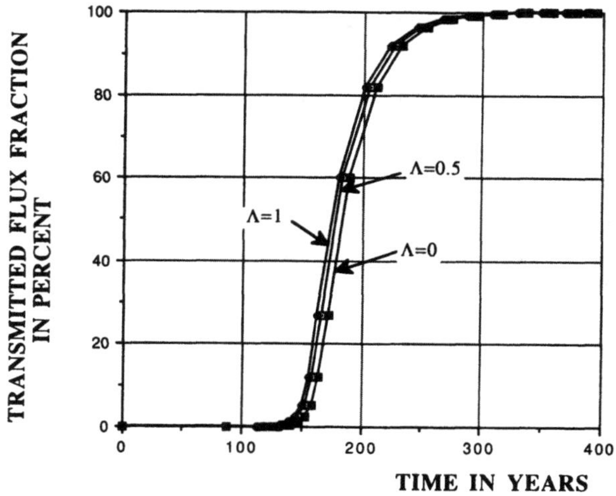

Fig. 8.7. Influence of numerical scheme parameter, λ, on the transmitted fraction

the suction reaches a limit of 50 mm, well within the capillary fringe of the fine soil but clearly outside that of the coarse soil.

8.6.3 Numerical Scheme Parameter Sensitivity

The influence of the value of the parameter λ is shown in Fig. 8.7, for the same data set of the Reference run. There is very little difference in the results. The value 0.5 is recommended since there is no problem of stability in this numerical procedure.

8.6.4 Influence of Effective Capillary Drive Magnitude

For the same data as in the Reference run, the effect of halving the value of the entry pressure in the fine soil h_{ceF} from 400 to 200 mm has been tested. The behaviour is as expected, though the influence is not a mere proportionality or translation as Fig. 8.8 would seem to suggest. The dependence of time in Eq. (20) on h_{ceF} is felt directly and indirectly through H_{cF} [See Eq. (7)] X_i [see Eq. (21)], A and B [Eqs. (19a) and (19b)]. It is a complex one.

8.6.5 Influence of Supply Rate

Again, the same data as in the Reference run are used, but the supply rate is doubled. Results are shown in Fig. 8.9. Naturally, the faster the supply rate, the faster the mound rise above the interface and the decline of suction at the interface. The dependence of t on r in Eq. (20) is again quite complex and is definitely not a simple proportionality.

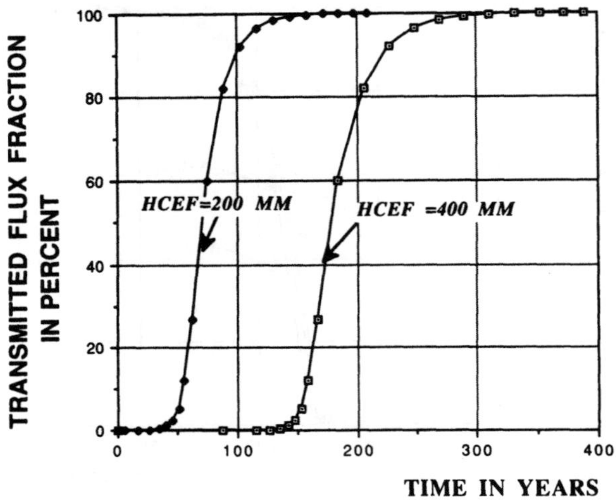

Fig. 8.8. Influence of effective capillary drive on the transmitted fraction

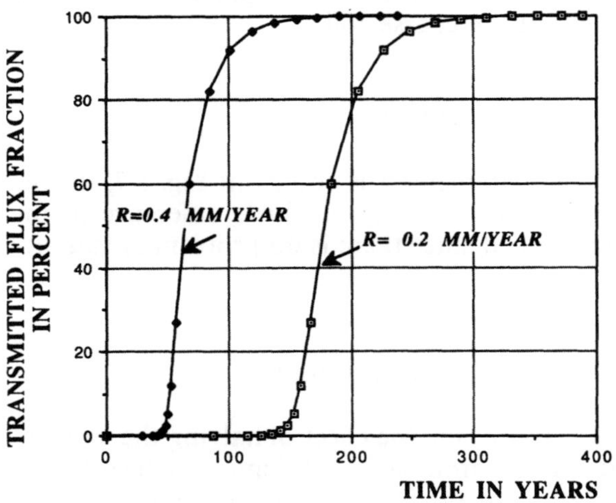

Fig. 8.9. Influence of supply rate on the transmitted fraction

8.7 Discussion

It is clear from the results that the influence of the various parameters combines to give the transmitted fraction curve a characteristic shape. At first, the curve hardly deviates from the time axis and does not do so for a significant length of time (130 years in Fig. 8.3). Then the process accelerates and the fraction evolves from 20 to 80% in roughly 40 years. It takes another 100 years to essentially reach the 100% value. As a rough approximation, one could reduce to response to a step function. In the case of Fig. 8.3, one could take the fraction to be zero up to year 180 and then one after that. If one were to do so, there would remain the need to determine a physical criterion to define the threshold level at which the transmission fraction changes rapidly from essentially zero to practically one.

The literature, particularly in textbooks, seems to indicate the belief that a threshold, before which *no* flux in transmitted at all, does exist. However, no rigorous treatment, based on physical principles, of this effect is presented. It is sometimes stated that water will not penetrate the coarse material until the suction at the interface is zero. It is shown here that this is not the case. It is also said that water will not penetrate the coarse soil until the mounding height, z_{rf}, exceeds the entry pressure of the fine soil. In the reference run, no significant flow occurs until z_{rf} reaches the value 400 mm, the entry pressure. However, prior to that, when the value of z_{rf} is 200 mm, the value of the transmitted flux is very small but is *not* zero. The various rules for the breakthrough of water into the coarse soil developed empirically from experimental evidence in the past. The results here suggest that, though these rules are not without merit, they fail to portray the full physical complexity of the "capillary barrier" effect. Such rules should not solely depend upon the entry pressure of the fine soil. The physical properties of the coarse soil also affect the solution. Such rules may have merit, especially in situations where the contrast between the two soils is extreme, but for moderate levels they may be quite misleading.

The depth of the interface has a role in the process of transmission. For example, the analysis in this chapter applies for the case when H_{cF} exceeds $Df_w(\theta_0)$ significantly, otherwise $h_{cc}(\theta_i^+)$ in Eq. (12) would be negative. This may occur if both D and r, and thus $f_w(\theta_0)$, are large. In such a situation, a nonzero flux would be transmitted at incipient time. There would be no capillary barrier effect at all. Physically, this means that because $Df_w(\theta_0)$ is large relative to H_{cF}, there is more weight of water in the soil at incipient time than the fine soil effective capillarity can hold under static equilibrium. Some of that water must drain immediately into the coarse soil. The analysis presented here does not apply in this case. The more complete analysis is presented in a separate document (Morel-Seytoux 1992).

8.8 Conclusions

The exploitation of the analytical solution presented in this chapter is quite simple compared with the use of numerical schemes such as finite differences or elements. It is also intrinsically more accurate.

The solution has many applications. It will be useful in the study of designs of buried toxic waste repositories and landfill disposal. For example, the value of the supply rate used in the numerical illustration corresponds to an estimated recharge rate at the Yucca Mountain project site.

The solution has applications in traditional hydrology for the proper description of runoff generation and, in particular, for the description of interflow. In hydrology, the case of a coarse soil overlying a fine one is more important because it is in this situation that a significant amount of rapid interflow can occur. Though not specifically treated in this chapter it is clear that the methodology is applicable to this situation with minor changes. In such a case, both sides of the interface are likely to be saturated.

The solution has applications for studies of contamination of groundwater by organic spills (oil tank, gasoline tank leakage, etc.). In this case, the displacement is that of water by a nonwetting phase. The capillary barrier effect will be stronger, everything else being the same, because the relevant capillary pressure curves are those of drainage, and the effective capillary drive is lager.

The approach presented here can be incorporated into numerical schemes. Since the solution's behaviour is naturally *singular* (meaning existence of very sharp gradients) at the interface, the traditional numerical schemes are not appropriate because their approximations presume a well-behaved solution. A better approximation is obtained by combining the numerical and analytical procedures.

Finally, because the solution is analytical and explicit, it is remarkably easy to incorporate the influence of uncertainty about some parameters on the behaviour of the solution. For example, suppose one was interested in knowing the time at which the transmitted fraction is of a given percentage, say 50%, but the value of K_C was uncertain. Considering this parameter to be a random variable with a known statistical distribution, then t in Eq. (20) for a fixed value of Z would be a random variable. Given the distribution of K_C, one can use Eq. (20) to derive the distribution of the time t for the 50% value of the transmitted fraction, or at least its expected value and variance. Naturally, one can test the sensitivity of that time to any change in all the many parameters of the problem, and sort out that way the relative importance of the various factors in shaping the solution.

Acknowledgements. A first version of this work was prepared while the author was on a NORCUS visiting assignment at the Battelle Pacific Northwest Laboratories in Richland, Washington, and was presented in a seminar there on June 4, 1991. The author wishes to thank especially Lance Vail and the Department of Geosciences, Rick Skaggs, and Marcel Bergeron for making this assignment possible and also thanks him and his colleagues Charlie Cole, Charlie Kincaid, Glendon Gee, Bob Lenhard, Steve Simmons and Steve Yabusaki for useful discussions on this particular subject and, in general, on hydrology, flow in porous media, solute transport and numerical techniques. That visit was most instructive and rewarding.

References

Corey AT (1977) Mechanics of heterogeneous fluids in porous media. Water Resources Publications, Fort Collins, Colorado

Morel-Seytoux HJ (1987) Multiphase flows in porous media. In: Novak P (ed) Developments in hydraulic engineering, vol. 4. Elsevier Applied Science, London, 103 p

Morel-Seytoux HJ (1992) The capillary barrier effect at the interface of two soil layers with some contrast in properties. Hydrology Days Publications, Report 92.4, 57 Selby Lane, Atherton, California, USA

Morel-Seytoux HJ, Billica JA (1984) A two-phase numerical model for prediction of infiltration: applications to a semi-infinite soil column. Water Resour Res J 21(4): 607–15

Morel-Seytoux HJ, Khanji J (1974) Derivation of an equation of infiltration. Water Resour Res J 10(4): 795–800

List of Symbols

a	composite parameter defined by Eq. (17a)
b	composite parameter defined by Eq. (17b)
c	composite parameter defined by Eq. (17d)
c	also as a subscript for "capillary"
f_w	little "f" function, defined by Eq. (4a)
g	acceleration of gravity
h_a, h_w	respectively air and water pressures, expressed as equivalent heights of water
h_c	capillary pressure expressed as an equivalent height of water
h_{ce}	entry pressure
h_{cF}, h_{cC}, h_{cCi}	capillary pressure respectively in the fine and coarse soil, at incipient time
k_{ra}, k_{rw}	relative permeability respectively to air and water
p	parameter in the power law for relative permeability
r	supply rate (length per time) at the soil surface
r^*	normalized supply rate
t, t^0, t^v	time, at beginning of interval, at end of interval
v	total velocity
v_w	water velocity in the Darcy sense
z	vertical coordinate oriented positive downward
z_f	depth of penetration of wetting front below the interface
z_{rf}	elevation of reflected front above the interface
A, B	composite parameters defined by Eqs. (19a) and (19b)
C	subscript for "coarse"
D	thickness of upper fine soil
H_c, H_{cM}	effective capillary drive (dimension of length), its maximum value
F	subscript for "fine"
K	hydraulic conductivity at natural saturation
L	composite parameter defined by Eq. (17c)
M	parameter in power law for capillary pressure
X, X_i, X_R	relative permeability to water in the coarse soil, its initial value at incipient time, value intermediate between that at beginning and at end of time interval
X^0, X^v	values of X at time t^0 and at time t^v
Z	$X - X_i$
ζ	vertical coordinate with origin at soil surface
θ	water content
θ^*	normalized water content, defined by Eq. (1)
$\tilde{\theta}$	water content at natural saturation
θ_r	residual water content
θ_0	uniform water content associated with flux r in fine soil
θ_i	initial value of water content just below the interface at incipient time
λ	parameter in numerical scheme
μ_w, μ_a	viscosity respectively of water and air
μ_r	total relative vicosity defined by Eq. (4b)
ρ	ratio of hydraulic conductivity in the fine soil over that in the coarse soil. If the contrast is sharp, ρ is small
$-$, $+$	superscripts referring to a location just above, just below the interface

9 Constant-Rainfall Infiltration on Hillslopes and Slope Crests

J.R. Philip[1]

9.1 Introduction

Eshel Bresler was single-minded in his pursuit of the problems of water flow in unsaturated soils and the associated transport of solutes. He was as adept in analytical methods as in numerical ones, and he worked with both linear and nonlinear models. An early paper (Bresler 1967) prophetically spanned the range of his concerns. That paper, which continues to receive its share of citations, made fruitful use of linear mathematics. In this contribution we also use both linear and nonlinear analysis to seek insight into fundamentally nonlinear processes of soil-water movement, with the detailed results based on linearization.

The nonlinear problem of ponded infiltration and downslope unsaturated flow in a long planar hillslope of homogenous isotropic soil at uniform initial moisture content was solved recently (Philip 1991a). Related nonlinear investigations of infiltration into divergent and convergent slopes (Philip 1991b) and into convex and concave topography (Philip 1991c) showed that the planar slope solution may be applied in such cases, so long as the reciprocal of the total curvature of the soil surface is of larger magnitude than the sorptive length of the soil. The later studies thus served to widen considerably the applicability of the planar hillslope solution.

The planar hillslope investigation (Philip 1991a) was much simplified by the assumption that, except in a small region immediately below the slope crest (or a point of slope angle change), the solution is independent of the downslope coordinate and is a function only of time and the space coordinate normal to the slope. Philip (1991a) included a primitive (and, in the event, very conservative) estimate of the magnitude of the region where the long-slope solution was inaccurate. The wide potential applicability of the long-slope solution made a more careful study desirable; and this was provided by a linearized study of the slope-crest effect (Philip 1992).

These various investigations were for ponded infiltration, and analogous studies of constant-rainfall infiltration on hillslopes and slope crests are clearly needed.

For horizontal topography the problems of constant-rainfall infiltration are, in general, mathematically less tractable than those of ponded

[1] CSIRO Centre for Environmental Mechanics, GPO Box 821, Canberra, Australian Capital Territory, 2601, Australia

infiltration. This greater difficulty carries over to slopes and crests. Whereas quasianalytical solution of long-slope ponded infiltration was relatively straightforward (Philip 1991a), nothing comparable can be done for the constant-rainfall case. Certain special nonlinear exact results are possible, but we limit ourselves here to linearized solutions.

9.2 Constant-Rainfall Slope-Crest Infiltration: Nonlinear Formulation

9.2.1 Flow Equation and Conditions

We begin with the formulation of the exact nonlinear problem for constant-rainfall slope-crest infiltration. We write the flow equation for water in unsaturated soil as:

$$\frac{\partial \theta_*}{\partial t_*} = \frac{\partial}{\partial x_*}\left(D\frac{\partial \theta_*}{\partial z_*}\right) + \frac{\partial}{\partial z_*}\left(D\frac{\partial \theta_*}{\partial z_*}\right) - \frac{\mathrm{d}K}{\mathrm{d}\theta}\cdot\frac{\partial \theta_*}{\mathrm{d}z_*}. \tag{1}$$

We seek the solution of Eq. (1) subject to the conditions:

$$\begin{aligned}
&t_* = 0, \quad -\infty \leqslant x_* \leqslant \infty, \quad z_* > |x_*|\tan\gamma, \quad \theta_* = \theta_{0*};\\
&t_* \geqslant 0, \quad -\infty \leqslant x_* \leqslant \infty, \quad z_* = |x_*|\tan\gamma,\\
&K + D\left(\left|\frac{\partial \theta_*}{\partial x_*}\right|\tan\gamma - \frac{\partial \theta_*}{\partial z_*}\right) = R_*.
\end{aligned} \tag{2}$$

Here, t_* is time, θ_* is volumetric moisture content, D the moisture diffusivity, and K the hydraulic conductivity. The horizontal Cartesian space coordinate is x_*, and the vertical one, taken positive downward, is z_*. The slope crest is at $(x_*, z_*) = (0, 0)$. The slope angle is $\gamma(>0)$, in both the positive and negative x_* directions (Fig. 9.1). θ_{0*} is the uniform initial θ_* value. The constant-rainfall rate, normal to the horizontal, is R_*. The second of conditions (2) equates, at the slope surface, the components normal to that surface of precipitation and of soil-water flow velocity.

9.2.2 Time-to-Ponding: The Two Cases

Equation (1) is limited to flow systems with the moisture potential $\Psi \leqslant 0$, and θ_* cannot exceed its value at saturation, θ_{1*}. We need to discriminate between two cases, depending on the relative magnitudes of R_* and of $K(\theta_{1*}) = K_1$, the saturated hydraulic conductivity.

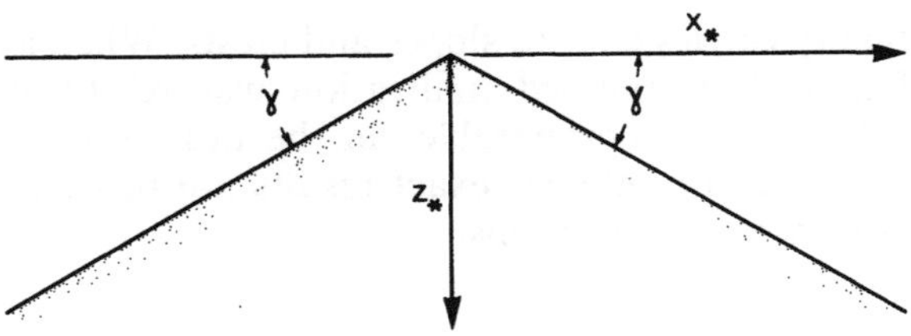

Fig. 9.1. Scheme illustrating physical Cartesian space coordinates (x_*, z_*)

9.2.2.1 Case 1: $R_* \leqslant K_1$

In this case the maximum permissible value of θ_*, θ_{1*}, is attained only in the case $R_* = K_1$, and only in the limit as $t_* \to \infty$, so the solution of Eq. (1) subject to conditions (2) applies for all values of $t_* \geqslant 0$.

9.2.2.2 Case 2: $R_* > K_1$

For these larger rainfall rates soil saturation and incipient ponding [at the slope crest $(x_*, z_*) = (0, 0)$] will occur at some finite value of t_*, T_*. In this case the solution given by Eq. (1) subject to conditions (2) is valid only up to this critical event of incipient ponding, and thus applies only for $0 \leqslant t_* \leqslant T_* < \infty$.

9.2.3 Remark

The solution of Eq. (1) subject to conditions (2) can, in principle, be obtained by machine computation using numerical methods. Exploring the various possibilities by the brute-force approach would be demanding of resources and labour. Here we seek, rather, to use analytical methods to develop a general understanding of crest and slope effects. One appropriate strategy is the linearization technique first used by Philip (1966a, 1969) and subsequently employed by Bresler (1973), Warrick (1974, 1975), Lomen and Warrick (1974), Warrick and Lomen (1976), Ben Asher et al. (1978), and Philip (1986).

9.3 Constant-Rainfall Slope-Crest Infiltration: Linearized Formulation

9.3.1 Flow Equation and Conditions

A convenient linearized form of Eq. (1) is (Philip 1986, 1992):

$$\frac{1}{\bar{D}}\frac{\partial \Theta}{\partial t_*} = \frac{\partial^2 \Theta}{\partial x_*^2} + \frac{\partial^2 \Theta}{\partial z_*^2} - \alpha \frac{\partial \Theta}{\partial z_*}, \tag{3}$$

where the Kirchhoff potential

$$\Theta = \int_{\theta_{0*}}^{\theta_*} D(\theta_*)\,\mathrm{d}\theta_* = \int_{\Psi_0}^{\Psi} K(\Psi)\,\mathrm{d}\Psi. \tag{4}$$

Here $\Psi_0 = \Psi(\theta_0)$. The constants $\bar{D}$ and α should be chosen to match optimally the nonlinear capillary and hydraulic properties of the soil over the appropriate θ_* range, which we denote by:

$$\theta_{0^*} \leqslant \theta_* \leqslant \theta_{R^*}. \tag{5}$$

For $R_* \leqslant K_1 = K(\theta_{1^*})$ with θ_{1^*} being the saturated moisture content and K_1 the saturated conductivity, θ_{R^*} is defined by the equation:

$$K(\theta_{R^*}) = K_R = R_*, \tag{6}$$

On the other hand, for $R_* > K_1$.

$$\theta_{R^*} = \theta_{1^*}. \tag{7}$$

Thus, since K is a monotonic increasing function of θ_*, we have that

$$\begin{aligned} &\text{for } R_* < K_1, \quad \theta_{R^*} < \theta_{1^*} \text{ and } K_R < K_1, \\ &\text{and for } R_* \geqslant K_1, \quad \theta_{R^*} = \theta_{1^*} \text{ and } K_R = K_1. \end{aligned} \tag{8}$$

We now use a matching procedure consistent with that of the linearized formulation of ponded infiltration on a slope crest (Philip 1992). We take:

$$\bar{D} = \frac{4}{\pi}\left[\frac{\Theta_R}{S_R}\right]^2, \tag{9}$$

where

$$\Theta_R = \Theta(\theta_{R^*}) = \int_{\theta_{0*}}^{\theta_{R^*}} D(\theta_*)\mathrm{d}\theta_* = \int_{\Psi_0}^{\Psi_R} K(\Psi)\mathrm{d}\Psi \tag{10}$$

and S_R is the sorptivity (Philip 1957, 1969) for the case with initial moisture content θ_{0^*} and applied surface moisture content θ_{R^*}. Ψ_R denotes $\Psi(\theta_{R^*})$. The constant α [dimensions $(\text{length})^{-1}$] is the sorptive number entering the exponential representation of $K(\Psi)$,

$$K(\Psi) = K_R \exp[\alpha(\Psi - \Psi_R)], \quad \Psi \leqslant \Psi_R, \quad \alpha > 0, \tag{11}$$

which yields the quasilinear analysis of multidimensional unsaturated flows (Philip 1966b, 1968, 1969). Adopting the value of α given by Eq. (11) ensures that, for flows approaching a steady state at large times, the solution of Eq. (3) in the limit as $t_* \to \infty$ is the corresponding steady quasilinear solution.

9.3.2 Dimensionless Forms

It is convenient to express Eq. (3) in the dimensionless form:

$$\frac{\partial\theta}{\partial t}=\frac{\partial^2\theta}{\partial x^2}+\frac{\partial^2\theta}{\partial z^2}-2\frac{\partial\theta}{\partial z}. \tag{12}$$

We normalize lengths with respect to the sorptive length, $2\alpha^{-1}$ (Philip 1987a, 1989). The various dimensionless quantities are then defined by:

$$\theta=\frac{\Theta}{\Theta_R},\quad \frac{x}{x_*}=\frac{z}{z_*}=\tfrac{1}{2}\alpha,\quad \frac{t}{t_*}=\frac{T}{T_*}=\tfrac{1}{4}\alpha^2\bar{D},\quad \frac{R}{R_*}=\frac{2}{\alpha\Theta_R}. \tag{13}$$

Note that R_*, and later various physical flow velocities, are normalized with respect to $\tfrac{1}{2}\alpha\Theta_R$. The dimensionless form of conditions (2) is then:

$$t=0,\quad -\infty\leqslant x\leqslant\infty,\quad z\geqslant|x|\tan\gamma,\quad \theta=0;$$

$$t\geqslant 0,\quad -\infty\leqslant x\leqslant\infty,\quad z=|x|\tan\gamma,\quad 2\theta+\left|\frac{\partial\theta}{\partial x}\right|\tan\gamma-\frac{\partial\theta}{\partial z}=R. \tag{14}$$

9.3.3 Time-to-Ponding: The Two Cases

Without significant loss of generality we take θ_{0*} small enough for $K(\theta_{0*})$ to be negligibly small. This is consistent with taking $\Psi_1=-\infty$. It then follows from (10) and (11) that $K_R=\alpha\Theta_R$, so that R_* (and various physical flow velocities) are normalized also with respect to $\tfrac{1}{2}K_R$. The dimensionless restatement of the two cases identified in Section 9.2.2 is then as follows:

Case 1: $R=2$, solution of Eq. (12) subject to conditions (14) valid for all values of $t\geqslant 0$.

Case 2: $R>2$, solution of Eq. (12) subject to conditions (14) valid only for $0\leqslant t\leqslant T<\infty$. (15)

9.3.4 Remark

Even though Eq. (12) and conditions (14) are linear, it does not appear possible to solve Eq. (12) subject to conditions (14) in useful simple closed form for γ arbitrary. Later (Sect. 9.7), we show that the special case $\gamma=45°$ is readily solved, but we first examine the constant-rainfall long-slope solutions.

9.4 Long-Slope Constant-Rainfall Infiltration: Nonlinear Formulation

Investigating ponded infiltration was greatly facilitated by the long-slope assumption (Philip 1991a), and the same advantages obtain here. It has been shown (Philip 1992) that, for ponded infiltration, at all points further than a distance of order one sorptive length downslope of the crest (or slope change), long-slope infiltration dynamics may be applied. Accordingly, here we proceed to study constant-rainfall infiltration on a planar slope at points far enough downslope for the long-slope assumption to apply; i.e., we take the solution independent of downslope distance and a function only of t_* and the space coordinate normal to the slope.

9.4.1 Flow Equation and Conditions

It follows immediately that:

$$\frac{\partial}{\partial x_*} = -\tan\gamma \cdot \frac{\partial}{\partial z_*}, \tag{16}$$

so that, as in Philip [1991a, Eq. (13)], Eq. (1) may be written as:

$$\frac{\partial \theta_*}{\partial t_*} = \sec^2\gamma \cdot \frac{\partial}{\partial z_{1*}}\left[D\frac{\partial \theta_*}{\partial z_{1*}}\right] - \frac{dK}{d\theta}\cdot\frac{\partial \theta_*}{\partial z_{1*}}, \tag{17}$$

where the vertical physical coordinate z_{1*} is chosen such that $z_{1*} = 0$ at the slope surface, i.e.,

$$z_{1*} = z_* - |x_*|\tan\gamma. \tag{18}$$

Using Eqs. (16) and (18) in Eq. (2) yields the conditions on Eq. (17):

$$\begin{aligned} &t_* = 0, \quad z_{1*} > 0, \quad \theta_* = \theta_{0*}; \\ &t_* \geqslant 0, \quad z_{1*} = 0, \quad K - D\sec^2\gamma \cdot \partial\theta_*/\partial z_{1*} = R_*. \end{aligned} \tag{19}$$

9.4.2 Remark

The solution of Eq. (17) subject to conditions (19) for arbitrary $K(\theta_*)$ and $D(\theta_*)$ may be obtained, with some trouble, through numerical methods and machine computation. On the other hand, analytical solutions may be had for the Broadbridge–White (Broadbridge and White 1988a, b), the Burgers–Knight (Philip 1974, 1987b), and the linear soils. We proceed here to the linearized analysis.

9.5 Long-Slope Constant-Rainfall Infiltration: Linearized Formulation

9.5.1 Flow Equation and Conditions, Dimensionless Forms

We apply the linearization and normalization techniques of Section 9.3. The linear dimensionless forms of Eq. (17) subject to conditions (19) are then:

$$\frac{\partial\theta}{\partial t} = \sec^2\gamma\frac{\partial^2\theta}{\partial z_1^2} - 2\frac{\partial\theta}{\partial z_1}. \tag{20}$$

$$t = 0, \quad z_1 > 0, \quad \theta = 0;$$

$$t \geqslant 0, \quad z_1 = 0, \quad 2\theta - \sec^2\gamma\,\partial\theta/\partial z_1 = R. \tag{21}$$

Note that $z_1 = \frac{1}{2}\alpha z_{1*}$. The further substitutions:

$$\tau = t\cos^2\gamma, \quad z_2 = z_1\cos^2\gamma, \tag{22}$$

reduce Eq. (20) and conditions (21) to the following forms, independent of γ:

$$\frac{\partial\theta}{\partial\tau} = \frac{\partial^2\theta}{\partial z_2^2} - 2\frac{\partial\theta}{\partial z_2} \tag{23}$$

$$\tau = 0, \quad z_2 > 0, \quad \theta = 0;$$

$$\tau \geqslant 0_1, \quad z_2 = 0, \quad 2\theta - \partial\theta/\partial z_2 = R. \tag{24}$$

9.5.2 Solution

The solution of Eq. (23) subject to conditions (24) is well known (Mason and Weaver 1924; Storm 1951; Braester 1973). It is:

$$\theta = \tfrac{1}{2}RG(2z_2, 4\tau), \tag{25}$$

where

$$G(X, Y) = \frac{1}{2}\left\{\operatorname{erfc}\frac{X - Y}{2Y^{1/2}} - (1 + X + Y)\exp(X)\operatorname{erfc}\frac{X + Y}{2Y^{1/2}}\right\} + \left(\frac{Y}{\pi}\right)^{1/2}\exp\left[-\left(\frac{X - Y}{2Y^{1/2}}\right)^2\right] \tag{26}$$

Various properties of the function $G(X, Y)$ are set out in the Appendix.

In Section 9.6 we explore the various physical implications of this long-slope solution, including:

1. The time course of the horizontal and downslope components of unsaturated flow velocity, and of their spatial integrals.
2. The variation with time of the flow velocity vector at the slope surface.
3. The dependence of time-to-ponding T_* on rainfall rate R_* and slope angle γ.

9.6 Physical Implications of Long-Slope Solutions

In this section we explore various implications of the constant-rainfall long-slope solutions and give illustrative examples. We defer until Section 9.9

a comparative discussion of these results and the analogous ones for ponded infiltration (Philip 1991a, 1992).

9.6.1 Distribution of Potential and Moisture Content

Evolution of the moisture content and potential profiles is described by $\theta(z_2, \tau)$. For a particular soil with $D(\theta)$ known, there is a one-to-one relation between θ and the volumetric moisture content θ_*; and, when the moisture characteristic $\Psi(\theta_*)$ is known, there is a one-to-one relation also between θ and Ψ (Cf. Philip 1992). Of special interest is $\theta(0, \tau)$, giving the time course of moisture content and potential at the slope surface. It follows from Eq. (25) that:

$$\begin{aligned} 2\theta(0,\tau)/R &= G(0,4\tau) \\ &= 1-(1+2\tau)\operatorname{erfc}\tau^{1/2}+2\left(\frac{\tau}{\pi}\right)^{1/2}\exp(-\tau). \end{aligned} \tag{27}$$

We note the expansions for small and large τ:

$$2\theta(0,\tau)/R = 4\left(\frac{\tau}{\pi}\right)^{1/2}\left[1-\frac{(\pi\tau)^{1/2}}{2}+\frac{\tau}{3}-0(\tau^{3/2})\right], \tag{28}$$

$$2\theta(0,\tau)/R = 1-(\pi\tau^3)^{-1/2}e^{-\tau}[1-3\tau^{-1}+0(\tau^{-2})]. \tag{29}$$

Figure 9.2 shows the time course of surface moisture content and potential in the reduced form $2\theta(0,\tau)/R$.

Figure 9.3 shows the evolution of the profiles in the form $2\theta(z_2,\tau)/R$ for $\tau = 0.01$, 0.25, and 1. The effect of slope angle on profile evolution is illustrated in Fig. 9.4, which compares profiles in the form $2\theta(z_1, t)/R$ for $\gamma = 0$, 25.842, and 36.870, and for $t = 0.25$.

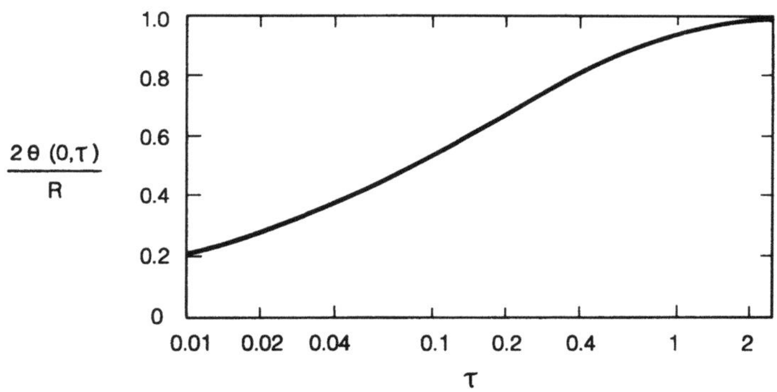

Fig. 9.2. Long-slope constant-rainfall infiltration. Time course of surface moisture content and potential in the dimensionless form $2\theta(0,\tau)/R$. θ and R are, respectively, normalized Kirchhoff potential and normalized rainfall rate. τ is time, normalized to be independent of slope angle γ

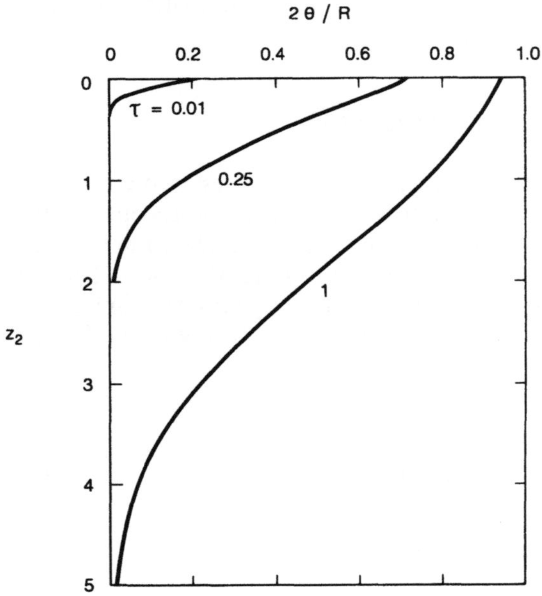

Fig. 9.3. Long-slope constant-rainfall infiltration. Evolution of moisture profiles in the dimensionless form $2\theta(z_2, \tau)/R$ for $\tau = 0.01, 0.25$, and 1. z_2 is the space coordinate, normalized to be independent of slope angle γ

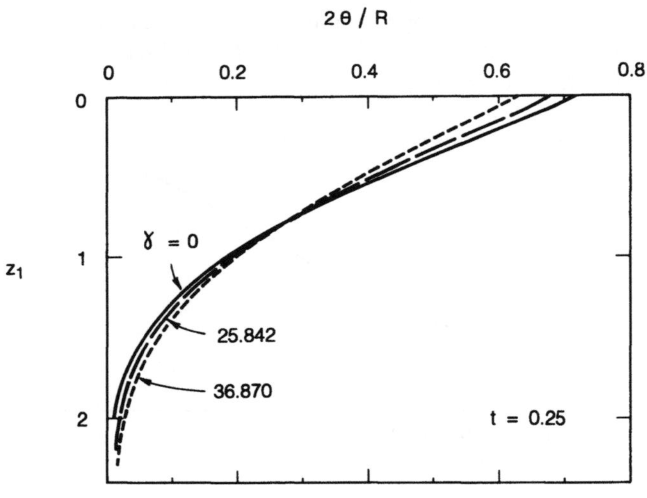

Fig. 9.4. Long-slope constant-rainfall infiltration. Effect of slope angle γ on moisture profile, shown in the dimensionless form $2\theta(z_1, t)/R$ for $t = 0.25$ and the indicated values of γ. z_1 and t are, respectively, the space coordinate and time, normalized without elimination of γ. Note that the γ values 0, 25.842, and 36.870 are, respectively, $\cos^{-1} 1$, $\cos^{-1} 0.9$, and $\cos^{-1} 0.8$

9.6.2 Standard and Rotated Flow Components

As well as standard Cartesian coordinates (x_*, z_{1*}), we use the rotated coordinates (η_*, ζ_*) with dimensionless form (η, ζ) (Fig. 9.5). We use the same flow velocity components, and the same symbols for them, as in Philip (1992) (Fig. 9.6). Here, however, we normalize the various flow components with respect to $\frac{1}{2}\alpha\theta_R$, so that:

$$\frac{u}{u_*} = \frac{u_d}{u_{d*}} = \frac{v}{v_*} = \frac{v_n}{v_{n*}} = \frac{2}{\alpha\theta_R}. \tag{30}$$

9.6.2.1 *Horizontal and Vertical Components*

The dimensionless components are u and v. Consistent with Philip (1991a, 1992), we take u positive into the slope (the negative x direction) and v positive downward (the positive z_1 direction). The equations for u and v are then:

$$u = \partial\theta/\partial x = -\partial\theta/\partial z_1 \tan\gamma, \quad v = 2\theta - \partial\theta/\partial z_1. \tag{31}$$

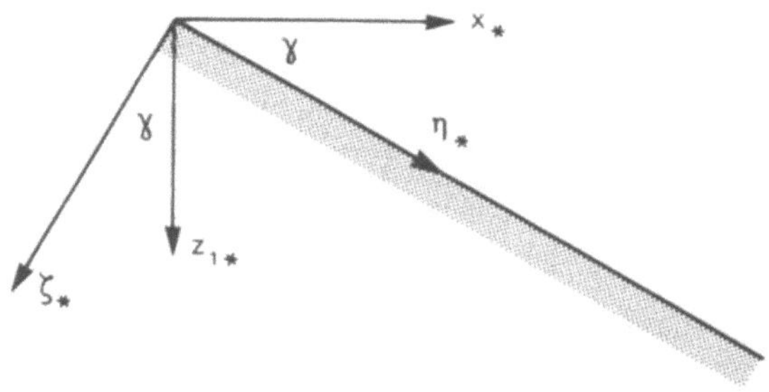

Fig. 9.5. Scheme illustrating Cartesian space coordinates (x_*, z_{1*}) and rotated coordinates (η_*, ζ_*)

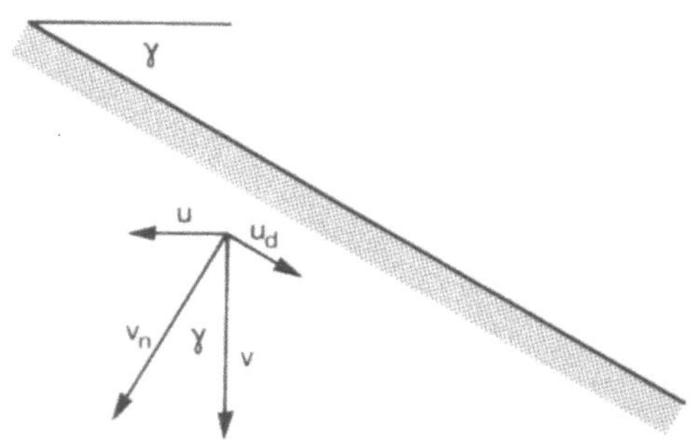

Fig. 9.6. Scheme illustrating various dimensionless flow velocity components: u, horizontal inslope; v, vertical; u_d, downslope; v_n, normal to slope

9.6.2.2 Horizontal Inslope Flow Velocity

It follows from Eqs. (22), (25) and (31) that:

$$\frac{u(z_2,\tau)}{R\sin\gamma\cos\gamma} = -G'(2z_2, 4\tau). \tag{32}$$

Combining Eq. (32) with Eq. (85) (see Appendix) then gives:

$$\frac{u(z_2,\tau)}{R\sin\gamma\cos\gamma} = (1+z_2+2\tau)\exp(2z_2)\operatorname{erfc}\frac{z_2+2\tau}{2\tau^{1/2}} - 2\left(\frac{\tau}{\pi}\right)^{1/2}\exp\left[-\left(\frac{z_2-2\tau}{2\tau^{1/2}}\right)^2\right], \tag{33}$$

with the special result:

$$\frac{u(0,\tau)}{R\sin\gamma\cos\gamma} = (1+2\tau)\operatorname{erfc}\tau^{1/2} - 2\left(\frac{\tau}{\pi}\right)^{1/2}\exp(-\tau). \tag{34}$$

Figure 9.7 shows profiles of horizontal inslope flow velocity in the dimensionless form $u(z_2,\tau)/(R\sin\gamma\cos\gamma)$ for $\tau = 0.01$, 0.25, and 1.

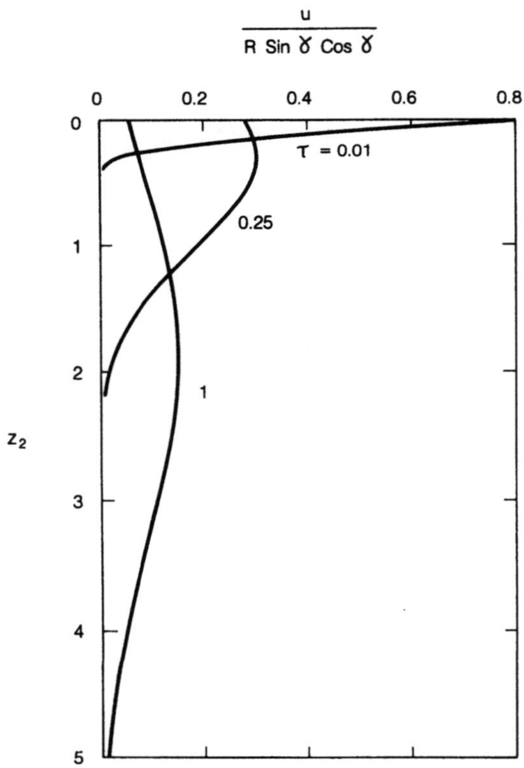

Fig. 9.7. Long-slope constant-rainfall infiltration. Evolution of profiles of inslope flow velocity in the dimensionless form $u(z_2,\tau)/(R\sin\gamma\cos\gamma)$ for $\tau = 0.01, 0.25$, and 1

9.6.2.3 *Vertical Flow Velocity*

Similarly, Eqs. (22), (25), and (31) give:

$$\frac{v(z_2,\tau)}{R} = G(2z_2,4\tau) - G'(2z_2,4\tau)\cos^2\gamma, \tag{35}$$

with the special result:

$$\frac{v(0,\tau)}{R} = 1 - \sin^2\gamma\left[(1+2\tau)\operatorname{erfc}\tau^{1/2} - 2\left(\frac{\tau}{\pi}\right)^{1/2}\exp(-\tau)\right]$$
$$= 1 - \frac{u(0,\tau)}{R}\tan\gamma. \tag{36}$$

9.6.2.4 *Downslope and Normal Components*

The dimensionless components are u_d and v_n. We take u_d positive in the downslope (η positive) direction and v_n positive into the slope (ζ positive direction). The equations for u_d and v_n are then:

$$u_d = 2\theta\sin\gamma,$$
$$v_n = (2\theta - \partial\theta/\partial z_1)\cos\gamma + \partial\theta/\partial z_1\tan\gamma\sin\gamma$$
$$= 2\theta\cos\gamma - \partial\theta/\partial z_1\sec\gamma. \tag{37}$$

9.6.3 Integrated Horizontal and Downslope Components

9.6.3.1 *Integrated Inslope Horizontal Flow*

The total inslope horizontal flow per unit cross-slope length, U_*, is given by:

$$U_* = \int_0^\infty u_*\,dz_{*1}, \tag{38}$$

with the dimensionless form

$$U = \int_0^\infty u\,dz_1 = -R\int_{z_1=0}^{z_1=\infty} d\theta\tan\gamma = R\theta(0,t)\tan\gamma. \tag{39}$$

The second equality of Eq. (39) uses Eq. (31). Note that it follows from Eqs. (38) and (39) that:

$$U_* = \Theta_R U. \tag{40}$$

We see from Eq. (39) that Fig. 9.2, as well as depicting the evolution of surface moisture content, represents also $U_*(t_*)$ in the dimensionless form $[2U(\tau)\cot\gamma]/R$.

9.6.3.2 *Integrated Downslope Flow*

The integrated downslope flow per unit cross-slope length, U_{d^*}, is defined by:

$$U_{d^*} = \int_0^\infty u_{d^*}\, d\zeta_*, \tag{41}$$

with the dimensionless form

$$U_d = \int_0^\infty u_d\, d\zeta = \int_0^\infty u_d\, dz_2 \sec\gamma, \tag{42}$$

with

$$U_{d^*} = \Theta_R U_D. \tag{43}$$

It follows from the first equality of Eq. (37) that:

$$U_d = 2R\int_0^\infty \theta dz_2 \tan\gamma = \frac{2\Theta R}{\bar{D}} Rt \sin\gamma \cos\gamma, \tag{44}$$

the final equality following from Eq. (90) (see Appendix). In Eq. (13) we normalized t_* with respect to $4(\alpha^2\bar{D})^{-1}$, and not to $4(\alpha^2\Theta_R)^{-1}$: this leads to the dimensionless factor $\Theta_R/\bar{D}$ in the last expression of Eq. (44). It is of interest that the integrated downslope flow rate is directly proportional to t_*.

9.6.4 Time Dependence of Surface Flow Velocity Vector

Both the magnitude and direction of the flow velocity vector at the slope surface, **v**, are time dependent. Its magnitude,

$$|\mathbf{v}(\tau)| = \{[u(0,\tau)]^2 + [v(0,\tau)]^2\}^{1/2}, \tag{45}$$

may be readily evaluated using Eqs. (34) and (36). Figure 9.8 shows $|\mathbf{v}(t)|/R$ for $\gamma = 0$, 15, 30, and 45. For $\gamma = 0$ $|\mathbf{v}(t)| = R$ for all values of $t \geqslant 0$. For $\gamma > 0$ $|\mathbf{v}(t)|$ increases monotonically from $R\cos\gamma$ at $t = 0$ to R as $t \to \infty$.

The direction of **v** is specified by the angle it makes with the downward vertical, β (degrees). We have:

$$\beta(\tau) = \tan^{-1}\frac{u(0,\tau)}{v(0,\tau)}, \tag{46}$$

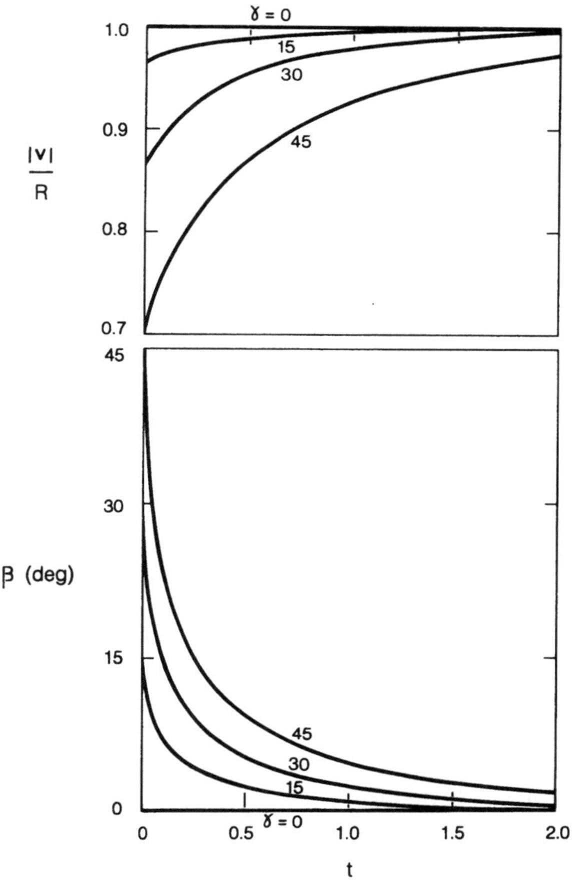

Fig. 9.8. Long-slope constant-rainfall infiltration. Time dependence of the magnitude and direction of the surface flow velocity vector, **v**. *Above* $|\mathbf{v}(t)|/R$, and *below* $\beta(t)$, for $\gamma = 0$, 15, 30, and 45. Note that $|\mathbf{v}(0)|/R = \cos\gamma$ and $\beta(0) = \gamma$. β is the angle specifying the direction of the vector. It is normal into the slope for $\beta = \gamma$ and is vertically downward for $\beta = 0$

and β may be evaluated from Eqs. (34) and (36). Figure 9.8 shows $\beta(t)$ for $\gamma = 0$, 15, 30, and 45. For $\gamma = 0$, $\beta = 0$ for all values of $t \geqslant 0$. For $\gamma > 0$, β decreases monotonically with increasing t from γ at $t = 0$ to 0 as $t \to \infty$.

9.6.5 Long-Slope Time-to-Ponding

In case 2, with $R > 2$, long-slope ponding is incipient at finite time $t_* = T_*$. Note that, in general, this T_* value is greater than the time to slope-crest ponding discussed in sections 9.2.2.2 and 9.3.3. Putting $\theta(0, \tau) = 1$ and $\tau = T\cos^2\gamma$ in Eq. (28) gives R as an explicit function of T.

$$R(T) = 2[G(0, 4T\cos^2\gamma)]^{-1}. \tag{47}$$

Expansion (38) gives the approximation, valid for $R \gg 2$:

$$T(R) \approx \frac{\pi \sec^2\gamma}{4R^2}. \tag{48}$$

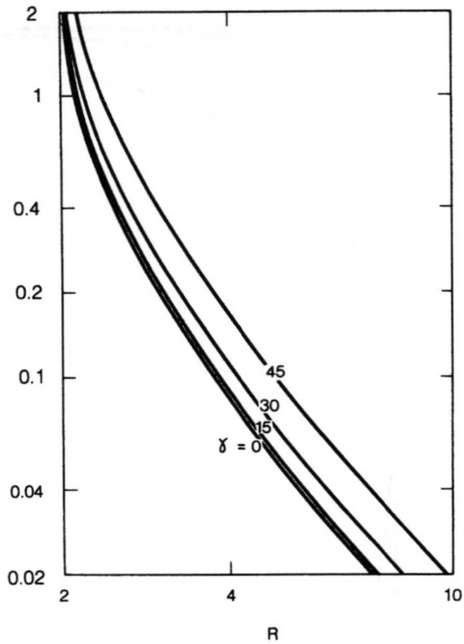

Fig. 9.9. Long-slope constant-rainfall infiltration. Dependence of time-to-ponding on rainfall rate in dimensionless form $T(R)$ for slope angle $\gamma = 0$, 15, 30, and 45

In physical quantities this becomes:

$$T_*(R_*) \approx \frac{\pi \sec^2 \gamma}{4\alpha^2 \bar{D}} \left(\frac{K_1}{R_*}\right)^2. \tag{49}$$

Figure 9.9 shows $T(R)$ for $\gamma = 0$, 15, 30, and 45.

9.7 Constant-Rainfall Slope-Crest Solution for $\gamma = 45$

For the case $\gamma = 45$ we introduce Cartesian coordinates (η, ζ) produced by rotating (x, z) clockwise through 45°, as in Philip (1992). The righthand slope is then the positive η axis and the lefthand slope the positive ζ axis (Fig. 9.10). Then:

$$\eta = \frac{1}{\sqrt{2}}(z + x),$$

$$\zeta = \frac{1}{\sqrt{2}}(z - x). \tag{50}$$

Using Eq. (50), we can rewrite Eq. (12) subject to conditions (14) as:

$$\frac{\partial \theta}{\partial t} = \frac{\partial^2 \theta}{\partial \eta^2} + \frac{\partial^2 \theta}{\partial \zeta^2} - \sqrt{2}\left(\frac{\partial \theta}{\partial \eta} + \frac{\partial \theta}{\partial \zeta}\right), \tag{51}$$

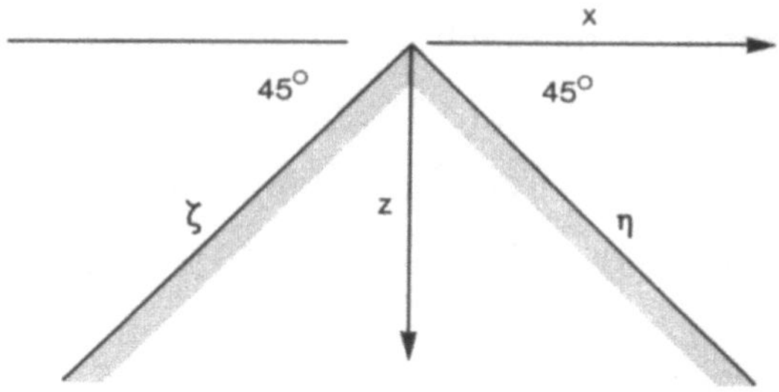

Fig. 9.10. Scheme illustrating dimensionless coordinates (x, z) and rotated coordinates (η, ζ) for slope angle $\gamma = 45°$

subject to:

$$t=0, \quad \eta>0, \quad \zeta>0, \quad \theta=0;$$
$$t \geqslant 0, \quad \eta=0, \quad \zeta>0, \quad \sqrt{2}\theta - \partial\theta/\partial\eta = R/\sqrt{2},$$
$$\zeta=0, \quad \eta>0, \quad \sqrt{2}\theta - \partial\theta/\partial\zeta = R/\sqrt{2}. \tag{52}$$

Knight (1992) showed that Eq. (51) subject to conditions (52) has a product solution, similar to that of Philip (1992), of the form:

$$\theta = \tfrac{1}{2}R[1 - H(\eta, t)\cdot Z(\zeta, t)]. \tag{53}$$

Substituting Eq. (53) in Eq. (51) subject to conditions (52) gives:

$$\frac{1}{H}\left[\frac{\partial H}{\partial t} - \frac{\partial^2 H}{\partial \eta^2} + \sqrt{2}\frac{\partial H}{\partial \eta}\right] = -\frac{1}{Z}\left[\frac{\partial Z}{\partial t} - \frac{\partial^2 Z}{\partial \zeta^2} + \sqrt{2}\frac{\partial Z}{\partial \zeta}\right]. \tag{54}$$

Equating each side of Eq. (54) to zero we get:

$$\frac{\partial H}{\partial t} = \frac{\partial^2 H}{\partial \eta^2} - \sqrt{2}\frac{\partial H}{\partial \eta}, \tag{55}$$

$$\frac{\partial Z}{\partial t} = \frac{\partial^2 Z}{\partial \zeta^2} - \sqrt{2}\frac{\partial Z}{\partial \zeta}. \tag{56}$$

Putting Eq. (53) in Eq. (52) then gives the conditions on Eqs. (55) and (56):

$$t=0, \quad \eta>0, \quad H=1; \quad t \geqslant 0, \quad \eta=0, \quad \sqrt{2}H - \partial H/\partial\eta = 0. \tag{57}$$

$$t=0, \quad \zeta>0, \quad Z=1; \quad t \geqslant 0, \quad \zeta=0, \quad \sqrt{2}Z - \partial z/\partial\zeta = 0. \tag{58}$$

We revert now to Eq. (23) subject to conditions (24). We see that replacing θ by $\frac{1}{2}R(1-H)$, z_2 by $\eta/\sqrt{2}$, and τ by $t/2$ in Eq. (23) and conditions (24) gives Eq. (55) subject to conditions (57). The same replacements in Eq. (25) thus give the solution of Eq. (55) subject to condition (57), namely:

$$H(\eta, t) = 1 - G(\sqrt{2}\eta, 2t). \tag{59}$$

Similarly, the solution of Eq. (56) subject to condition (58) is:

$$Z(\zeta, t) = 1 - G(\sqrt{2}\,\zeta, 2t). \tag{60}$$

It follows from Eq. (53) that that required solution is therefore:

$$\theta(\eta, \zeta, t) = \tfrac{1}{2}R\{1 - G(\sqrt{2}\,\eta, 2t)][1 - [1 - G(\sqrt{2}\,\zeta, 2t)]\}. \tag{61}$$

9.8 Physical Implications of Slope-Crest Solution

The solution (61) enables us to explore various aspects of constant-rainfall slope-crest infiltration. Inter alia we use it to establish limits to the applicability of long-slope constant-rainfall solutions and to develop the comparison of ponded and constant-rainfall infiltration on slopes and crests. Note that, because of the symmetry of the problem, we limit the treatment to the right-hand slope, i.e., the region $\eta \geqslant 0$, $0 \leqslant \zeta \leqslant \eta$ (Fig. 9.10). Results for the lefthand slope follow by simple transposition of η and ζ.

9.8.1 Evolution of Moisture Content and Potential Distributions

The evolution of the distribution of θ as infiltration proceeds is described by Eq. (61). We have the following special results: on the plane of symmetry through the slope crest, $x = 0$, $\eta = \zeta = z/\sqrt{2}$, and Eq. (61) becomes:

$$\theta = \tfrac{1}{2}R\{1 - [1 - G(z, 2t)]^2\}. \tag{62}$$

Also, since [cf. Eq. (83) of the Appendix] $\lim_{\eta \to \infty} G(\sqrt{2}\,\eta, 2t) = 0$,

$$\lim_{\eta \to \infty} \theta(\eta, \zeta, t) = \tfrac{1}{2}R\,G(\sqrt{2}\,\zeta, 2t). \tag{63}$$

We thus establish that, far enough down the slope, θ is independent of η and depends only on ζ and t. The solution (63) is, of course, the long-slope solution (25) for the special case $\gamma = 45$.

Figure 9.11 illustrates the evolution of the solution as t increases, showing the distribution of θ for $t = 0.02$, 0.50, and 2. The contours of θ are horizontal at $x = 0$, $z > 0$, i.e., they intersect the plane of symmetry normally. The slope of each contour increases downslope. Contours of θ greater than the long-slope surface value $\theta(\infty, 0, t)$ intersect the soil surface. Contours of θ less than this approach parallelism to the slope asymptotically. The depth of moisture penetration from the surface is greatest on $x = 0$, decreasing downslope until it approaches the depth given by the long-slope solution. The process has many similarities to that of ponded infiltration, except that here values of θ at the slope surface decrease with increasing η from a maximum at the crest, and, of course, increase with t (cf. Fig. 4 of Philip 1992).

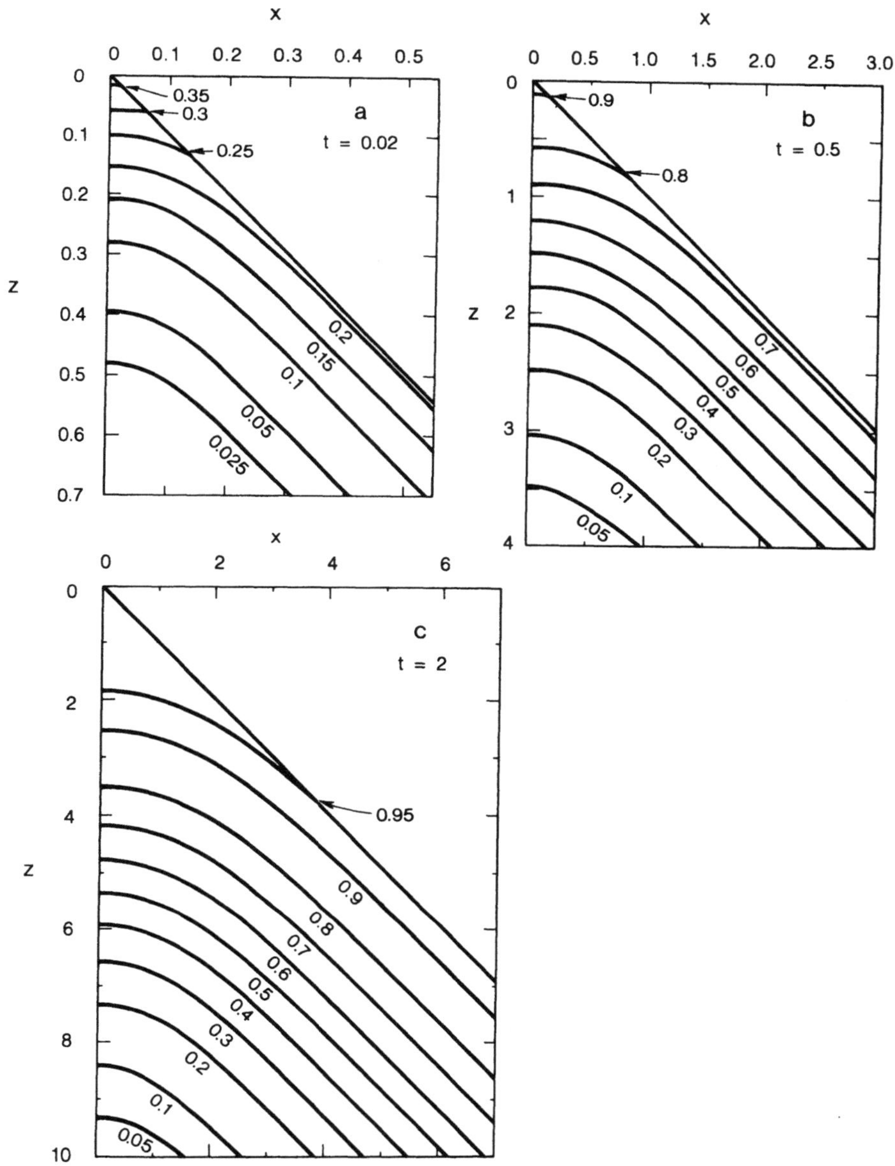

Fig. 9.11. Slope crest constant-rainfall infiltration. The distribution of the dimensionless potential in the form $2\theta/R$ for dimensionless time, $t = 0.02$, 0.5, and 2. Note that x and z are the horizontal and vertical dimensionless space coordinates. The scale of the figures decreases with increasing t

9.8.2 The Time Course of Surface Moisture Content

At the slope surface:

$$\theta(\eta, 0, t) = \tfrac{1}{2}R\{1 - [1 - G(\sqrt{2}\eta, 2t)][1 - G(0, 2t)]\}, \tag{64}$$

with two particular results. At the crest:

$$\theta(0, 0, t) = \tfrac{1}{2}R\{1 - [1 - G(0, 2t)]^2\}$$
$$= \tfrac{1}{2}R\left\{1 - \left[(1+t)\operatorname{erfc}\left(\frac{t}{2}\right)^{1/2} - \left(\frac{2t}{\pi}\right)^{1/2}\exp\left(-\frac{t}{2}\right)\right]^2\right\}; \tag{65}$$

and the long-slope result is:

$$\theta(\infty, 0, t) = \tfrac{1}{2}RG(0, 2t)$$
$$= \tfrac{1}{2}R\left[1 - (1+t)\operatorname{erfc}\left(\frac{t}{2}\right)^{1/2} + \left(\frac{2t}{\pi}\right)^{1/2}\exp\left(-\frac{t}{2}\right)\right], \tag{66}$$

which is simply Eq. (27) for $\gamma = 45$. Combining Eqs. (65) and (66) gives the relation:

$$\frac{\theta(0, 0, t)}{\theta(\infty, 0, t)} = 2 - \frac{2\theta(\infty, 0, t)}{R}. \tag{67}$$

We see that in the limit $t \to 0$ and $\theta(\infty, 0, t) \to 0$, $\theta(0, 0, t)/\theta(\infty, 0, t) = 2$; that the ratio decreases with increasing t; and that in the limit as $t \to \infty$ and $\theta(\infty, 0, t) \to R/2$, $\theta(0, 0, t)/\theta(\infty, 0, t) = 1$. Downslope differences in moisture content thus become relatively less as t increases. This behaviour is illustrated in Fig. 9.12, which shows $2\theta(\eta, 0, t)/R$ for the indicated values of t.

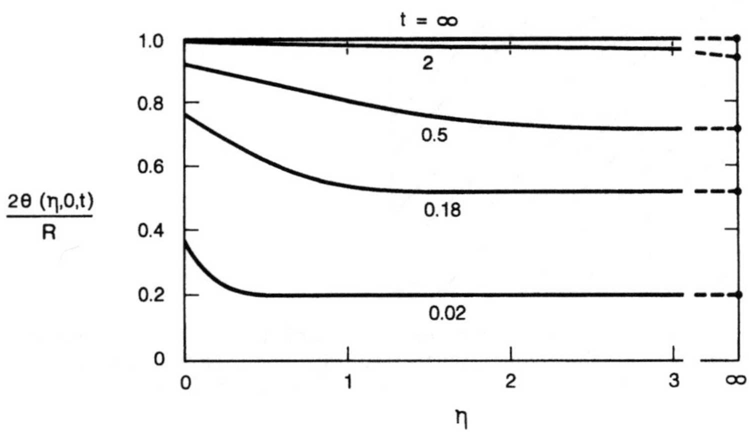

Fig. 9.12. Slope crest constant-rainfall infiltration. Variation of surface moisture content with time and downslope distance in dimensionless form $2\theta(\eta, 0, t)/R$. η is dimensionless downslope distance. The *dots* shown for $\eta = \infty$ represent the long-slope solution

9.8.3 Slope-Crest Time-to-Ponding

Note that for case 1, with $R_* \leqslant K_1$, $R=2$, the solution applies for all values of $t \geqslant 0$, just as for the long slope. Case 2, with $R_* > K_1$, $R>2$, however, requires further consideration. Since, for $t < \infty$, θ at the surface is a maximum at the crest, incipient ponding must occur there first, i.e., at $\eta = 0$. For case 2 our analysis is strictly valid only for $0 \leqslant t \leqslant T(0)$, where we take η as the argument of T. It follows from Eq. (65) that the relation between R and $T(0)$ is:

$$R = 2\{1 - [1 - G(0, 2T\{0\})]^2\}^{-1}. \tag{68}$$

In the slope-crest context we need to address the question of just what ponding implies. Does ponding at the crest produce runoff down the slope? Or do we assume, on the other hand, that surface rugosity is such that ponding is local and ponded water stays at its point of arrival? Either or some mixture of these alternatives may happen in practice. If downslope surface runoff occurs instantly on ponding, $T(0)$ is the one time-to-ponding which our analysis yields with any assurance.

On the other hand, if water ponds without runoff an a rugose surface, more can be deduced. For example, Eq. (66) gives the long-slope time-to-ponding relation as:

$$R = 2[G(0, 2T\{0\})]^{-1}. \tag{69}$$

This is simply Eq. (47) for $\gamma = 45$. One can argue that the effect of local ponding near the crest [and therefore local failure of the second of conditions (52)] will be negligible far enough downslope; and that therefore, far down the slope, the long-slope time-to-ponding $T(\infty)$ has real physical significance.

More problematic is the concept of $T(\eta)$ varying continuously downslope. The difficulty is that, although it is clear, from both previous hillslope studies (Philip 1991a, b, c, 1992) and the present results, that the boundary conditions upslope by more than a few sorptive lengths have a negligible effect on the solution at a given point, at the instant of incipient ponding the boundary condition has altered *everywhere* upslope of the point. Nevertheless $T(\eta)$ has some interest as being indicative of how local ponding will spread downslope. In fact, physical arguments show that $T(\eta)$ represents a lower bound on the actual time at which local ponding will reach the slope point specified by η.

Without prejudice to these reservations, we give the relation between R and $T(\eta)$, which follows from Eq. (64):

$$R = 2\{1 - [1 - G(\sqrt{2}\eta, 2T\{\eta\})][1 - G(0, 2T\{\eta\})]\}^{-1}. \tag{70}$$

These various matters are illustrated in Fig. 9.13, which shows T as a function of R for $\eta = 0$, $2^{-1/2}$, $2^{1/2}$, $2^{3/2}$, and ∞.

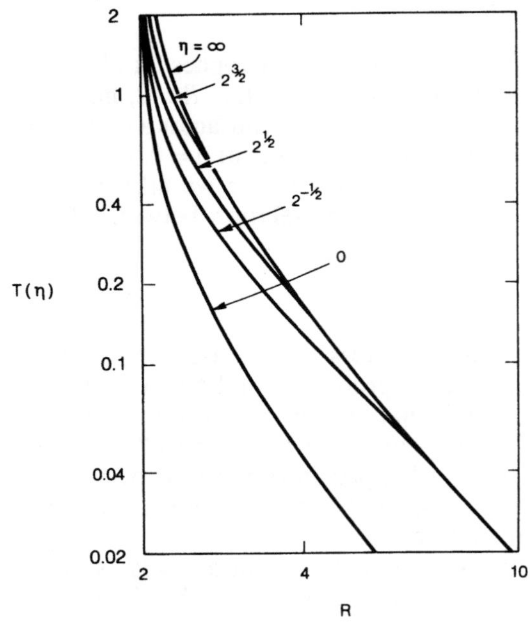

Fig. 9.13. Slope crest constant-rainfall infiltration. Dependence of time-to-ponding on rainfall rate in dimensionless form $T(\eta)$ as a function of R. $\eta = 0$ represents the slope crest and $\eta = \infty$ represents the long-slope solution. The intermediate curves for $\eta = 2^{-1/2}$, $2^{1/2}$, and $2^{3/2}$ represent lower bounds on T for these η values on a rugose slope with local ponding

9.8.4 Criterion for Validity of Long-Slope Solution

Philip (1992) gave a criterion for the downslope distance from a crest, beyond which the long-slope solution for ponded infiltration was valid. The present study yields an analogous criterion for constant-rainfall infiltration.

A primary descriptor of the constant-rainfall infiltration process is the surface moisture content; and, accordingly, we propose the criterion that:

$$\theta(\eta, 0, t) \leqslant 1.01\theta(\eta, 0, \infty). \tag{71}$$

Inequality (71) implies that the surface value of θ should not exceed its long-slope value by more than 1%. Putting Eq. (64) into (71) gives:

$$G(\sqrt{2}\eta, 2t) \leqslant \frac{0.01G(0, 2t)}{1 - G(0, 2t)}. \tag{72}$$

We denote by $\eta_{0.01}$ the value of η for which $G(\sqrt{2}\eta, 2t) = 0.01\, G(0, 2t)/[1 - G(0, 2t)]$. Then, (72) may be rewritten as:

$$\eta \geqslant \eta_{0.01}(t). \tag{73}$$

Figure 9.14 shows $\eta_{0.01}(t)$. It increases approximately as $t^{1/2}$ at small t, has a maximum of about 4.8 at $t \approx 2.4$ and then decreases rapidly to 0 at $t \approx 4.2$.

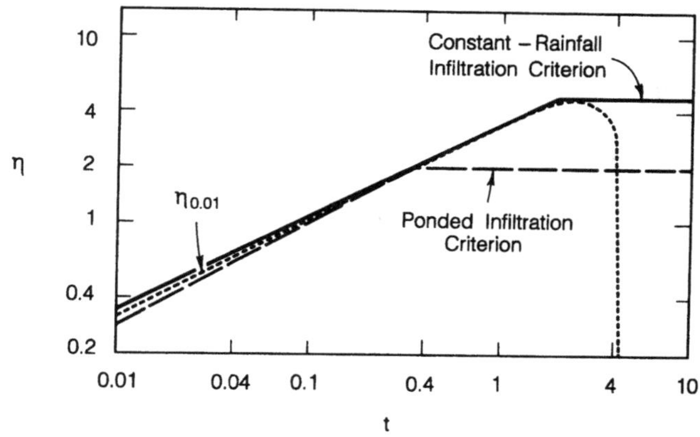

Fig. 9.14. Slope crest constant-rainfall infiltration. *Short-dashed curve* represents $\eta_{0.01}(t)$. *Solid curve* indicates practical criterion for validity of long-slope constant-rainfall solution. *Long-dashed curve* represents analogous criterion for long-slope ponded infiltration

The bold curve on Fig. 9.14 represents the suggested practical criterion,

$$0 \leqslant t \leqslant 2, \quad \eta \geqslant 3.4t^{1/2};$$

$$t > 2, \quad \eta \geqslant 4.8. \tag{74}$$

Also shown in Fig. 9.14 is the analogous criterion developed (Philip 1992) for the validity of long-slope ponded infiltration solutions. It is of interest that the two criteria essentially agree for $0 \leqslant t \leqslant 0.33$, but the constant-rainfall lower bound on η continues to increase until $t = 2$.

9.8.5 Limits on Long-Slope Solution for Arbitrary γ

The preceding criterion [Eq. (74)] applies only to the special case of slope angle $\gamma = 45$. We recognize frankly that 45° slopes are wildly improbable in nature, but this case does supply us with a datum against which we may estimate the limits of validity of the long-slope solution for arbitrary γ.

As noted elsewhere (Philip 1992) the term of the flow equation [Eq. (4) of Philip (1991a)] involving downslope variation is proportional to $\sin\gamma$, suggesting that the slope-crest effect should increase with $\sin\gamma$; and this was confirmed in the primitive criterion for validity of the long-slope ponded infiltration solution found in Philip (1991a). Analogous to Philip (1992), we conjecture here that we may extend Eq. (74) to arbitrary γ as follows:

$$0 \leqslant t \leqslant 2, \quad x \geqslant 3.4t^{1/2} \sin\gamma;$$

$$t > 2, \quad x \geqslant 4.8 \sin\gamma. \tag{75}$$

Note that Eq. (75) is an inequality on x, the dimensionless horizontal distance from the crest.

9.9 Concluding Discussion

Our concluding discussion centres on a comparison of the properties of hillslope and slope-crest infiltration and downslope unsaturated flow in the two cases, ponded and constant-rainfall.

9.9.1 Inslope Horizontal Flow

The long-slope nonlinear solution for ponded infiltration (Philip 1991a) showed that the integrated horizontal flow rate into the slope, U_*, was constant in time and equal to $\Theta_1 \tan\gamma$, where $\Theta_1 = \Theta(\theta_1)$. For constant-rainfall infiltration we find, on the other hand, that [cf. Eq. (39)]:

$$U_* = \Theta(0, t_*) \tan\gamma. \qquad (76)$$

The time dependence of U_* thus follows that of Θ at the slope surface. This result holds generally in the nonlinear case, and not just in the linear one. It follows from Eqs. (28) and (29) that

$$\text{for small } t_*, \quad U_* \propto t_*^{1/2},$$

and

$$\text{for large } t_*, \quad U_* \approx \text{constant} = t_*^0. \qquad (77)$$

It can be shown that (77) holds generally and not only for the linearized solution.

A significant point is that, just as for ponded infiltration, inslope horizontal flow occurs during constant-rainfall infiltration into a homogeneous isotropic soil. We do not need to invoke soil layering or anisotropy (e.g., Zaslavsky and Rogowski 1969; Zaslavsky and Sinai 1981; McCord and Stephens 1987) to explain the horizontal component: the fact of capillarity suffices.

9.9.2 Downslope Flow

The long-slope nonlinear ponded-infiltration solution (Philip 1991a) indicated that U_{d*}, the integrated downslope flow rate, was roughly proportional to the cumulative infiltration, behaving like $t_*^{1/2}$ at small t_* and approaching proportionality to t_* at large t_*. We find also for constant-rainfall infiltration proportionality between U_{d*} and the cumulative infiltration. In this case [cf. Eq. (44)]:

$$U_{d*} = K_R R_* t_* \sin\gamma \cos\gamma. \qquad (78)$$

This exact linearity of $U_{d*}(t_*)$ is a consequence of our linearization of the problem. Nevertheless, it is demonstrable that, in sufficiently regular nonlinear cases, U_{d*} is proportional to t_* in the limits of both small and large t_*, with differing constants of proportionality in the two limits. It may further be shown that Eq. (78) is the limiting form of $U_{d^*}(t_*)$ as $t_* \to \infty$ in the general nonlinear case.

Here also, as for the ponded case, we find downslope flow during constant-rainfall infiltration into a homogeneous isotropic soil. There is no need to call on soil layering or anisotropy (Zaslavsky and Rogowski 1969; Zaslavsky and Sinai 1981; McCord and Stevens 1987) to account for it: the fact of a downslope component of gravity suffices.

9.9.3 The Surface Flow Velocity Vector

Requirements of brevity limit us to a summary exposition of the behaviour of the surface flow velocity vector.

9.9.3.1 Ponded Infiltration

This vector was not discussed specifically in Philip (1991a, 1992), though the relevant information for ponded infiltration is implicit in the solutions of those papers. In general terms, for ponded infiltration on slope crests, the magnitude of the vector decreases as t_* increases, and its direction shifts gradually from normal into the slope at $t_* = 0$ to vertically downward as $t_* \to \infty$. This behaviour is explained physically by the decrease of capillary effects as t_* increases. The exception to this behaviour is at the slope crest, where the vector has magnitude K_1 and direction vertically downward for all values of $t_* > 0$. This is also the final state of the vector as $t_* \to \infty$ at all points downslope of the crest. The approach to this state is rapid at small downslope distances, and decreases as distance increases.

9.9.3.2 Constant-Rainfall Infiltration

For constant-rainfall infiltration with $R_* \leqslant K_1$ the direction of the vector behaves similarly, being normal to the slope at $t_* = 0$ and shifting towards vertically downward as $t_* \to \infty$. On the other hand, in this case the magnitude of the vector *increases* from $R_* \cos\gamma$ at $t_* = 0$ to R_* as $t_* \to \infty$. Figure 9.8 shows the long-slope result. In this case also, the slope crest is an exceptional point where the vector has magnitude R_* and is directed vertically downward for all values of $t_* > 0$; and here also the rate of approach to this final state decreases as distance downslope of the crest increases.

9.9.4 Comparing the Dynamics of Ponded and Constant-Rainfall Long-Slope Infiltration

The foregoing analysis yields, inter alia, the limiting power dependence on t_* of various processes associated with constant-rainfall infiltration for t_*

Table 9.1. Limiting power dependence on t_* of various quantities of hillslope ponded and constant-rainfall infiltration

	Powers of t_* in the limit			
	Ponded		Constant rainfall	
Quantity	Small t_*	Large t_*	Small t_*	Large t_*
Infiltration rate	$-\frac{1}{2}$	0	0	0
Cumulative infiltration	$\frac{1}{2}$	1	1	1
Integrated horizontal inslope flow	0	0	$\frac{1}{2}$	0
Integrated downslope flow	$\frac{1}{2}$	1	1	1

both small and large. Table 9.1 gives a useful overall picture of the dynamics of the various processes, and shows for comparison the analogous results for ponded infiltration (Philip 1991a). For all processes the index for small t_* is greater by exactly $\frac{1}{2}$ for constant rainfall than for the ponded case, whereas the large t_* indices are the same in both cases.

9.9.5 The Slope-Crest Effect

For ponded infiltration the primary effect of the slope crest is to increase θ in its neighbourhood (Philip 1992). Within the region of increased θ, moisture gradients, and hence capillary effects, tend to be reduced. The region of increased θ and reduced capillarity propagates downstream with increasing t_*.

The present study shows that the slope-crest effect for constant-rainfall infiltration is similar in character (Fig. 9.11). The difference is that, whereas for ponded infiltration the slope crest was the site of the minimum infiltration capacity for $0 \leqslant t_* < \infty$, here it is the site of the maximum surface moisture content.

For ponded infiltration it was found (Philip 1992) that the slope-crest effect became unimportant for large t_*, so that the long-slope solution was valid provided that the downslope distance exceeded a few sorptive lengths at most. This result was attributed to the diminution of capillary effects downslope as t_* increases. In the present work we have found a similar result for constant-rainfall infiltration, and its physical explanation again lies with the decrease of capillary effects as t_* increases. The practical criterion proposed for validity of the long-slope solution in this case gives values up to 2.4 times that suggested for ponded infiltration. In an order-of-magnitude sense the criterion is unaltered: the minimum downslope distance is of the order of the sorptive length.

9.9.6 The Downslope Propagation of Ponding from a Slope Crest

The analysis of ponded infiltration on slope crests showed that the infiltration capacity was least at the crest and increased downslope (Philip 1992). This led to the expectation that, for constant-rainfall infiltration onto a rugose slope crest, with $R_* > K_1$, ponding would occur first at the crest and then propagate downslope. As we have seen, the present study fully confirms this expectation.

This remarkable behaviour, suggested by earlier results (Philip 1992) and confirmed here, will occur independent of linearity and for γ arbitrary. I believe this interesting result was not obvious before these linearized studies for a special slope angle were carried out and this illustrates how simplified analysis may give surprising new insights.

Acknowledgements. I am grateful to Drs. I. White and P.W. Ford for helpful comments and criticisms, and especially to Dr. J.H. Knight for setting me right on an important issue.

Appendix: Properties of $G(X, Y)$

We set out briefly significant properties of the function G defined by:

$$G(X, Y) = \frac{1}{2}\left\{\operatorname{erfc}\frac{X - Y}{2Y^{1/2}} - (1 + X + Y)\exp(X)\operatorname{erfc}\frac{X + Y}{2Y^{1/2}}\right\} + \left(\frac{Y}{\pi}\right)\exp\left[-\left(\frac{X - Y}{2Y^{1/2}}\right)^2\right]. \tag{79}$$

Special results for G include:

$$0 \leqslant X \leqslant \infty, \quad G(X, 0) = 0; \tag{80}$$

$$0 \leqslant Y \leqslant \infty, \quad G(0, Y) = 1 - \left(1 + \frac{Y}{2}\right)\operatorname{erfc}\frac{Y^{1/2}}{2} + \left(\frac{Y}{\pi}\right)^{1/2}\exp\left(-\frac{Y}{4}\right); \tag{81}$$

$$G(0, \infty) = 1; \tag{82}$$

$$0 \leqslant Y < \infty; \quad G(\infty, Y) = 0; \tag{83}$$

$$\begin{aligned} \lim_{Y \to \infty} G(\lambda Y, Y) &= 1 \text{ for } 0 < \lambda < 1, \\ &= \tfrac{1}{2} \text{ for } \lambda = 1, \\ &= 0 \text{ for } \lambda > 1. \end{aligned} \tag{84}$$

For the derivative $\partial G(X, Y)/\partial X = G'(X, Y)$ we have:

$$G'(X, Y) = \left(\frac{Y}{\pi}\right)^{1/2} \exp\left[-\left(\frac{X - Y}{2Y^{1/2}}\right)^2\right] - \left(1 + \frac{X+Y}{2}\right)\exp(X)\operatorname{erfc}\frac{X+Y}{2Y^{1/2}}, \tag{85}$$

with the special results:

$$0 \leqslant X \leqslant \infty, \quad G'(X, 0) = 0; \tag{86}$$

$$0 \leqslant Y \leqslant \infty, \quad G'(0, Y) = \left(\frac{Y}{\pi}\right)^{1/2} \exp\left(-\frac{Y}{4}\right) - \left(1 + \frac{Y}{2}\right)\operatorname{erfc}\frac{Y^{1/2}}{2}; \tag{87}$$

$$G'(\infty, Y) = 0; \tag{88}$$

$$G'(0, \infty) = 0. \tag{89}$$

Finally, we note that:

$$\int_0^\infty G(X, Y)\,dX = Y. \tag{90}$$

References

Ben-Asher JD, Lomen DO, Warrick AW (1978) Linear and nonlinear models of infiltration from a point source. Soil Sci Soc Am J 42: 3–6

Braester C (1973) Moisture variation at the soil surface and the advance of the wetting front during infiltration at constant flux. Water Resour Res 9: 687–694

Bresler E (1967) A model for tracing salt distribution in the soil profile and estimating the efficient combination of water quality and quantity under varying field conditions. Soil Sci 104: 226–233

Broadbridge P, White I (1988a) Constant rate rainfall infiltration: a versatile nonlinear model. 1. Analytic solution. Water Resour Res 24: 145–154

Broadbridge P, White I (1988b) Constant rate rainfall infiltration: a versatile nonlinear model. 2. Applications of solutions. Water Resour Res 24: 155–162

Knight JH (1992) A linearized solution of the slopecrest constant flux infiltration problem. Water Resour Res (submitted)

Lomen DO, Warrick AW (1974) Time-dependent linearized infiltration. II. Line sources. Soil Sci Soc Am Proc 38: 568–572

McCord JT, Stephens DB (1987) Lateral moisture flows beneath a sandy hillslope without an apparent impeding layer. Hydrol Processes 1: 225–238

Mason M, Weaver W (1924) The settling of small particles in a fluid. Phys Rev 23: 412–426

Philip JR (1957) The theory of infiltration. 4. Sorptivity and algebraic infiltration equations. Soil Sci 84: 257–264

Philip JR (1966a) A linearization technique for the study of infiltration. In: Rijtema RE, Wassink H (eds) Water in the unsaturated zone: proceedings of IASH/UNESCO symposium, vol 1. UNESCO, Paris, pp 471–478

Philip JR (1966b) Absorption and infiltration in two- and three-dimensional systems. In: Rijtema RE, Wassink H (eds) Water in the unsaturated zone: proceedings of IASH/UNESCO symposium, vol 1. UNESCO, Paris, pp 503–525
Philip JR (1968) Steady infiltration from buried sources and spherical cavities. Water Resour Res 4: 1039–1047
Philip JR (1969) Theory of infiltration. Adv Hydrosci 5: 215–296
Philip JR (1974) Recent progress in the solution of nonlinear diffusion equations. Soil Sci 117: 257–264
Philip JR (1986) Linearized unsteady multidimensional infiltration. Water Resour Res 22: 1717–1727
Philip JR (1987a) The quasilinear analysis, the scattering analog, and other aspects of infiltration and seepage. In: Fok Y-S (ed) Irrigation development and application. Water Resource Res Center, Honolulu, pp 1–27
Philip JR (1987b) The infiltration joining problem. Water Resour Res 23: 2239–2245
Philip JR (1989) The scattering analog for infiltration in porous media. Res Geophys 27: 431–448
Philip JR (1991a) Hillslope infiltration: planar slopes. Water Resour Res 27: 109–117
Philip JR (1991b) Hillslope infiltration: divergent and convergent slopes. Water Resour Res 27: 1035–1040
Philip JR (1991c) Infiltration and downslope unsaturated flows in concave and convex topographies. Water Resour Res 27: 1041–1048
Philip JR (1992) A linearized solution of the slopecrest infiltration problem. Water Resour Res 28: 1128–1132
Storm ML (1951) Heat conduction in simple metals. J Appl Phys 22: 940–951
Warrick AW (1974) Time-dependent linearized infiltration. I. Point sources. Soil Sci Soc Am Proc 38: 383–386
Warrick AW (1975) Analytical solutions to the one-dimensional linearized moisture flow equation for arbitrary input. Soil Sci 120: 79–84
Warrick AW, Lomen DO (1976) Time-dependent linearized infiltration. III. Strip and disc sources. Soil Sci Soc Am J 40: 639–643
Zaslavsky D, Rogowski AS (1969) Hydrological and morphological implications of anisotropy and infiltration in soil profile development. Soil Sci Soc Am Proc 33: 594–599
Zaslavsky D, Sinai G (1981) Surface hydrology, 1. Explanation of phenomena. J Hydraul Div Am Soc Civ Eng 107: 1–16

10 The Transport of Sorbed Chemicals in Eroded Sediment

C.W. Rose[1]

10.1 Introduction

In countries which depend extensively on groundwater as a source of drinking water, and where monitoring the quality of such water is carried out, there has been significant concern as to the origin of contaminants found. Eutrophication of lakes and rivers with plant nutrients has also drawn attention to the loss of such nutrients from agricultural land.

Chemicals can move by a variety of transport processes. One characteristic of chemicals which strongly affects the type of transport processes of dominant importance is the strength with which the chemical is sorbed to soil. The strength of the sorption process, along with the equilibrium between the aqueous and vapour phases of the chemical, determines the relative fraction of chemical in each of the phases (solid, aqueous, and vapour) (Weber and Miller, 1989).

For lightly sorbed chemicals, transport in a dissolved state is commonly of dominant importance; the movement of saline salts and leaching of nitrate are typical examples (White, 1988). For chemicals more strongly sorbed to organic matter, clays, and mineral surfaces, movement is more strongly linked to the movement of the soil in erosion processes, such as those caused by water, wind, or mass movement of soil.

This chapter concentrates on the transport of sorbed chemicals due to erosion by water, though some attention will also be given to wind erosion. Particular attention will be given to reasons for the common observation that eroded sediment is richer in sorbed chemicals then the soil from which the sediment is derived. Explanations of this observation are sought in terms of processes which appear to be of general application to all sorbed chemicals. These explanations will be described in terms of the processes involved, and will be illustrated by experimental data on a variety of chemicals, both natural and human-applied.

10.1.1 Overview of Chemical Enrichment Mechanisms

The enrichment ratio (E_R) for any chemical (Massey and Jackson 1952; Menzel 1980) is the ratio of the chemical concentration in the eroded

[1] Centre for Catchment and In-Stream Research, Faculty of Environmental Sciences, Griffith University, Nathan Campus, Brisbane, Queensland 4111, Australia

sediment to that of the original soil, given as:

$$E_R = \frac{\text{kg chemical/kg soil in eroded sediment}}{\text{kg chemical/kg soil in original soil}}. \quad (1)$$

The value of E_R is commonly greater than unity, and when this is so it commonly declines with the magnitude of soil lost during an erosion event (Menzel 1980; Knisel 1980). For water erosion of plant nutrients, Foster et al. (1985) and Rose and Dalal (1988) provide illustrations of this relationship for E_R which is of the form:

$$E_B = AM^{-B}, \quad (2)$$

where M is the accumulated mass per unit area of soil lost during a water erosion event and A and B are empirically fitted parameters which can vary appreciably. The loss of any chemical N (kg/m^2) is then given by:

$$N = E_R M c_N \quad (\text{kg/m}^2), \quad (3)$$

where c_N is the concentration of chemical N in the original soil (kg N/kg soil).

It is clear from Eq. (3) that if no enrichment mechanisms are operative (i.e. $E_R = 1$), then the erosive loss of any soil-sorbed chemical is given by the product of the amount of sediment lost from the given area of soil and the concentration of the chemical of concern in the eroded layer. This statement is true for soil erosion by either fluid, water or wind (air).

Recent work has shown that at least two quite different erosion processes result in the enrichment of sorbed chemicals. Before going into detail, these two enrichment processes will now be briefly outlined.

The first process applies both to water and wind erosion. In both forms, the erosion mechanisms appear to be non-selective for all but quite large aggregates and gravel-sized particles. However, once dislodged into the moving fluid, the rate of falling through the fluid, and hence the rate of deposition, is strongly size dependent. The size selectivity of the deposition processes, combined with the commonly non-uniform concentration of sorbed chemical with particle size, is one reason for enrichment of chemicals in the eroded sediment (Rose 1988).

A second type of process which can lead to chemical enrichment occurs when rainfall impact is an important erosion mechanism and where soil aggregates are stable and do not immediately collapse under such impact. This can occur in some soils whose stable aggregates remain intact for a number of direct hits by raindrops. Such collision of the drop with soil leads to very high-speed lateral water flow, and the resultant high stresses strip off the outer layer of the aggregate, producing much finer particles which are then preferentially eroded. This process is termed "raindrop stripping" (Ghadiri and Rose 1991a) and, especially when depths are shallower than the size of larger stable aggregates, this enrichment process can be dominant.

For water erosion, Rose et al. (1989) and Palis et al. (1990a) have shown that it is when rainfall detachment dominates runoff entrainment as an erosion mechanism that enrichment effects are likely to be more significant.

Hence, in Section 10.2, sediment transport when rainfall impact is the main erosive agent will be described. When this is so, the sediment concentration at equilibrium depends on the balance between the rate at which impacting raindrops add sediment to the water layer that is transporting sediment, and the rate at which sediment is removed from that layer by the process of deposition or sediment return to the soil surface.

10.2 Sediment Transport with Rainfall Impact the Dominant Erosive Agent

10.2.1 Deposition

After sediment has been detached by rainfall impact, the majority of it returns to the soil surface after being carried some distance downslope in the overland flow responsible for the sediment flux. Sediment falling in water quickly achieves a terminal or settling velocity, and, on reaching the soil surface again, is said to have been deposited. During an erosion event, the continuous addition of sediment to the water layer by raindrop impact implies that deposition is also a continuous process, except for fine sediment which forms the "suspended load". The suspended load consists of sediment with a settling velocity less than the typical velocity of whirling eddies in flow made turbulent by surface roughness or raindrop impact; this velocity is called the "friction velocity". Such sediment essentially remains in suspension, is removed in the overland flow, does not deposit unless non-turbulent waters are met, and so may continue in suspension until it is discharged into a stream.

The term c_i denotes the sediment concentration of the class of soil particles or aggregates whose settling velocity is v_i. Then, by definition of these terms, the rate of deposition, d_i, defined as the mass of sediment reaching the soil surface per unit area per second, is given by:

$$d_i = v_i c_i \quad (\mathrm{kg/m^2/s}). \tag{4}$$

The distribution by mass of settling velocities v_i for a soil can be measured in a bottom withdrawal tube (Lovell and Rose 1988). This relationships between the mass of sediment in any settling velocity class and settling velocity is referred to as a "settling veolcity characteristic".

10.2.2 Rainfall Detachment and Redetachment

Rainfall detachment is defined as the removal of sediment from the original uneroded soil bed by raindrop impact. Deposited sediment forms a layer which can also be removed by rainfall impact, and this is referred to as redetachment. There is a need to distinguish between detachment and redetachment of deposited sediment, firstly, because for cohesive soil the sediment forming the deposited layer is very much weaker (and so more readily removed) than the original soil matrix. Secondly, sediment forming the deposited layer can differ substantially in size or settling-velocity distribution from the original soil, at least until a steady-rate or equilibrium situation is achieved.

Theory describing the sediment concentration achieved at equilibrium between deposition and rainfall detachment and redetachment is given by Rose and Hairsine (1988) and Hairsine and Rose (1991). This theory assumes that at equilibrium some fraction (H) of the original soil surface is effectively shielded from raindrop impact by the layer of sediment formed by deposition. Thus, fraction (1-H) of the original soil surface is exposed to raindrops. The rates of rainfall detachment and redetachment are assumed to be proportional to the rainfall rate raised to some power which is found experimentally to be rather close to unity. Where water depths are sufficient for most soil to be immersed, rates of detachment and redetachment are also assumed not to be size specific (i.e. the rate of removal of sediment in any given size-class range is taken to be proportional to the fraction of sediment present in that size range).

Theory is development by quantitatively expressing these assumption on erosion rates, and by combining these expressions together with Eq. (4) in an equation expressing mass conservation of sediment (Hairsine and Rose 1991), and an equation describing the hydrology of overland flow (Rose et al. 1983). These equations can then be solved to provide a solution for sediment concentration (Hairsine and Rose 1991).

10.2.3 Changes in Time of Sediment Settling-Velocity Distributions During Erosion in the Presence of a Water Layer

A controlled rainfall experiment was carried out with soil in a flume set at very low slope to ensure that rainfall, and not runoff, was the dominant erosive agent. Simulated rainfall of a constant rate of 100 mm/h and drop size 2.24 mm fell through 9 m onto a slightly dispersive sandy clay loam soil, classified as an Aridisol (Soil Survey Staff 1975) or as solonchak (Stace et al. 1968). The soil was pre-wetted and a mean water depth of 5 mm was maintained throughout the experiment. Since aggregate breakdown due to rainfall can occur, a reference settling-velocity distribution for the soil was obtained for soil which had been wet and rained on for the same period of time as the experiment duration (40 min), with the same depth of surface water.

Figure 10.1 shows the changes with time of the settling-velocity characteristics of sediment leaving the 6 m-long flume. This figure shows that the settling-velocity distribution of eroded sediment at the first sampling time of 1.5 min contained a much greater proportion of slow settling (finer) particles than did sediment sampled after 40 min of rainfall. Thus, Fig. 10.1 shows that the sediment became progressively coarser with time. After 40 min exposure the eroded sediment was similar to, but still a little finer than, the

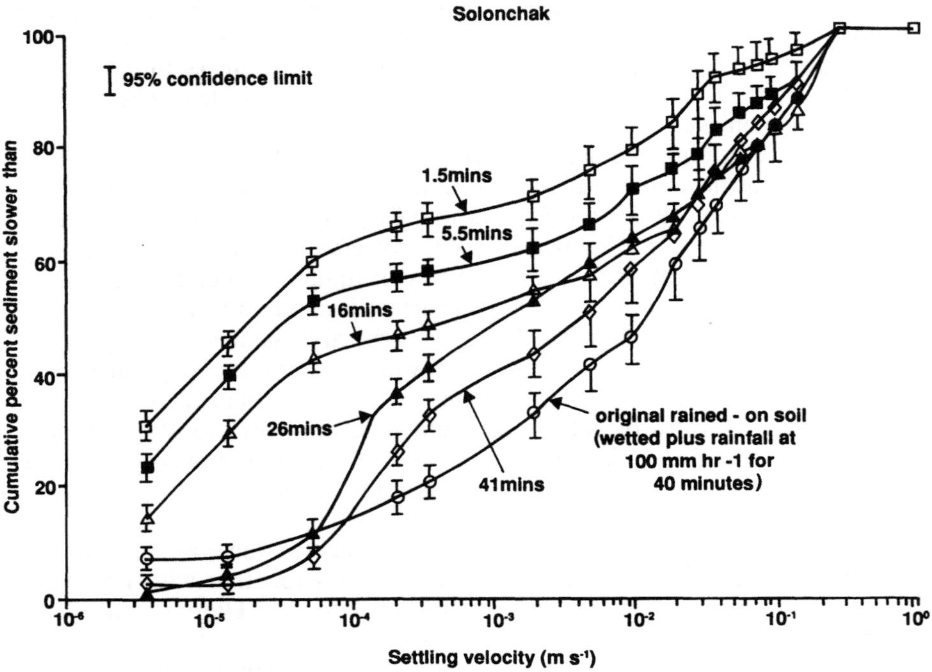

Fig. 10.1. Mean settling-velocity distributions of the original rained-on Aridisol (or solonchak) soil and of eroded sediment at selected sampling times. (After Proffitt et al. 1991)

rained-on soil which had not suffered soil loss due to erosion. Whilst the details vary with soil type, rainfall rate and other factors, the type of behaviour illustrated in Fig. 10.1 appears to be quite general (Proffitt et al., 1991). This behaviour, and associated changes in sediment concentration, can be understood in terms of changes with time in the character of the deposited layer.

If a sorbed chemical is not of uniform concentration across all soil-size ranges, but is higher in concentration for finer particles, then the changes in fineness of eroded sediment, illustrated in Fig. 10.1, will lead to enrichment of the sorbed chemical which is initially high and decreases with time (or with the amount of soil eroded). Thus, this general behaviour of sediment eroded by rainfall provides a basis for understanding one form of chemical enrichment, which will be outlined in Section 10.3.

10.2.4 Sediment Detachment and Transport Under Rainfall with Small Water Depth on the Soil

Erosion with significant consequences for sediment and sorbed chemical transport can take place in different situations where the depth of water on the soil surface is small. Two examples are: (1) on the tops of furrows produced by cultivation and (2) on the slightly higher elements of even relatively flat land. Restricted slope lengths of cultivated land can also keep water depths below the size of many aggregates. When an initially flat soil surface was exposed to simulated rainfall in a flume, Ghadiri and Rose (1991a, b) described the resulting overland flow as taking place dominantly in small ponds and transient microrills. Raindrop impact appeared to be associated with the formation of these transient micro-features which, especially as the slope was increased, could be readily observed to be the major vehicle of sediment transport.

Thus, in general, it is likely that at least some fraction of a land surface where erosion is dominantly due to rainfall impact will experience only small depths of water. Hence, the sedimentation processes described in Section 10.2.3 are restricted to that subset of soil aggregates and particles whose sizes are less than even the small depth of water flowing over the soil surface.

Ghadiri and Payne (1979, 1981) have shown in detail how the impact of raindrops causes a breakdown of soil aggregates and the formation of finer sediment. When raindrops fell on stable aggregates, these authors showed that the high stresses induced by rapid flows following impact could strip fine material from the outer surfaces of the aggregates.

Figure 10.2 shows the size and mass distribution of the aggregates of a clay soil of stable aggregation, and the same distribution for eroded sediment collected in a series of experiments following exposure to stimulated rainfall

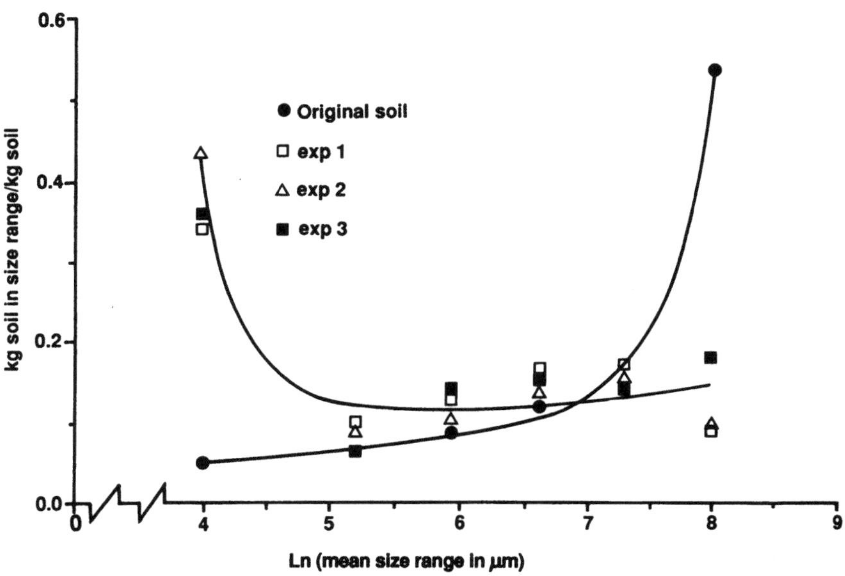

Fig. 10.2. Size fractionation of the original soil and of the eroded sediment collected after 46 min of exposure to simulated rain in three similar experiments. Data hand fitted. (After Ghadiri and Rose 1991b)

of a rate of 100 mm/h and drop size 2.2 mm (Ghadiri and Rose 1991b). Figure 10.2 shows that, in all experiments, eroded sediment had a higher proportion of fine particles and a lower proportion of coarse particles compared with the parent soil. In all three experiments water depths were so small that the effect of size selectively of sedimentation, described in Section 10.2.3, is unlikely to be an important reason for this result. In any case, the data in Fig. 10.2 refer to the end of the experiment when such effects are least, even with a significant depth of water (Fig. 10.1). Whether data such as those shown in Fig. 10.2 simply indicate the breakdown of large aggregates into fine sediment, or represent the result of raindrop stripping is not clear from Fig. 10.2 itself, and this issue will be discussed when data on sorbed chemicals are considered in Section 10.4.

With the shallow water layers produced in these experiments there may also have been difficulty in transport of the larger aggregates, and this may also have contributed to the paucity of large aggregates shown in Fig. 10.2 for the erosion experiments 1 to 3.

10.3 Sorbed Chemical Enrichment with a Significant Water Depth

10.3.1 The Effect of Erosion Process and Surface Contact Cover on the Settling-Velocity Distribution of Eroded Sediment

Figure 10.1 illustrates the type of change with time which occurs in the settling velocity characteristics of eroded sediment when rainfall detachment and redetachment are the erosion processes taking place in the presence of a sufficient depth of water for sedimentation to occur. Palis et al. (1990a, b) present similar information to that shown in Fig. 10.1 for a range of water erosion mechanisms, and for a range of degree and type of contact over in close contact with the soil surface. This study was carried out in the same 6-m-long tilting flume with simulated rainfall as that referred to above. The outcome of this extensive study may be summarized as follows.

In the absence of rainfall, when the erosion agent was only overland flowing water (introduced at the top of the 6-m flume), the settling-velocity distribution of eroded sediment very quickly becomes similar to that of the parent soil. In this type of experiment the only erosion processes involved are entrainment and re-entrainment of deposited sediment (Rose and Hairsine 1988). This result contrasts with the type of behaviour illustrated in Fig. 10.1. The rates of erosion with entrainment and re-entrainment are generally much higher than with rainfall, as the sole agent of erosion, and it appears that the rate of reworking of the deposited layer is proportionaly faster. It is this reworking which results in change in the settling-velocity distribution, and such rapid reworking leads to the early achievement of an equilibrium or steady-rate situation in which the settling-velocity distributions of eroded sediment and the parent soil are expected to be theoretically the same (Hairsine and Rose 1992).

Palis et al. (1990a, b) also investigated, for the same sandy loam soil, the concentration and settling-velocity distribution of sediment eroded by a combination of rainfall and runon at the top of the soil in the flume (the latter being the sole source of overland flow in no rainfall experiments). The rainfall and runon rates were arranged so that the streampower of overland flow at exit from the flume was the same in both types of experiment. As might be expected, these "combination" experiments resulted in eroded sediment whose settling-velocity characteristics were intermediate between those produced by rainfall alone, on the one hand, and by runon alone, on the other hand (Palis et al. 1990a, b).

The effect of cover by corn stalks and simulated leaves placed on the soil surface at cover levels of 10, 30, and 50% was also investigated by Palis et al. (1990a, b). Whilst cover resulted in very considerable reductions in the concentration of eroded sediment, as is well known, cover tended in general to produce eroded sediment finer in its settling-velocity distribution than that produced by bare soil under the same experimental conditions. This has an impact on the enrichment ratio of loaded sediment, as will be discussed later.

10.3.2 The Effect of Erosion Process and Surface Contact Cover on the Enrichment Ratio of Nitrogen

The consequences of the experiments of Palis et al. (1990a, b), described in Section 10.3.1, for the enrichment ratio of total nitrogen [Eq. (1)] will now be described. For the soil investigated most of the total nitrogen (1.15 g/kg) was in the organic form, organic carbon being 19 g/kg, and so the nitrogen may be regarded as sorbed to or incorporated in the soil.

Figure 10.3 illustrates the variation with time during the experiment in the enrichment ratio for total nitrogen for the three types of erosion described as detachment (i.e. rainfall alone), entrainment (i.e. runon alone), and detachment and entrainment (the combination experiments). The figure also shows the effect of percentage cover by corn stalks laid across the slope on the soil surface.

Whilst increasing surface contact cover by corn stalks substantially reduced the concentration of eroded sediment, Fig. 10.3 shows that, in general, increased cover leads to higher values of enrichment ratio, though this was not so when runon (entrainment) was the cause of erosion.

Considering together the results described in Section 10.3 so far leads to the conclusion that higher values of enrichment ratio are associated with the degree to which finer eroded sediment is produced by the erosion mechanisms at work, or by the degree of cover. A mechanistic understanding of why this is so will be given in terms of a conceptual model in the next section.

Results reported in quite a wide section of the literature can be interpreted as supporting the conclusion given in the preceeding paragraph. For example, the commonly found decline in E_R with mass of soil eroded,

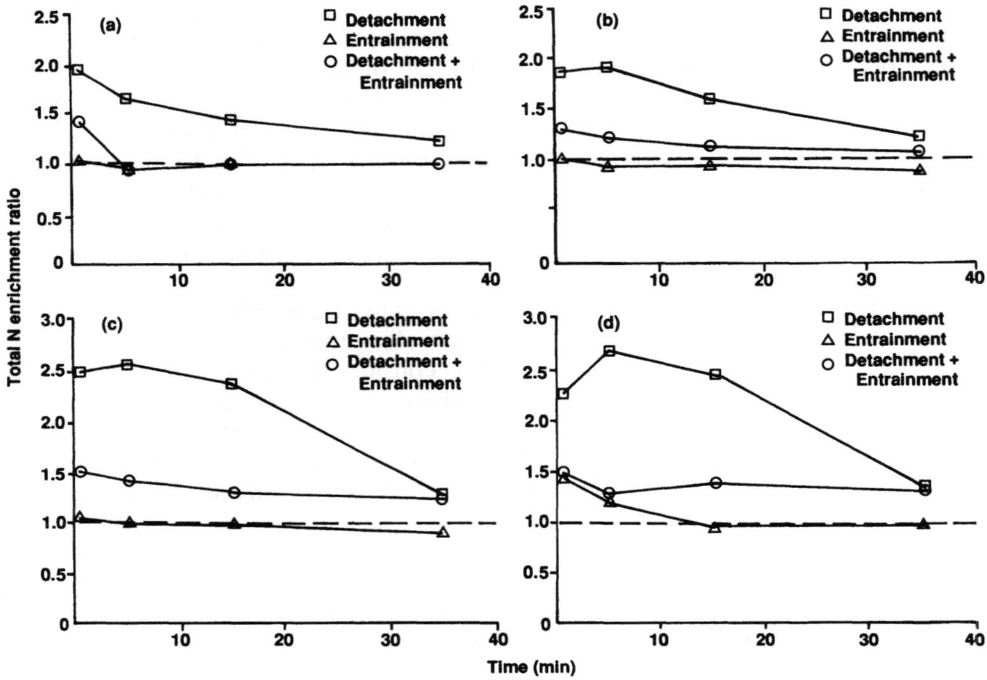

Fig. 10.3a–d. Variation with time of E_R for the two types of erosion processes and their combination (referred to as detachment and entrainment) under **a** 0%, **b** 10%, **c** 30% and **d** 50% cover by corn stalks for soil in a flume at 3% slope. (After Palis et al. 1990b)

represented in Eq. (2), can be understood in terms of the type of changes in settling-velocity characteristics illustrated in Fig. 10.1, if the finer soil fractions are also richer in nutrients, which is certainly the case for sandy soils. However, these matters will be discussed further following presentation of the conceptual model of enrichment in Section 10.3.3.

10.3.3 Theoretical Framework for Interpreting Enrichment Effects with a Significant Depth of Overland Flow

Rose and Dalal (1988) presented a conceptual framework which is helpful in understanding the reason for enrichment of sorbed chemicals. This framework is illustrated schematically in Fig. 10.4. As has been illustrated in Fig. 10.1 and discussed in Section 10.1, the settling-velocity characteristics of eroded sediment depends, for any given soil, on the type of erosion mechanism and time (or mass of soil eroded), as well as on the degree and type of cover, if present on the soil surface. This is illustrated in the distribution *A* of Fig. 10.4, where the concentration of eroded sediment in different size or settling-velocity classes is shown by distribution (a) at a

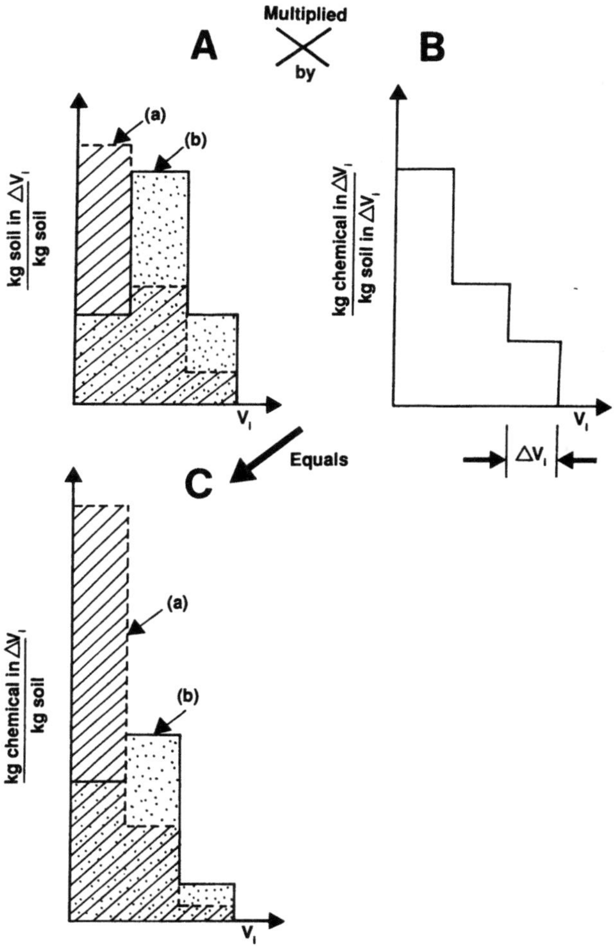

Fig. 10.4A–C. Schematic representation of the analytical framework. The distribution in **C**, from which E_R is calculated, depends on the product of the two distributions illustrated in **A** and **B**. In all distributions, the x axis is the size or settling velocity (v_i). In **A** the y axis is the concentration of sediment for various size ranges. The histogram distribution labelled a in **A** refers to the eroded sediment, which varies with the type or mix of erosion mechanisms and with time. Distribution b in **A** and **C** refers to the original soil. The sorbed chemical distribution in **B** is assumed to be the same for both the original soil and eroded sediment. The value of the enrichment ratio is obtained from the ratio of the *two differently hatched areas* shown in **C**. (After Palis et al. 1990a)

particular time. That such a distribution can change with time is illustrated in Fig. 10.1. The abscissa, shown as the settling velocity v_i, in Fig. 10.4 could be in v_i or particle size. Distribution (b) refers to the original soil.

Distribution B in Fig. 10.4 illustrates the experimentally measured distribution of the particular sorbed chemical with size or settling-velocity range. The single distribution B in Fig. 10.4 could be the distribution for the original soil. Palis et al (1990a) showed that this type of chemical concentration distribution changed little during an erosion event.

The final distribution concerned with evaluating the enrichment ratio [Eq. (1)] is distribution C in Fig. 10.4; this is obtained as the product of the component distributions in A and B, since:

$$\frac{\text{kg soil in size range}}{\text{kg soil}} \times \frac{\text{kg } N \text{ in size range}}{\text{kg soil in size range}} = \frac{\text{kg } N \text{ in size range}}{\text{kg soil}}, \tag{5}$$

where N in Eq. (5) can stand for any sorbed chemical.

Suppose the histogram (b) in distribution C is for the original uneroded soil (or, very closely, for eroded sediment in which entrainment and re-entrainment are the dominant erosion mechanisms). Then, the value of E_R for eroded sediment at the time at which the histogram (a) in A and C apply is given by the ratio of the area for (a) to that for (b) in distribution C. Clearly, the value of E_R will change with time as the histogram for mechanism (a) in distribution A alters with time.

Palis et al (1990a, b) showed that the conceptual framework illustrated in Fig. 10.4 provided a quantitatively useful way of interpreting the reason for change with time in the directly measured value of E_R for total nitrogen with a particular soil. The same authors showed that the interpretation remained very good even if the somewhat minor changes with experimental duration in distribution B were ignored, and the distribution measured for the original uneroded soil was used to calculate E_R. This finding indicates that if a theory were developed to predict the change with time in the distribution A (a current challenge to research), then with the distribution B measured for the soil of interest, E_R could be predicted as a function of time.

Even though such a prediction is not yet fully possible, the framework facilities a theoretical consideration of this mechanism for enrichment. For example, it shows that for enrichment to occur by this mechanisms, the chemical concentration distribution B in Fig. 10.4 must be non-uniform. Furthermore, if nutrients are distributed with a bias towards the fine fraction, as illustrated in B of Fig. 10.4, then changes in settling velocity or size distribution with time, as illustrated in Fig. 10.1, must lead to a decrease in E_R with time or with accumulated mass of sediment eroded; thus giving a theoretical basis for the empirically derived Eq. (2). The more rapid the change with time in settling-velocity characteristics (cf. Fig. 10.1), the greater will be the value of exponent B in Eq. (2). As shown by Proffitt et al (1991), this rate of change depends on the soil type, all other things being equal. Thus, as has been found experimentally, exponent B will depend on soil type.

The common finding that sediment originating from interrill regions is higher in E_R than sediment from rills (e.g. Rose and Dalal 1988) can also be understood in these terms, since erosion in rills will be dominated by runoff processes, whereas in interrill regions rainfall will often play an important role.

Foster et al. (1965) showed that enrichment ratios tend to be highest for sandy soils and lowest for clay soils. For sandy soils most chemicals are sorbed to the clay fraction, which leads to a distribution *B* in Fig. 10.4 strongly biased towards the finer fraction, thus leading to higher values of E_R. With clay soils, even larger aggregates contain a high clay content, so that the distribution *B* is much more uniform, and much less enrichment occurs in eroded sediment.

Foster et al. (1985) also showed that, for any given soil, the value of E_R for nitrogen measured in the sediment leaving plots declined with an increase in the ratio of sediment leaving the plot to sediment eroded within the plot (the sediment delivery ratio). Again, since it is the finer fractions of sediment which are preferentially delivered, this result can be understood if distribution *B* has a bias of the type illustrated in Fig. 10.4.

This theoretical framework associates the major reason for sorbed chemical enrichment with changes in time in the distribution *A* of Fig. 10.4. This change involves the processes of sedimentation and deposition, processes which are constrained to smaller aggregates as the depth of water in overland flow decreases. Thus, this conceptual model will become less applicable, or enrichment due to the processes it describes will become less important, if the depth of water on the soil surface is very shallow. Also, a shallow water layer leaves aggregates more exposed to raindrop impact, and collapse of less stable aggregates may be expected to lead to a more significant change in distribution *B* (Fig. 10.4) during any erosion event.

There will generally be some proportion of any soil surface which has only a shallow water layer during an erosion event, whether this is due to cultivation effects, other surface irregularities, or the surface-draining effect of rill incision. An enrichment process which can operate even with very shallow water depths will be described in Section 10.4.

10.4 Sorbed Chemical Enrichment with Shallow Water Depths

The characteristics of sediment eroded by rainfall with shallow water depths was discussed in Section 10.2.4. Here, this discussion will be extended to consider resultant enrichment effects.

10.4.1 Enrichment due to Differential Sorbed Chemical Concentration Within Soil Aggregates

There are reasons why the concentration of sorbed chemicals may not be uniform throughout the volume of aggregates, especially if these are stable, so that the processes of aggregate breakdown and reaggregation occur only over long time periods. The inner regions of stable aggregates may remain anaerobic for a sufficient time to reduce the nitrogen content, for example. More obviously, the concentration of applied pesticides is likely to be greater in the outer coatings of soil aggregates.

Ghadiri and Rose (1991a) report the results of analysing separately outer aggregate coatings removed either by careful hand peeling of larger stable

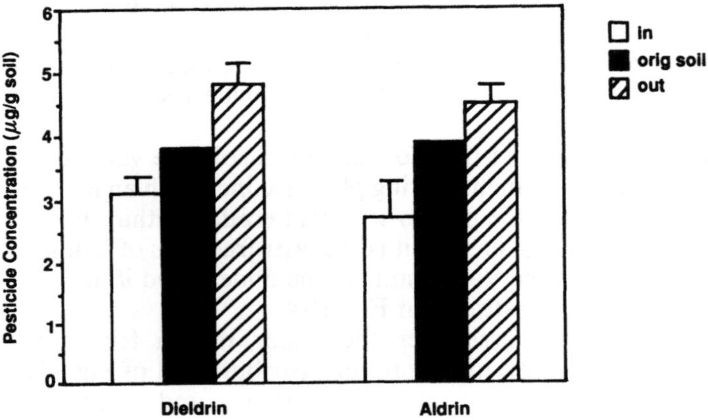

Fig. 10.5. Pesticide concentration in the outer layers and inner parts of aggregates compared with the concentration in the original whole aggregate, for dieldrin and aldrin. (After Ghadiri and Rose 1991a)

aggregates, or by dry sieving. For a soil with 55% clay and which had significant concentrations of pesticides, including dieldrin, aldrin and endrin, the outer coating of aggregates was shown to have significantly higher concentrations of these chemicals than the remaining interior part of the aggregates. Pesticide concentration of the original whole soil aggregate was intermediate between these two levels. This is illustrated in Fig. 10.5.

Evidently, the chemical applied to the soil (or the polymers formed before or after application) are sorbed onto the aggregate surface, and diffusion, which will tend over time to produce a uniform distribution throughout the aggregate, is sufficiently slow that this was not yet complete at the time of measurement.

The term enrichment ratio refers to the original soil or eroded sediment as a whole. The term "concentration ratio" was introduced by Ghadiri and Rose (1991a) as the enrichment ratio for any fraction of the soil.

Based on the results obtained from the aggregate peeling and sieving experiments referred to above, the outer layer of aggregates was found to be richer in the chemicals investigaged. As described in Section 10.2.4, Ghadiri and Payne (1979, 1981) showed raindrops to be capable of stripping away the outer coating of stable aggregates. If this is so, then the fine particles removed by this raindrop stripping should be richer in the sorbed pesticides than particles of the same size in the original soil. Figure 10.6 for dieldrin confirms this expectation since the concentration ratios of finer particles are greater than one. Similar results were obtained for other chemicals and also for organic matter and total nitrogen which were also of a slightly richer concentration on the outer coatings of aggregates (Ghadiri and Rose 1991a).

The theory given by Rose and Hairsine (1988), Hairsine and Rose (1991), and the experimental results of Palis et al. (1990a, b) and Proffitt et al. (1991)

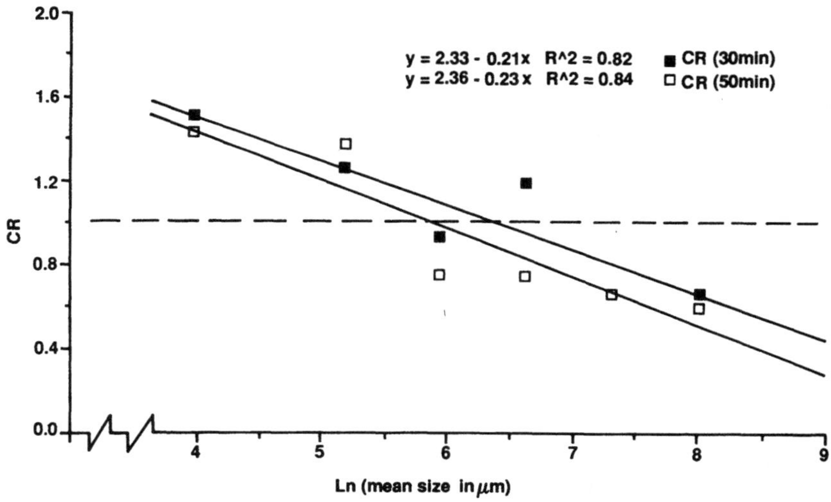

Fig. 10.6. Change in the concentration ratio (CR) of dieldrin with mean aggregate size for durations of 30 and 50 min under rainfall of rate 100 mm/h. Data are the mean of three and six sampling positions along the 6-m flume for 30 and 50 min sampling times, respectively. (After Ghadiri and Rose 1991a)

indicate that quite soon after it is exposed to rainfall, the soil surface becomes partly, but substantially, covered by a deposited layer of previously detached sediment. This could be a reason why the decoating or surface stripping process appears to occur dominantly on the larger aggregates for which the concentration ratio is less than unity (Fig. 10.6).

10.4.2 Relationship Between Enrichment Ratio and Cumulative Loss of Eroded Sediment

The form suggested by the literature for this relationship is given in Eq. (2). Since it follows from Eq. (2) that:

$$\ln E_R = \ln A - R \ln M, \quad (6)$$

then Eq. (2) can be conveniently tested by plotting $\ln E_R$ against $\ln M$. For the clay soil described in Ghadiri and Rose (1991a), these log–log plots were well fitted by linear relationships, both for total nitrogen and organic carbon, as shown in Ghadiri and Rose (1991b). Hence, Eq. (2) fits this set of experimental data, with the values of the constants $\ln A$ and B either within the general range suggested by Menzel (1980) and Sharply (1980) or a little outside it, varying somewhat with the experimental arrangement.

The concentrations of both total nitrogen and organic carbon were also found to be greater on the outside coatings of aggregates, though the difference between the outer coating and inner part of the aggregate was not as great as for applied pesticides (Ghadiri and Rose 1991a). Thus, the cause of enrichment and its decline with cumulative mass of soil eroded may be understood in terms of raindrop stripping as at least one operative process, and probably the dominant process. Equation (2) has been used to fit data in situations where raindrop stripping is both likely and unlikely to be the dominant cause of chemical enrichment. Thus, the ability of Eq. (2) to describe experimental data does not appear to be restricted to one type of enrichment mechanism.

10.5 General Discussion and Conclusions

Under situations where high rates of soil loss are produced by the dominance of runoff over rainfall erosion processes (Flanagan and Foster 1989; Palis et al. 1990a, b), then little, if any, enrichment of sorbed chemicals in the eroded sediment takes place. When gravitationally induced and mass movement of soil takes place, i.e. when soil is not carried in runoff water, the same conclusion would be expected to hold.

However, the extensive literature on this topic, of which only a small subset is reviewed here indicates that enrichment of chemicals or organic matter, which is sorbed to or intimately associated with soil, is a common occurrence. Whilst recognizing that other enrichment mechanisms can play a role, it is suggested as likely that the two mechanisms considered in this chapter are the most common and important reasons for enrichment.

When erosion by water is of significance, it is common for some parts of the soil surface to be covered by a significant depth of water. This is so in furrows, preferred pathways of water movement, and in rills or channels. In these water-covered parts, and provided that erosion is not totally dominated by the processes of entrainment and re-entrainment (Palis et al. 1990a, b), then chemical enrichment due to the sedimentation-related reasons described in Section 10.3 is expected to provide the dominant explanation.

However, during erosion, that fraction of the soil surface that is higher in the micro-relief will commonly have quite a shallow depth of water running over it, of depth more affected by "grain roughness" than by "form roughness" (Hairsine et al. 1992). This shallow water depth restricts the sedimentation process to sediment of sizes smaller than this shallow depth, and therefore restricts the enrichment process described in Section 10.3, and for which Fig. 10.4 provides a conceptual model. The enrichment of sorbed chemicals in this segment of the land surface with shallow water is expected to be dominated by the process of raindrop stripping, provided that the aggregates of soil are reasonably stable under wetting and raindrop impact.

In general, it is possible for both the described enrichment mechanisms to make a contribution to chemical enrichment. The very shallow water from higher micro-relief areas tends to be captured by pathways of preferred flow in transport from a field. Thus, there is the possibility for water with sediment chemically-enriched by raindrop stripping to enter channels of deeper flow where such stripping is suppressed, but where sedimentation-enrichment processes can operate. Some addition or even synergism between the two enrichment mechanisms, operating in series in this way, is possible.

As discussed in Section 10.3.3, the theoretical framework for interpreting enrichment effects illustrated in Fig. 10.4 provides an explanation of why E_R would be expected to decline with the duration or mass eroded during an erosion event. Another reason for this common observation, expressed in Eq. (2), is available in terms of the raindrop stripping mechanism. If the concentration of sorbed chemical increases continuously towards the outside of the aggregates, then continuous stripping of this outer coating by raindrops will lead to stripped sediment of a concentration which decreases with duration of the erosion event, and thus with accumulated soil lost by erosion.

The loss of sediment in wind erosion also leaves source areas of sediment with less plant nutrients, and with potentially lower productivity (Rose 1988). In the case of wind erosion there is no analogous process to the solution and transport of soluble forms of plant nutrients. However, for nutrients sorbed to soil or associated with its organic matter (e.g. organic nitrogen) enrichment of eroded sediment has been measured (Hagen and Lyles 1985). This enrichment can be understood in terms of the preferential long-distance transport of the finer sediment fraction lost in the suspended load, leaving behind the coarser sediment involved in saltation and creep. This remaining sediment, coarsened by the erosion event, will also be lower in plant nutrients if they are sorbed preferentially to sediment forming the finer soil fraction. There is some analogy between this and the preferential sedimentation cause of enrichment described for water erosion is Section 10.3.

Acknowledgements. Some of the work reviewed in this chapter has been supported by the Land and Water Research and Resource Development Council (LWRRDC, incorporating AWRAC) of the Australian Government, and by the Centre for Catchment and In-Stream Research of the Division of Environmental Sciences, Griffith University. Other work reviewed has been supported by the Australian Centre for International Agricultural Research (ACIAR). The support of these organizations, and discussion with my colleague Dr. H. Ghadiri, is much appreciated.

References

Flanagan DC, Foster GR (1989) Storm pattern effect on nitrogen and phosphorous losses in surface runoff. Trans Am Soc Agric Eng 32(2): 535–544

Foster GR, Young RA, Neibling WH (1985) Sediment composition for nonpoint source pollution analyses. Trans ASAE 29: 133–139, 146

Ghadiri H, Payne D (1979) Raindrop impact and soil splash. In: Greenland DJ, Lal R (eds) Soil physical properties and crop production in the tropics. John Wiley, New York, 95 pp

Ghadiri H, Payne D (1981) Raindrop impact stress. J Soil Sci 32: 41–49

Ghadiri H, Rose CW (1991a) Sorbed chemical transport in overland flow. Part 1. A nutrient and pesticide enrichment mechanism. J Environ Qual 20: 628–633

Ghadiri H, Rose CW (1991b) Sorbed chemical transport in overland flow. Part 2. Enrichment ratio variation with erosion processes. J Environ Qual 20: 634–641

Hagen LJ, Lyles L (1985) Amount and nutrient content of particles produced by soil aggregate abrasion. In: Erosion and soil productivity. Am Soc Agric Eng, St Joseph, MI, pp 117–129
Hairsine PB, Rose CW (1991) Rainfall detachment and deposition: modelling the physical processes. Proc Soil Sci Soc Am 55: 320–324
Hairsine PB, Rose CW (1992) Modelling water erosion due to overland flow using physical principles. I. Sheet flow. Water Resour Res 28: 237–243
Hairsine PB, Moran CJ, Rose CW (1992) The influence of surface soil characteristics on overland flow and erosion. Aust J Soil Res 30: 249–264
Knisel WG (1980) CREAMS: a field scale model for chemicals, runoff, and erosion from agricultural management systems. USDA, Cons Res Rep No 26, Washington, DC
Lovell CJ, Rose CW (1988) Measurement of soil aggregate settling velocities. I. A modified bottom withdrawal tube. Aust J Soil Res 26: 55–71
Massey HF, Jackson ML (1952) Selective erosion of soil fertility constituents. Soil Sci Soc Am J 16: 353–356
Menzel RG (1980) Enrichment ratios for water quality modelling. In: Knisel WG (ed) CREAMS: a field-scale model for chemicals, runoff, and erosion from agricultural management systems. USDA Cons Res Rep No 26, Washington, DC
Palis RG, Okwach G, Rose CW, Saffigna PG (1990a) Soil erosion processes and nutrient loss. I. The interpretation of enrichment ratio and nutrient loss in runoff sediment. Aust J Soil Res 28: 623–639
Palis RG, Okwach G, Rose CW, Saffigna PG (1990b) Soil erosion processes and nutrient loss. II. The effect of surface contant cover and erosion processes on enrichment ratio and nitrogen loss in eroded sediment. Aust J Soil Res 28: 641–658
Profit APB, Rose CW, Hairsine PB (1991) Rainfall detachment and deposition. Experiments with low slopes and significant water depth. Soil Sci Soc Am J 55: 325–332
Rose CW (1988) Water and wind erosion processes and strategies for soil land water conservation. In: Unger PW, Sneed TV, Jordan WR, Jensen RW (eds) Challenges in dryland agriculture – A global perspective. Proc Int Conf Dryland farming. Aug 15–19, 1988, Amarillo/Bushland, TX, Texas Agricultural Experimental Station pp 65–73
Rose CW, Dalal RC (1988) Erosion and runoff of nitrogen. In: Wilson JR (ed) Advances in nitrogen cycling in agricultural ecosystems. Commonwealth Agric Bureau Int Wallingford, UK, pp 212–235
Rose CW, Hairsine PB (1988) Processes of water erosion. In: Steffen WL, Denmean OT (eds) Flow and transport in the Natural environment. Springer Berlin Heidelberg New York, 312 pp
Rose CW, Parlange J-Y, Sander GC, Campbell SY, Barry DA (1983) Kinematic flow approximation to runoff on a plane: an approximate analytical solution. J Hydrol 62: 363–369
Rose CW, Saffigna PG, Hairsine PB, Palis RG, Okwach EG, Proffitt APB Lovell CJ (1989) Erosion processes and nutrient loss. Proc 5th Int Soil Conservation Conf on Land conservation for future generations. Bangkok, Thailand, vol 1, pp 163–175
Sharpley AN (1980) The enrichment of soil phosphorus in runoff sediments. J Environ Qual 9: 521–526
Soil Survey Staff (1975) Taxonomy: a basic system of soil classification for making and interpreting soil surveys. USDA Agric Handb 436, US Government Printing Office, Washington, DC
Stace HCT, Hubble GD, Brewer R, Northcote KH, Sleeman JR, Mulcahy MJ, Hallsworth EG (1968) A handbook of Australian soils. Rellim Tech Publ, Glenside, South Australia
Weber JB, Miller CT (1989) Organic chemical movement over and through soil. In : Sawhney BL, Brown K (eds) Reactions and movement of organic chemicals in soils. Soc Sci Soc Am Spec Publ No 22, Madison, Wisconsin
White RE (1988) Leaching. In: Wilson JR (ed) Advances in nitrogen cycling in agricultural ecosystmes. Commonwealth Agricultural Bureau, Wallingford, UK

Part III
Experimental Techniques and Application of Statistical Methods to Field Problems

11 Field Measurement of Water and Solute Transport Parameters in Soils

D.E. Elrick[1], *W.D. Reynolds*[2], *R.G. Kachanoski*[1], *and G.W. Parkin*[1]

11.1 The Use of Permeameters and Infiltrometers to Determine Hydraulic Properties

The transient and steady rate of infiltration of water into soil reflects the arrangement of soil particles and aggregates into the resultant maze of pore voids which conduct water. The size, distribution and continuity of pores is reflected in changes in both the saturated hydraulic conductivity and the unsaturated hydraulic conductivity components, flux potential and sorptivity. These hydraulic components can be measured in the field using well permeameters for subsurface measurements, and pressure and tension infiltrometers for surface measurements. Note that pressure and tension infiltrometers, which give more detailed information on hydraulic properties, can also be used at depth if a pit is dug. In this chapter we focus on procedures using the Guelph permeameter and its pressure and tension infiltrometer attachments.

11.2 Hydraulic Properties

The hydraulic properties applicable to flow from well permeameters and pressure or tension infiltrometers involve both the saturated and unsaturated components of the hydraulic conductivity (Figs. 11.1 and 11.2).

In Fig. 11.1a–c, the extent of the saturated bulbs, the unsaturated wetting zones, and the position of the wetting fronts are shown for some arbitrary times which are sufficiently large that steady flow has essentially been established. In Fig. 11.1c, where water is supplied at the surface under a negative potential, only an unsaturated wetting zone occurs. At true steady state, which occurs as time approaches infinity, the wetting front would theoretically extend to infinity (or to the boundaries established for a numerical solution). The size of the saturated bulbs, however, remain finite and essentially the same as those depicted in Fig. 11.1a and b at true steady state. There is no saturated bulb under the tension infiltrometer (Fig. 11.1c).

[1] Department of Land Resource Science, University of Guelph, Guelph, Ontario N1G 2W1 Canada

[2] Centre for Land and Biological Resources Research, Research Branch, Agriculture Canada, Ottawa, Ontaric K1A 0CA, Canada

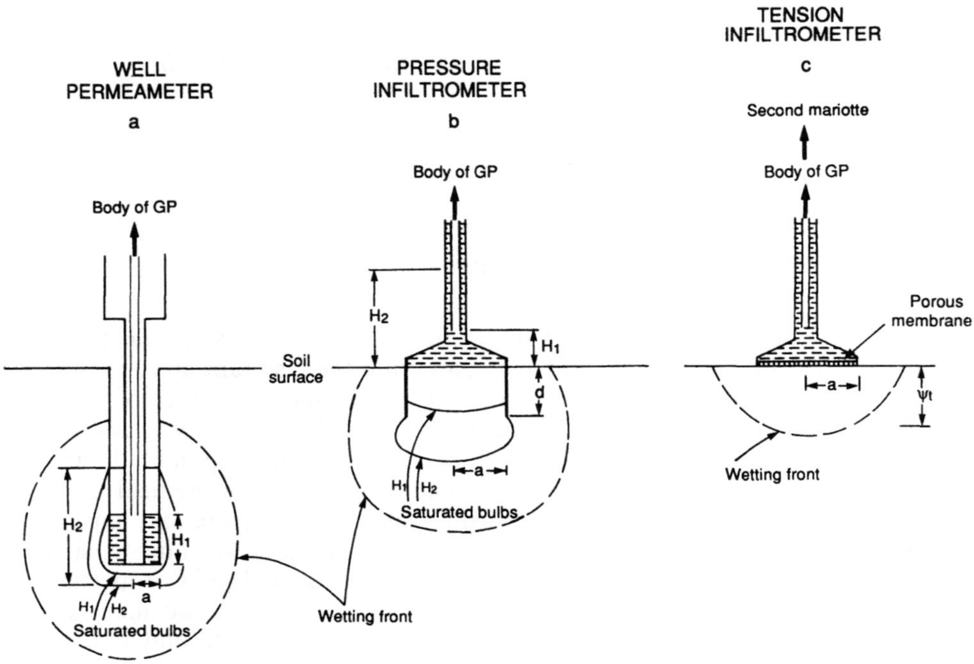

Fig. 11.1a–c. Field techniques for measuring the hydraulic conductivity of soils. **a** The constant head well permeameter; **b** the pressure infiltrometer; and **c** the tension infiltrometer. H_1 and H_2 are constant positive heads, Ψ_t is the constant negative head, a is the radius of the well, pressure infiltrometer or tension infiltrometer and d is the depth of insertion of the pressure infiltrometer. *GP* represents Guelph Permeameter

The relationship between hydraulic conductivity, $K(LT^{-1})$ and water potential, $\Psi(L)$, is shown in Fig. 11.2. In our approximate analytical analysis, it is the integral properties of the $K(\Psi)$ relationship that we show later to be of importance. We define the driving flux potential, $\phi_d(L^2T^{-1})$, as follows:

$$\phi_d = \int_{\Psi_i}^{H} K(u)du = \int_{\Psi_i}^{0} K(u)du + \int_{0}^{H} K(u)du = \phi_m + \phi_v, \tag{1}$$

where $\Psi_i(L)$ is the initial or background soil water pressure head and $H(L)$ is the depth of ponding in a well or pressure infiltrometer ring. The matric flux potential, $\phi_m(L^2T^{-1})$, is defined as:

$$\phi_m = \int_{\Psi_i}^{0} K(u)du, \tag{2}$$

and $\phi_v(L^2T^{-1})$, the velocity (flux) potential (Kirkham and Powers 1972) is

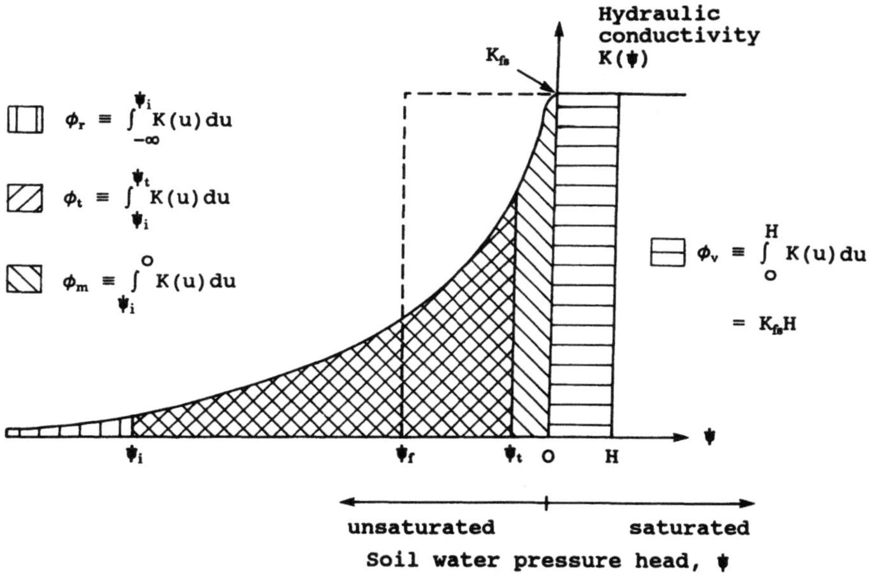

Fig. 11.2. The relationship between hydraulic conductivity and soil water pressure head

defined as:

$$\phi_v = \int_0^H K(u)du = K_{fs}H, \tag{3}$$

where $K_{fs}(LT^{-1})$ is the field-saturated hydraulic conductivity.

In a soil where Ψ_i corresponds to an unsaturated water content, $\theta_i(L^3L^{-3})$, the driving flux potential is given by $\phi_d = \phi_m + \phi_v$ (the sum of the areas shown in Fig. 11.2). The ϕ_d value describes the driving flux potential at the soil surface inside a ponded ring, and at the bottom surface of a ponded test well.

An empirical representation of the unsaturated $K(\Psi)$ relationship is given by Gardner (1958):

$$K(\Psi) = K_{fs}e^{\alpha\psi}; \quad \Psi \leqslant 0. \tag{4}$$

Note that the α parameter $(0 \leqslant \alpha \leqslant \infty)$ is the slope of $ln\,K$ vs. Ψ. We may set

$$\lambda_c = \alpha^{-1}, \tag{5}$$

where $\alpha(L^{-1})$ is called the sorptive number, and $\lambda_c(L)$ the macroscopic capillary (or sorptive) length scale (White and Perroux 1987).

Substituting Eq. (4) into Eq. (2) and carrying out the integration gives:

$$\alpha = (K_{fs} - K_i)/\phi_m, \tag{6}$$

where K_i is $K(\Psi)$ at $\Psi = \Psi_i$.

For this analysis we are more interested in α^* (L^{-1}), the ratio of K_{fs} to ϕ_m, where

$$\alpha^* = K_{fs}/\phi_m = \alpha/(1 - K_i/K_{fs}). \tag{7}$$

Thus, the relationship between α^* and α depends upon Ψ_i. For soils when Ψ_i is at or below the field capacity value (i.e. at the field capacity water content or drier, $\Psi_i \leqslant -1$ m) then $K_i \ll K_{fs}$ and $\alpha^* \approx \alpha$. However, in soils where Ψ_i approaches zero (soils very close to saturation), which can occur in very wet and in poorly drained soils, then α^* can be larger than α.

It is noted that α^* can also be interpreted as the suction at the wetting front of the Green-Ampt model (see Fig. 11.2):

$$|\Psi_f| = 1/\alpha^* \approx \lambda_c, \tag{8}$$

where $\Psi_f(L)$ is the effective wetting front potential (Mein and Farrell 1974).

11.3 Flow Equations and Analyses

The steady-state flow of water from the constant head well permeameter, the pressure infiltrometer and the tension infiltrometer can be approximated by equations of the form (Philip 1985; Reynolds et al. 1985; Reynolds and Elrick 1990, 1991):

$$\begin{aligned} Q &= f(H, a)K_{fs} + g(H, a)\phi_m \\ &= f(\Psi_t, a)K_t + g(\Psi_t, a)\phi_t, \end{aligned} \tag{9}$$

where $Q(L^3T^{-1})$ is the *steady* intake rate of water, $a(L)$ is the radius of the well or infiltrometer, $K_t = K(\Psi)$ at $\Psi = \Psi_t$ (see Fig. 11.2), and ϕ_t is the flux potential at Ψ_t, i.e. $\phi_t \equiv \int_{\Psi_i}^{\Psi_t} K(u)du$.

Note from Eq. (9) that it is the integral properties of the $K(\Psi)$ relationship that are of importance when determining the steady-state flow. From Fig. 11.2 it is shown that ϕ_m and ϕ_v represent, respectively, the unsaturated and saturated areas under the $K(\Psi)$ curve. It is easy to obtain K_{fs} directly from ϕ_v, however, because the rectangular area is given by $K_{fs}H$ and H is known. Thus, the important integral properties are K_{fs}, ϕ_m and their ratio α^*. For a more detailed characterization of unsaturated (near-saturated) flow, values of K_t, ϕ_t and the ratio K_t/ϕ_t can be determined for decreasing values of Ψ_t (such as $\Psi_t = -0.01$, -0.05, and -0.1 m) using the tension infiltrometer.

The following equations apply to the permeameter and infiltrometer analyses:

1. Constant Head Well Permeameter – Steady-State Measurements

$$Q = (2\pi H^2/C + \pi a^2)K_{fs} + (2\pi H/C)\phi_m \tag{10}$$

Two-height technique

$$= [2\pi H^2/C + \pi a^2 + 2\pi H/(\alpha^* C)K_{fs} \tag{11}$$

$$= [(2\pi H^2/C + \pi a^2)\alpha^* + 2\pi H/C]\phi_m. \tag{12}$$

(11), (12): One-height technique with α^* site estimated or determined independently

Note that H is the constant height of ponded water in the well and that a is the well radius (see Fig. 11.1a).

2. Pressure Infiltrometer – State Measurements

$$Q = (aH/G + \pi a^2)K_{fs} + (a/G)\phi_m \quad (13) \text{ Two-height technique}$$

$$= [aH/G + \pi a^2 + a/(\alpha^* G)]K_{fs} \quad (14)$$

$$= [(aH/G + \pi a^2)\alpha^* + a/G]\phi_m. \quad (15)$$

(14), (15): One-height technique with α^* site estimated or determined independently

Note that H is the constant height of ponded water in the ring and that a is the ring radius (see Fig. 11.1b).

In the above equations C is a dimensionless shape parameter primarily dependent on the H/a ratio (and to a lesser extent on $K_{fs}/\phi_m = \alpha^*$) and G is given by:

$$G = 0.316d/a + 0.184, \quad (16)$$

where $d(L)$ is the depth of insertion of the ring and a is the ring radius.

3. Tension Infiltrometer – Steady-State Measurements

$$Q = \pi a^2 K_t + (a/G_t)\phi_t \quad (17) \text{ Two-tension technique}$$

$$= [\pi a^2 + a(\alpha^* G_t)]K_t \quad (18)$$

$$= [\pi a^2 \alpha_t^* + a/G_t]\phi_t, \quad (19)$$

(18), (19): One-tension technique with α_t^* site estimated or determined independently

where $G_t = 0.25$.

4. Pressure and Tension Infiltrometer Measurements – Early-Time Transient Measurements

Early-time measurements are based on the formula:

$$I = S_H t^{1/2} \quad (20) \text{ Pressure infiltrometer}$$

$$= S_t{}^{1/2}, \quad (21) \text{ Tension infiltrometer}$$

where $I(L)$ is the cumulative infiltration and t is time (Philip 1958). S_H is the sorptivity at a ponded height, H, and S_t is the sorptivity at the negative potential, Ψ_t.

The experimental procedures for using the above techniques will not be presented here but can be obtained from the following references:

1. Constant head well permeameter: Reynolds and Elrick 1985; Talsma 1987; Elrick et al. 1989; Elrick and Reynolds 1992.
2. Pressure infiltrometer: Reynolds and Elrick 1990.
3. Tension infiltrometer: Ankeny et al. 1990, 1991; Reynolds and Elrick 1991; Perroux and White 1988.

11.4 Results

There have been numerous reports in the literature concerning the constant head well permeameter technique (e.g. Stephens and Neuman 1982; Talsma 1987; Elrick et al. 1989; Gallichand et al. 1990; Jabro and Fritton 1990; Elrick and Reynolds 1992). We report no data here but summarize the results of several studies as follows. Use of the simultaneous equations approach by taking steady-state Q measurements at two different heights [based on Eq. (10)] often results in calculated values of K_{fs} or ϕ_m that are invalid, i.e. negative. Negative values, however, can be avoided by calculating K_{fs} and ϕ_m using a one-height technique based on Eqs. (11) and (12) in which the parameter α^* is site estimated or measured independently. Suggested estimated values of α^* are: $1\,m^{-1}$ for compacted clays (caps, liners, sediments, etc.); $4\,m^{-1}$ for unstructured fine textured soils; $12\,m^{-1}$ for most structured soils from clays to clay loams and for unstructured medium and fine sands and sandy loams (first choice if little is known about the soil); and $36\,m^{-1}$ for coarse and gravelly sands and for highly structured soils with macropores (Elrick et al. 1989; Elrick and Reynolds 1992). The single-height technique in which α^* is site estimated gives results that are sufficiently accurate for most purposes.

There have also been a number of studies carried out with the tension infiltrometer (e.g. Clothier and White 1981; Chong and Green 1983; Watson and Luxmoore 1986; Ankeny et al. 1988, 1990, 1991; Reynolds and Elrick 1991). In particular, it should be noted that the tension infiltrometer supplies water at a negative potential and thereby eliminates the effects of highly conducting macropores (if present) such as surface cracks, root passageways and worm channels. This technique is useful for soil management studies where tillage or other practices affect the soil structure (Ankeny et al. 1990a, b).

We report briefly here on some results obtained with the pressure infiltrometer, as little has been published on this new procedure. An experiment was carried out on a 30 × 30 m field plot located at CEMAGREF-Montpellier in France. Four rows, each 4 m apart, were selected and eight pressure infiltrometer measurements (Reynolds and Elrick 1990) were carried out on each row at 4-m intervals giving a total of 32 measurements. At each site, early-time data were obtained at 10-s intervals, beginning at 20 s and continuing for approximately 2 min, and thereafter each 20 or 30 s for the next few minutes. Steady-state Q values were then obtained at $H = 6$, 16 and 25.5 cm. All three steady-state Q values were obtained within about 35 min. All measurements at a site were generally obtained within 1 hour, taking refilling etc. into account. The multiple-height technique based on Eq. (13) and using all three Q determinations was employed to obtain K_{sf}, ϕ_m and α^*. Equation (20) was used to determine S_H from the early-time measurements. The sorptivity at $H = 0$, i.e. S_0, was then obtained from (Philip 1958):

$$S_0 = S_H / \sqrt{1 + \alpha^* H}. \tag{22}$$

The geometric means of K_{fs}, ϕ_m and α^* were $2.59 \times 10^{-5}\,\mathrm{ms}^{-1}$, $2.20 \times 10^{-6}\,\mathrm{m}^2\,\mathrm{s}^{-1}$ and $11.8\,\mathrm{m}^{-1}$, respectively, with associated standard deviation factors of 1.40, 1.73 and 1.40. Note that the standard deviation factor is a multiplication/division factor. The geometric mean of S_0, obtained by fitting the early time data to Eq. (20) and then using Eq. (22), was $1.56 \times 10^{-3}\,\mathrm{ms}^{-1/2}$ with a standard deviation factor of 1.55. For the loam textured field soil these numbers are appropriate and the standard deviation factors for $n = 32$ were surprisingly small. In addition, the multiple-height technique based on Eq. (13) gave no negative values of either K_{fs} or ϕ_m. The pressure infiltrometer has both physical and theoretical advantages over the constant head well permeameter technique as evidenced by the absence of negative K_{fs} or ϕ_m values obtained at the Montpellier site, and by the small standard deviation factors.

11.5 Discussion

During the past several decades the trend in soil physics has been to field experimentation. Many of the physical principles of water and solute transport had been thoroughly tested in the laboratory during the previous decades, primarily using air-dried, sieved soil. At times, "undisturbed"soil cores were used. However, applying these laboratory-determined hydraulic and solute transport parameters to field situations often underestimated the travel times of water and solutes to various depths (and ultimately to groundwater). What was once perceived to be one-dimensional vertical flow appeared to actually be a complex two-(or three-) dimensional flow under field conditions. Macropores, consisting of cracks, fractures, decayed root passageways, wormholes, etc., can lead to preferential vertical flow, particularly under conditions of surface ponding. Fingering effects, brought about by textural/structural discontinuities, can also lead to two-(or three-) dimensional flow. So what was once perceived to be a rather straightforward, one-dimensional vertical flow has turned out, in many instances, to be rather complex in nature. As a result, many researchers have resorted to a stochastic approach in order to describe water and solute flow under field conditions.

The constant head well permeameter and newly designed pressure and tension infiltrometer techniques were developed as a means for in situ determination of the field-saturated and near-saturated hydraulic properties of soils. These hydraulic properties are required to measure and characterize the important effects of soil macrostructure in tillage, infiltration, drainage and solute transport.

11.6 The Use of TDR to Determine Hydraulic and Solute Transport Properties

11.6.1 Hydraulic Properties

A new TDR-based technique for determining soil hydraulic properties using early-time transient measurements of soil water storage has been proposed by Parkin et al. (1992). The soil dielectric constant is measured over the length of vertically installed TDR (time domain reflectometry) probes and converted to volumetric water content (Topp et al. 1980). Water storage to the depth of the TDR probes is simply the measured volumetric water content, θ, multiplied by the length, L, of the TDR probes. Recently, analytical solutions of transient water contents in soils under constant rainfall infiltration have been published by Broadbridge and White (1988) and Sander et al. (1988). The parametric solutions form the basis for the integral calculations of soil water storage, W, over the depth L as follows:

$$W(L,t) = \int_0^L \theta(z,t)dz. \tag{23}$$

Integration of Eq. (23) yields an analytical solution for cumulative water storage under constant rainfall infiltration and contains three independent soil hydraulic parameters; K_{fs}, α^* and C (Broadbridge and White 1988) or ν (Sander et al. 1988). $C = \nu^{-1}$, governs the shape of the advancing wetting fronts. Figure 11.3 shows an example plot of storage vs. time for several K_{fs} values using the Sander et al. (1988) form as the basis for solution. A nonlinear least squares fitting procedure, or other procedures, may be used with field data to solve for the unknown parameters such as K_{fs} and α^*. A constant water content boundary condition at the surface (e.g. saturated or just slightly flooded) may produce conditions that may be more applicable in slowly permeable soils.

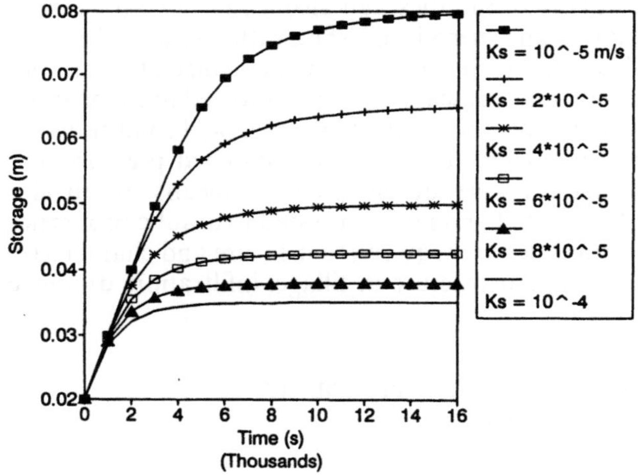

Fig. 11.3. Effect of saturated hydraulic conductivity on cumulative storage. Note that the steady surface rainfall rate $= 10^{-5}\,\mathrm{ms}^{-1}$, $\alpha^* = 12\,\mathrm{m}^{-1}$, $\nu = 0.67$, $\theta_s = 0.4$, $\theta_i = 0.1$ and the TDR probe length, L, is 0.2 m

11.6.2 Solute Transport Parameters

One of the limiting steps in our analyses of dispersion problems in heterogeneous soils is the lack of experimental data, not models to fit the data (Knight 1988). Unsaturated transport parameters have significant spatial variability (Bigger and Nielsen 1976; Van der Pol et al. 1977; Schulin et al. 1987), and their measured values depend on the scale of the measurement area (Butters and Jury 1989; Van Wesenbeeck and Kachanoski 1991). Thus, large numbers of samples are needed to adequately characterize a site, with the consequence that intensive field studies are often impractical because of labour and financial constraints.

The movement of a tracer such as bromide (Br^-) or chloride (Cl^-) through unsaturated field soils is usually measured by soil coring or by using vacuum solution samplers. Soil sampling gives resident concentration and mass distribution of the tracer with depth. However, the method is time-consuming, expensive, and very destructive. Large sample numbers can quickly render experimental sites useless for future work. In addition, it is only possible to obtain a sinlge distribution of solute at a particular time and location.

Problems with using solution samplers and with the representative nature of the water sample are well known (Hansen and Harris 1975). To the authors' knowledge, the question of whether the solution sample concentration represents the flux or resident concentration has never been answered. The physical disturbance caused by installing the samplers is less than that caused by soil coring, but can still be significant. In addition, the solute travel-time density function used to characterize solute transport is related to the mass flux of the solute tracer and not to the concentration (Jury et al. 1986). Thus, assumptions or information about the water flux density at the location of the solution cup are also necessary. Both the soil coring and solution sampler methods result in large sample numbers for subsequent laboratory determination of the tracer concentration.

A new procedure to measure travel-time density functions has been proposed by Kachanoski et al. (1992). The method uses vertically installed TDR probes. At constant θ, the TDR estimate of bulk electrical conductivity is related to the average pore-water concentration of an added solute tracer which, in turn, is related to the total specific mass ($g\,m^{-2}$) of the tracer between the TDR probes. The change in bulk electrical conductivity with time is shown to be linearly related to the mass flux of the solute past the ends of the TDR transmission rods. Laboratory experiments showed that solute mass flux out of a column was very similar to that measured using the TDR method. Excellent experimental agreement was also obtained between solute mass flux measurements in the field (0.2 m depth) using ceramic solution samplers, and the new TDR method (Fig. 11.4). The method measures mass flux of the solute at a particular depth and not solute concentration. Thus, no assumptions about local water flux density are necessary in estimating the travel-time density functions. The method also has the advantage that θ can be measured at the same location (soil volume) where the solute mass

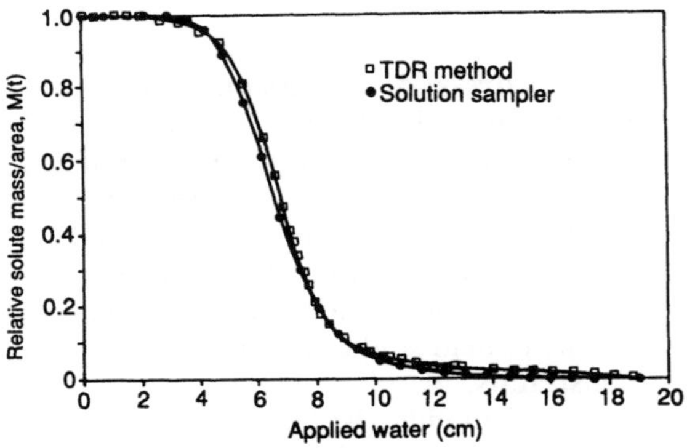

Fig. 11.4. Relative solute mass remaining in the 0- to 0.2-m depth in the field plot as a function of time, as measured using the TDR and solution sampler methods

flux is measured. The procedure is, however, limited to soils where the addition of a solute spike will cause a measurable reduction in the TDR signal.

In order to be able to predict the movement of chemicals to other depths, two steady-state models, the convective dispersion equation and the convective log-normal transfer function, were chosen along with a moment analysis to represent the transport process (Elrick et al., 1992). The reduced TDR measurements for a delta function (spike) input of solute are shown to be equivalent to the solution of a step function input of pure water at the surface to a soil containing a reduced concentration of unity to infinite depth, i.e., the complement of the flux-averaged reduced concentration for a step function input of solute. Nonlinear curve-fitting techniques were used to obtain the transport parameters. Fitting the measured TDR readings to the two models for both field and laboratory data gave good estimates of the transport parameters. The moment analysis, however, reflected the tailing of the solute, resulting in a doubling of the variance. The field TDR procedure is fast, inexpensive and non-destructive; the analysis yields needed parameters for model predictions.

Acknowledgments. We would like to thank our colleagues, J. L. Thony, G. Vachaud and M. Vauclin at the Institut de Mecanique de Grenoble and Ph. Revol and P. Ruelle of CEMAGREF, the Irrigation Division in Montpellier, for helping to obtain the data.

References

Ankeny MD, Kaspar TC, Horton R (1988) Design for an automated tension infiltrometer. Soil Sci Soc Am J 52: 893–895

Ankeny MD, Kaspar TC, Horton R (1990) Characterization of tillage and traffic effects on unconfined infiltration measurements. Soil Sci Soc Am 54: 837–840

Ankeny MD, Ahmed M, Kaspar TC, Horton R (1991) A simple field method for determining unsaturated hydraulic conductivity. Soil Sci Soc Am J 55: 467–470

Biggar JW, Nielsen DR (1976) Spatial variability of the leaching characteristics of a field soil. Water Resour Res 12: 78–84

Broadbridge P, White I (1988) Constant rate rainfall infiltration. A versatile nonlinear model. 1 Analytic solution. Water Resour Res 24(1): 145–154
Butters GL, Jury WA (1989) Field scale transport of bromide in an unsaturated soil. 1. Experimental methodology and results. Water Resour Res 25: 1595–1582
Chong SK, Green RE (1983) Sorptivity measurement and its application. In: Proc. Natl Conf Adv Infiltration, Chicago, IL. 12–13 Dec. Am Soc Agric Engr Publ 11–83. St Joseph, MI pp 82–91
Clothier BE, White I (1981) Measurement of sorptivity and soil water diffusivity in the field. Soil Sci Soc Am J 45: 241–245
Elrick DE, Reynolds WD (1992) Analysis of constant head well permeameter data. Soil Sci Soc Am J 56: 320–323
Elrick DE, Reynolds WD, Tan KA (1989) Hydraulic conductivity measurements in the unsaturated zone using improved well analyses. Ground Water Monit Rev 9: 184–193
Elrick DE, Kachanoski RG, Pringle EA, Ward A (1992) Parameter estimates of field solute transport models based on TDR measurements. Soil Sci Soc Am J (in press)
Gallichand J, Madramootoo CA, Enright P, Barrington SF (1990) An evaluation of the Guelph permeameter for measuring saturated hydraulic conductivity. Trans Am Soc Agric Eng 33: 1179–1184
Gardner WR (1958) Some steady-state solutions of the unsaturated moisture flow equation with application to evaporation from a water table. Soil Sci 85: 228–232
Hansen EA, Harris AR (1975) Validity of soil water samples collected with porous ceramic cups. Soil Sci Soc Am J 39: 528–536
Jabro JD, Fritton DD (1990) Simulation of water flow from a percolation test hole in layered soil. Soil Sci Soc Am J 54: 1214–1218
Jury WA, Sposito G, White RE (1986) A transfer function model of solute transport through soil. 1. Fundamental concepts. Water Resour Res 22: 243–247
Kachanoski RG, Pringle EA, Ward A (1992) Field measurement of solute travel times using solute time domain reflectometry. Soil Sci Soc Am J 56: 47–52
Kirkham D, Powers WL (1972) Advanced soil physics. Wiley-Interscience, New York
Knight JH (1988) Solute transport and dispersion. In: Steffen WL, Denmead OT (eds) Flow and transport in the natural environment: advances and applications. Springer Berlin Heidelberg New York, pp 17–29
Mein RG, Farrell DA (1974) Determination of wetting front suction in the Green-Ampt equation. Proc Soil Soc Am 38: 872–876
Parkin GW, Elrick DE, Kachanoski RG (1992) Cumulative storage of water under constant flux infiltration: analytical solution. Water Resour Res (in press)
Perroux KM, White I (1988) Designs for disc permeameters. Soil Sci Soc Am J 52: 1205–1215
Philip JR (1958) The theory of infiltration: 7. Soil Sci 85: 333–337
Philip JR (1985) Approximate analysis of the borehole permeameter in unsaturated soil. Water Resour Res 21: 1025–1033
Reynolds WD, Elrick DE (1985) In situ measurement of field-saturated hydraulic conductivity, sorptivity, and the α-parameter using the Guelph permeameter. Soil Sci 140: 292–302
Reynolds WD, Elrick DE (1990) Ponded infiltration from a single ring. I. Analysis of steady flow. Soil Sci Soc Am J 54: 1233–1241
Reynolds WD, Elrick DE (1991) Determination of hydraulic conductivity using a tension infiltrometer. Soil Sci Soc Am J 55: 633–639
Reynolds WD, Elrick DE, Clothier BE (1985) The constant head well permeameter: effect of unsaturated flow. Soil Sci 139: 172–180
Sander GC, Parlange J-Y, Kuhnel V, Hogarth WL, Lockington D, O'Kane JPJ (1988) Exact nonlinear solution for constant flux infiltration. J Hydrol 97: 341–346
Schulin R, Van Genucten MT, Fluhler H, Ferlin P (1987) An experimental study of solute transport in a stony field soil. Water Resour Res 23: 1785–1794
Stephens DB, Neuman SP (1982) Free surface and saturated-unsaturated analysis of borehole infiltration tests above the water table. Adv Water Resour 5: 111–116
Talsma T (1987) Re-evaluation of the well permeameter as a field method for measuring hydraulic conductivity. Aust J Soil Res 25: 361–368

Topp GC, Dais JL, Annan AP (1980) Electromagnetic determination of soil water content: measurements in coaxial transmission lines. Water Resour Res (13): 574–582
Van der Pol RM, Wierenga PJ, Nielsen DR (1977) Solute movement in a field soil. Soil Sci Soc Am J 41: 10–13
Van Wesenbeeck IJ, Kachanoski RG (1991) Spatial scale dependence of in situ solute transport. Soil Sci Soc Am J 55: 3–7
Watson KW, Luxmoore RJ (1986) Estimating macroporosity in a forest watershed by use of a tension infiltrometer. Soil Sci Soc Am J 50: 578–582
White I, Perroux KM (1987) Use of sorptivity to determine field soil hydraulic properties. Soil Sci Soc Am J 51: 1093–1101

12 Estimation of Regional Effective Soil Hydraulic Parameters by Inverse Modeling

R.A. Feddes[1], *G.H. de Rooij*[1], *J.C. van Dam*[1], *P. Kabat*[2], *P. Droogers*[1], *and J.N.M. Stricker*[1]

12.1 Introduction

Many environmental problems are manifest at the regional scale. To solve these problems it is necessary to have a proper description of the hydrological processes occurring at such scales. On the other hand, formulations for and the possibility of verifying hydrological processes are often available only at a local scale, i.e., at the field-plot scale. Hence, a procedure for scaling up from the local to the regional level is required to describe properly hydrological phenomena on such large scales. To meet that goal one can generally apply one of three types of approach: scaling, stochastic analysis or inverse modeling. The purpose of this chapter is to explore the inverse modeling approach.

Generally speaking, one may state that a physically-based model for water movement is required that reacts to changes in boundary conditions in a similar way as does a large area. Soils have nonlinear flow properties which transform sudden changes in boundary conditions into gradual changes in water movement. These macroscopic flow properties consist of the relationships between pressure head, soil water content and hydraulic conductivity. If we now assume that existing small-scale soil hydraulic property formulations can adequately describe the large-scale hydraulic properties, then traditional small-scale soil physics may be employed to estimate the areal vertical water movement, i.e., by application of the one-dimensional Richards' flow equation.

Based on this reasoning, a new approach is explored to efficiently calculate average vertical soil water movement for large areas. If successful, the approach will eliminate the need to investigate the hydrological behavior on (many) separate plots within the area. The basic idea is to derive by inverse modeling effective soil hydraulic properties for an area as a whole, using areal soil water flow data (e.g., areal evaporation, areal drainage, areal storage changes, and averages of measured moisture profiles of several plots within the area). In principle, the inverse method allows the area to be modeled for each boundary condition by one single simulation of a one-dimensional model that is based on the Richards' equation. In this way,

[1] Department of Water Resources, Wageningen Agricultural University, 6709 PA Wageningen, The Netherlands
[2] DLO The Winand Staring Centre for Integrated Land, Soil and Water Research (SC-DLO), 6709 PA Wageningen, The Netherlands

spatial variability is averaged before running the model (instead of averaging the results of many simulations, as occurs in traditional scaling methods).

The method extends the use of inverse modeling from plots to large areas. The soil hydraulic properties that are fitted to areal water movement in combination with Richards' equation data are thus assumed to describe effectively the hydrological behavior of that area. Until now, inverse modeling of unsaturated flow experiments has only been applied to laboratory columns, lysimeters or small field plots.

12.2 Methods of Regionalization

12.2.1 Scaling Approaches

Previously, soil physicists attempted to describe the hydrological behavior of large areas by scaling techniques (e.g., Miller and Miller 1955a,b, 1956; Warrick et al. 1977a; Tillotson and Nielsen 1984; Sposito and Jury 1985; Hopmans 1987). To arrive at areal values, the results of many simulations of one-dimensional soil water flow, each representing part of the area in a statistical manner, have to be averaged. The soil hydraulic properties are generated using the Monte-Carlo technique as demonstrated by, e.g., Hopmans (1988) and Hopmans and Stricker (1989). Warrick and Amoozegar-Fard (1979) introduced a technique that requires only one simulation of the flow problem, from which the solution for any particular site can be derived from its scaling factor. To arrive at areal values, the results of a sufficiently large number of sites still need to be averaged.

The effect of soil spatial variability has been studied by Sharma and Luxmoore (1979), who simulated the water balance components of a grassland watershed. The authors compared the integrated response of scale-different vertical soil columns with the results for "average" scale factors (mean, mode, and median). They found that the water balance simulated by the "average" scaling factors was significantly different from the integrated response as predicted by either the normal or log-normal distribution.

A study that considered not only spatial variability of the soil hydraulic functions but also the variability of the flux boundary condition at the bottom was performed by Hopmans and Stricker (1989). These authors simulated the water balance of a catchment and compared it with measured data. They showed that under relatively wet conditions the variation in water balance components, as caused by a stochastic lower flux boundary conditions, was of the same order of magnitude as the variation caused by the spatially variable soil hydraulic properties.

12.2.2 Non-Scaling Approaches

To investigate the effect of spatial variability, Bresler and Dagan (1983) applied a simple one-dimensional flow model. They argued that estimation errors caused by spatial variability may be much larger than errors originating from the approximate nature of some models. In the authors' view, actual flow phenomena under spatially variable field conditions are represented realistically by stochastic models in which the main statistical moments (averages and variances) of the flow variables are provided by simplified flow models. Bresler and Dagan showed effective soil hydraulic properties to be meaningful only in the case of very restricted flow conditions. Therefore, the authors concluded that the use of the traditional deterministic one-dimensional simulation approach cannot be justified in solving flow problems in spatially variable fields.

The failure to describe heterogeneous soil systems using average parameters will, however, be less marked in three dimensions than in one, because flow will bypass regions of low conductivity (Philip 1980). Yeh et al. (1985) applied a three-dimensional steady-state stochastic model approach, and showed that the effective hydraulic functions describing the mean large-scale behavior of the system were anisotropic. Mantoglou and Gelhar (1987) extended the analysis of Yeh et al. to transient unsaturated flow and showed the effective conductivity functions to be not only anisotropic but also hysteretic. Mishra et al. (1990) started from the same analysis and determined the effective moisture capacity and effective hydraulic conductivity numerically. They found that large-scale hysteresis and anisotropy did not significantly affect the flow. The authors suggest that for heterogeneous media with small spatial variability and correlation lengths, effective properties can indeed be useful for modeling transient unsaturated flow.

Hence, the conclusion may be that, although theoretically no justification for the existence of effective parameters on the field scale can be given, effective parameters may still simulate water balance components in a satisfactory way. It may be expected, however, that, when not only the soil hydraulic properties are variable but also boundary conditions, rooting depth, etc. vary, the use of "effective" soil hydraulic properties at catchment size may be warranted more.

The main goal of the present investigation is to find out whether field-effective parameters on a catchment scale do exist, and if so, how these field-effective parameters can be derived by means of inverse modeling techniques.

12.2.3 Inverse Modeling Approaches of Transient Flow

Laboratory Experiments. For many years laboratory outflow experiments have been used to determine the soil water retention curve $\theta(h)$ and the hydraulic conductivity curve $k(\theta)$. Kool et al. (1985) applied the inverse modeling approach to estimate $\theta(h)$ and $k(\theta)$ from outflow experiments. A key question to the inverse modeling approach is whether the measurements are accurate enough and contain sufficient information to estimate the effective hydraulic parameters with the accuracy required. To achieve this goal for outflow experiments, either the outflow data have to be supplemented with independent retention data or the boundary conditions should change more gradually (van Dam et al. 1992). Russo et al. (1991) investigated infiltration experiments in order to determine for which conditions the inverse problem is well posed. It appeared that for more complicated soil hydraulic functional relationships, such as the van Genuchten (1980) functions, some shape parameters were not simultaneously identifiable. This finding corresponds with the findings of van Dam et al. (1992).

Lysimeter Experiments. Only a few studies have been published to derive by inverse modeling soil hydraulic functions at the lysimeter scale. Dane and Hruska (1983) studied gravity drainage of a clay loam soil. They obtained good agreement between the predicted and independently measured $\theta(h)$ function, but overestimated the $k(\theta)$ curve. Kool and Parker (1988) showed by numerical simulation that the parameters in the extended van Genuchten (1980) model for hysteretic k-θ-h relations can be estimated by inverse modeling from lysimeter infiltration-drying experiments using error-free measurements of water content and pressure head. For the case of ponded infiltration, parameter estimates seem to be very sensitive to a correct identification of the position of the wetting front. During the stage of drainage, having at the same time a relatively high evaporative flux at the soil surface, a correct approximation of the top boundary conditions (in particular the flux) is paramount for an accurate estimation of the soil hydraulic parameters.

Field Experiments. The requirements for inverse modeling with respect to optimal design of the field experiment and optimal input data have not been investigated in a systematic manner (Kool et al. 1987). Attention should be given to the selection of the most suitable measurements (water contents, heads, and fluxes) and to optimal distributions of measurements in space and time. Consideration of optimal initial and boundary conditions for the field and the sensitivity to uncertainties in boundary and initial conditions are important points to consider in the development of field calibration methods.

12.2.4 Outline of the Present Study

Numerical infiltration-evaporation experiments were performed for 51 days on 32 artificial bare-soil columns having different soil hydraulic properties. The latter were taken from 32 samples of the top 30-cm layer of an existing soil in the watershed Hupselse Beek in the Netherlands. The output of the 32 simulation experiments as then averaged to obtain daily areal values, i.e., daily areal soil water contents at all depths and daily areal infiltration/evaporation fluxes.

Next, "effective" soil hydraulic paramters were generated by an inverse modeling approach for several subsets of these areal data for one single "effective" soil column (Fig. 12.a).

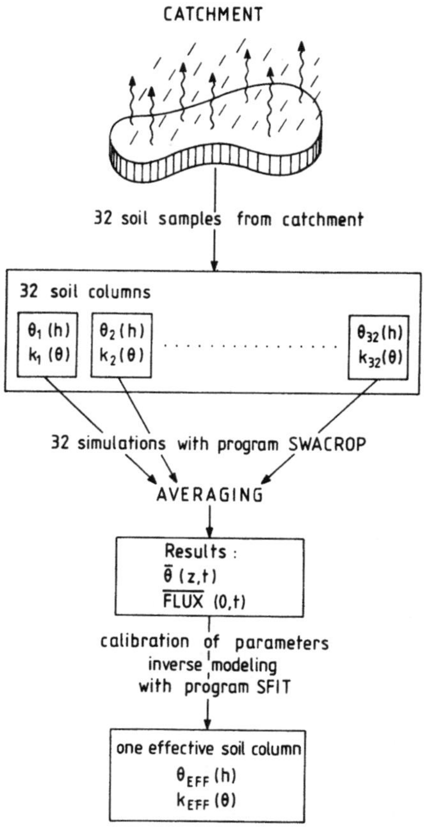

Fig. 12.1a Scheme of the numerical simulation and the inverse modeling procedure followed for the fitting case.

Three types of these areal data subsets were selected:

1. Water contents at several depths on selected days.
2. Cumulative infiltration/evaporation data (50 daily values).
3. Cumulative infiltration/evaporation data supplemented with the water contents at 0.5-cm depth for selected days.

Apart from all this, the measured soil hydraulic data of the 32 soil samples that were taken from the field were scaled in order to obtain a set of reference soil hydraulic properties.

To evaluate the resulting soil hydraulic parameter sets, three simulations were carried out for each parameter set for different boundary conditions. The first simulation had the initial and boundary conditions of the numerical infiltration-evaporation experiment. The second simulation was run under relatively wet boundary conditions and the third under relatively dry conditions. Then, the 32 soil columns were subjected to these three different sets of boundary conditions, after averaging the yielded areal values of water

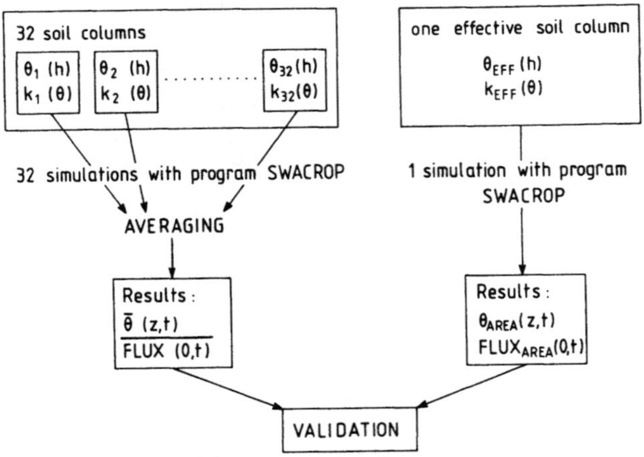

Fig. 12.1b Scheme of the validation procedure followed for the fitting, wet and dry cases

content and fluxes again. The outputs of the runs employing "effective" and reference soil hydraulic parameters were subsequently compared with the areal output resulting from the simulations on the 32 soil columns (Fig. 12.1b).

Next, the reference parameters resulting from the scaling procedure and the inversely fitted parameter set that performed best, were tested on the soil covered with grass. The water movement during two growing seasons was simulated by employing these two parameter sets, and again the results were compared with the average result of the 32 columns subjected to the same initial and boundary conditions.

12.3 Simulation Experiments

12.3.1 Numerical Infiltration-Evaporation Experiment on 32 Bare-Soil Columns

Simulations on 32 bare-soil columns were performed using the one-dimensional model SWACROP (Feddes et al. 1978; Belmans et al. 1983; Wesseling et al. a,b 1989).

To create a "soil" of known spatial variability, the hydraulic properties of 32 samples taken from the A-horizon in the Hupselse Beek watershed were assigned to 32 bare homogeneous columns. Of these samples, the soil water retention curves were derived from measurements on both slowly drying samples and samples put on a sandbox apparatus (Stakman et al. 1969). Unsaturated hydraulic conductivity values were determined for the wet range by means of the crust method (Bouma, 1977) and for the dry range by means of the hot air method of Arya et al. (1975) and the sorptivity method (Dirksen 1979). The soil hydraulic conductivity curves were parameterized by fitting the Mualem–van Genuchten functions (van

Genuchten 1980) to the measured data:

$$S_e = [1 + (\alpha|h|)^n]^{-m} \quad (1)$$

$$k = K_s S_e^l [1 - (1 - S_e^{1/m})^m]^2, \quad (2)$$

where $S_e = (\theta - \theta_r)/(\theta_s - \theta_r)$, $m = 1 - 1/n$, and K_s is the hydraulic conductivity at saturation (LT^{-1}). The subscripts r and s denote the residual and saturated values of the volumetric water content $\theta(-)$, while $n(-)$, $\alpha(L^{-1})$, and $l(-)$ are shape parameters. The measurement and fitting procedures are given in more detail by Hopmans and Stricker (1987). The homogeneous bare-soil columns had a zero flux lower boundary at $1-m$ depth.

For the first numerical simulation experiment, the boundary conditions were chosen in such a way that a reasonably wide range of pressure heads and water contents was covered. Initially, the soil was at static equilibrium with the pressure head at the soil surface at -4 m. At the start of the simulation, 3 cm of water was added at the soil surface. During the first day no evaporation was allowed. Then, for a period of 50 days a constant pressure head of -50 m was set as the boundary condition at the soil surface. The output of the simulation runs on the 32 columns resulted in averaged daily evaporation rates as well as in areally averaged daily soil moisture profiles.

12.3.2 Inverse Modeling Approach on Bare-Soil Columns

For the inverse modeling approach we took the finite element inverse model SFIT of Kool and Parker (1987). By this inverse model 32 effective soil hydraulic parameters of the closed-form $\theta(h)$ and $k(\theta)$ relationships of Eqs. (1) and (2) were generated. To maintain a proper areal mass balance, the initial condition was given not as a static equilibrium with a pressure head of -4 m at the soil surface, but as the areally averaged initial soil moisture profile that resulted from the original initial conditions being imposed on the 32 columns. Because of implicit differences between the deterministic model SWACROP and the inverse model SFIT, in the inverse approach the infiltration of water did not proceed at its maximum rate until 3 cm had infiltrated, but proceeded at a rate of 3 cm day^{-1} for the first day. Test runs of SWACROP and SFIT did not, however, reveal significant differences in moisture and pressure head profiles.

The least squares optimization process that is applied in SFIT is based on the Levenberg–Marquardt maximum neighbourhood method (Marquardt 1963). For details see Kool and Parke (1988). The method requires an initial estimate of all parameters that are to be optimized. When the optimization problem is ill posed (as in our case) the method is known to be quite sensitive to these initial estimates. By using the reference parameter set of Hopmans and Stricker (1989), obtained by scaling (see Table 12.1, set 6), we avoided this problem. (In all cases, we also included runs with two alternative initial estimates, the parameter values of which were either systematically smaller

Table 12.1. Soil hydraulic parameters resulting from several optimization and scaling procedures

Parameter set[a]	θ_r (–)	θ_s (–)	α (cm^{-1})	n (–)	k_s (cm day^{-1})	l (–)
1	0	0.419	0.0286	1.348	124.6	0.970
2	0[b]	0.406[b]	0.0300	1.333	106.4	0.807
3	0.0183	0.398	0.00788	1.386	33.7	0.500
4	0.00363	0.265	0.0201	1.228	37.8	0.410
5	0[b]	0.406[b]	0.00642	1.348	34.8	0.481
6	0	0.406	0.0190	1.593	33.7	0.500
7	0	0.4063	0.0274	1.4072	125.6	0.8799

[a] The number of the parameter set refers to the method by which the parameter values were obtained:
1. Fit on soil moisture profiles, all parameters optimized.
2. As 1, but with θ_r and θ_s fixed.
3. Fit on infiltration/evaporation, all parameters optimized.
4. Fit on evaporation and water content at 0.5-cm depth, all parameters optimized.
5. As 4, but with θ_r and θ_s fixed.
6. Reference curves obtained from scaling by Hopmans and Stricker (1989).
7. Reference curves obtained from scaling using soil hydraulic data from a wider range of water contents than applied in set 6.

[b] Value was fixed during the optimization process.

or larger than the values of set 6; the optimizations starting from set 6, however, always yielded by far the best results.)

The effective soil hydraulic parameters were then derived from three different subsets of areally averaged data mentioned in the outline. Subset 1 involves soil moisture contents taken at several depths. Alternatively, parameter estimation was also based on data taken at or just below the soil surface, i.e., at 0.5-cm depth. The parameters were fitted on cumulative infiltration/evaporation data only (subset 2) and on a combination of infiltration data and averaged water contents at 0.5-cm depth (subset 3).

Optimizations on average pressure heads were not considered. Unlike average fluxes and water contents, the average of the pressure head lacks a clear physical meaning.

Parameterizations Based on Soil Moisture Profile Data. For the fit on soil moisture profiles, the water content data were chosen in such a way that the rang of water contents covered was as wide as possible. The average water contents at five depths (1.5, 3.5, 6, 32.5, and 85 cm below the soil surface) on ten days (day 1, 2, 3, 4, 5, 10, 15, 20, 30, and 51) were used as input to the inverse model. As a first optimization, all six parameters which the program allows to be optimized were fitted to the data. The resulting parameter values are presented as set 1 in table 12.1 and as the soil water retention and hydraulic conductivity curves 1 in Fig. 12.2 and 12.3,

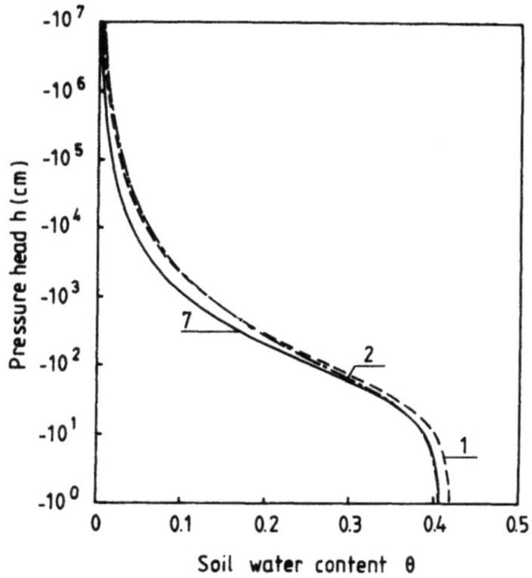

Fig. 12.2. Soil water retention curves resulting from the optimization of soil water contents. The *numbers* refer to the parameter sets described in Table 12.1. *Curve* 7 is shown for comparison

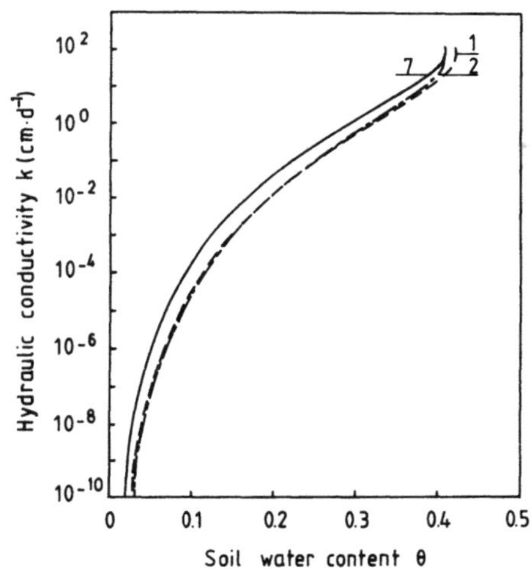

Fig. 12.3. Hydraulic conductivity curves resulting from the optimization of soil water contents. The *numbers* refer to the parameter sets described in Table 12.1. *Curve* 7 is shown for comparison

respectively. As a second optimization, the position of the soil moisture retention was pinned down by fixing the values of $\theta_r = 0$ and $\theta_s = 0.406$. The latter value represents the areal average, so sufficient water storage in the profile was guaranteed. Results are shown as set 2 of table 12.1 and curves 2 in Figs. 12.2 and 12.3.

Parameterizations Based on Infiltration/Evaporation and Water Content Data at 0.5-cm Depth. For the fits on data taken at or near the soil surface, all six parameters were first fitted to daily cumulative infiltration/evaporation data only. The results are presented as set 3 of Table 12.1 and curves 3 in Figs. 12.4 and 12.5.

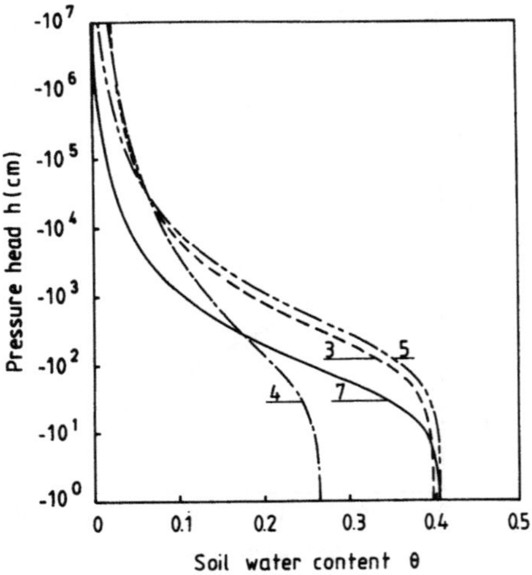

Fig. 12.4. Soil water retention curves resulting from the optimization of infiltration/evaporation and soil water contents at 0.5-cm depth. The *numbers* refer to the parameter sets described in Table 12.1. *Curve* 7 is shown for comparison

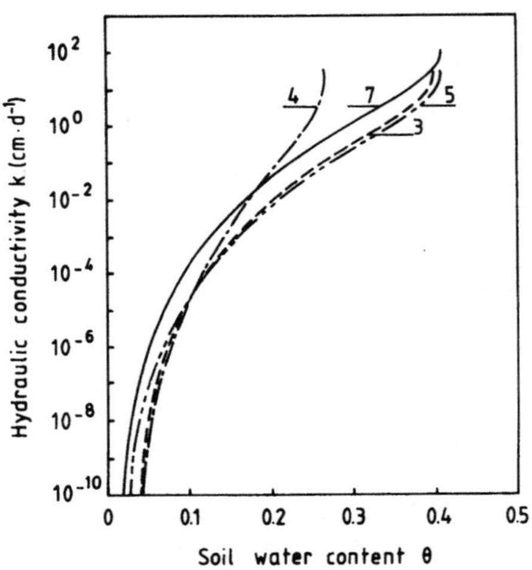

Fig. 12.5. Hydraulic conductivity curves resulting from the optimization of infiltration/evaporation and soil water contents at 0.5-cm depth. The *numbers* refer to the parameter sets described in Table 12.1. *Curve* 7 is shown for comparison

To avoid possible shifting of the soil water retention curve along the water content axis, an additional optimization was performed based on infiltration values, and supplemented with 0.5-cm-depth soil water content data. In this case, a value for cumulative infiltration/evaporation was included only if the difference from the previous value included was at least 0.1 cm. As a result, 23 daily values were excluded. These were partly replaced by the averaged water content at 0.5-cm depth on days 1, 2, 4, 8, 16, 28, and 51. The results are presented as set 4 in Table 12.1 and curves 4 in Figs. 12.4 and 12.5. Another procedure was to fit on these infiltration and water content data only four parameters, keeping θ_r fixed at 0 and θ_s at 0.406 (Similar to set 2 of Table 12.1). The results are shown as set 5 in Table 12.1 and curves 5 in Figs. 12.4 and 12.5.

Reference Curves as Obtained from Scaling Procedures. The data of the 32 samples comprising the artificial "soil" have been scaled by Hopmans and Stricker (1989) to yield a reference soil water retention and hydraulic conductivity curve (set 6 of Table 12.1 with corresponding curves in Figs. 12.6 and 12.7). Another set of reference curves was obtained from the same data set but using soil hydraulic data from a wider range of water contents (set 7 of Table 12.1 and curves 7 in Figs. 12.6 and 12.7). As mentioned earlier, the parameters of set 6 (Table 12.1) were used as initial estimates to start all the optimizations. Parameter set 7 (Table 12.1) was used as a comparison with the fitted parameter sets 1–5.

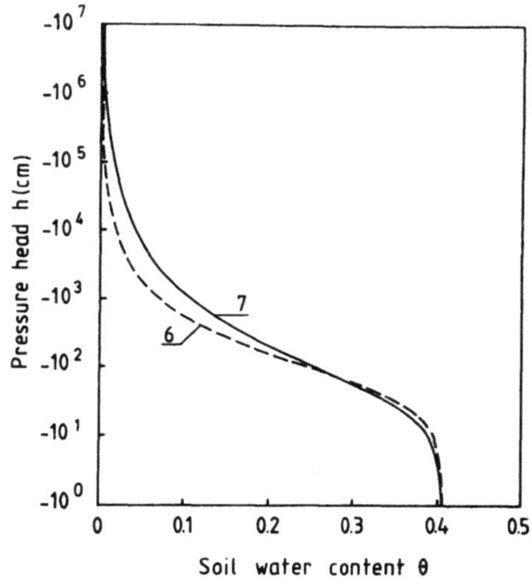

Fig. 12.6. Soil water retention curves resulting from the scaling procedure

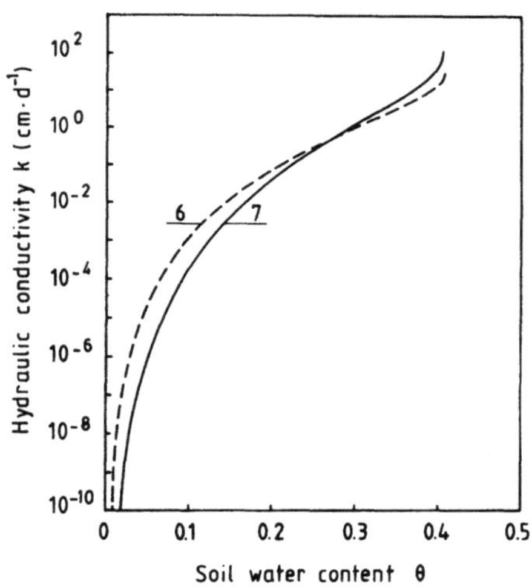

Fig. 12.7. Hydraulic conductivity curves resulting from the scaling procedure

12.4 Validation

The parameter sets 1–5 of Table 12.1, which were generated by the optimization program, and the reference parameter sets (6 and 7 of Table 12.1), were tested on two problems representing wet and dry conditions. The range of pressure heads and water contents covered by the test problems were partly beyond the range to which the parameters were fitted. The initial condition for the runs, employing the different parameter sets in both cases, was the average of the initial soil moisture profiles of the 32 artificial soil columns.

The initial pressure head distribution in the *dry case* corresponded to static equilibrium with the pressure head being −20 m at the soil surface. These conditions yielded in an average water content ranging from 0.085 at the soil surface to 0.086 at 1-m depth. Again, zero flux imposed at that depth. At day 1 no evaporation was allowed, while at day 5 0.5 cm of water was added to the soil surface and at day 15 1 cm of water was added. From day 1 through to day 14, the imposed pressure head at the soil surface decreased gradually from −20 to −85 m. At day 15 it increased to −20 m again, but since on that day it was raining, no evaporation was allowed. From day 15 till day 30, the imposed pressure head at the soil surface decreased again in a linear fashion, ending up at a value of −105 m.

For the *wet case*, the initial soil moisture profile represented static equilibrium with the pressure head being 1 m at the soil surface. This profile resulted in an average θ of 0.259 at the soil surface and 0.406 (saturation) at 1-m depth. At this 1-m depth a zero flux was imposed. At the soil surface, on days 1 and 5 1.2 cm of water was added, and on day 15 1 cm was added. On all other days the pressure head at the soil surface was kept at −20 m.

In both the dry and wet case a period of 30 days was simulated according to the procedure depicted in Fig. 12.1b. Hence, using the different $\theta(h)$ and $k(\theta)$ relationship of the 32 soil columns, areal values of water content $\bar{\theta}(z,t)$ and evaporation rates FLUX (o,t) were obtained by averaging the 32 SWACROP simulation results. In addition, one SWACROP simulation was performed for one effective soil column using the soil hydraulic data of sets 1–7 of Table 12.1, yielding $\theta_{\mathrm{AREA}}(z,t)$ and $\mathrm{FLUX}_{\mathrm{AREA}}(o,t)$.

The results of the runs on te fitting, (51 days) dry and wet cases are summarized in Tables 12.2 and 12.3. Table 12.2 presents the soil water contents at 1.5- and 32.5-cm depths for a dry day (during which evaporation was minimal over the entire simulation) and for a wet day (during which the water added to the soil surface caused evaporation to the maximal on the next day). The water contents in Table 12.2 adequately illustrate the range of water contents encountered during the simulations. Table 12.3 characterizes the daily evaporation of all tested sets for the three cases. The days presented are days with high and with low evaporation. The next category includes the days with evaporation peaks (after irrigation days), and the first day after each of these peaks. The dry days are the days just before irrigation and, simply, the last day of the simulation (day 30). Table 12.3 thus gives the range of daily evaporation covered by the simulations.

Table 12.2. Areal average moisture content $\bar{\theta}(z,t)$ obtained from 32 SWACROP simulations of the 32 different soil columns as compared with the moisture content $\theta_{AREA}(z,t)$ obtained from one SWACROP simulation of one effective soil column using the soil hydraulic data sets 1–7 of Table 12.1. Results are shown at two soil depths on days with low and high evaporation for the fitting, dry and wet case, respectively

	Depth (cm)	Volumetric soil water content $\bar{\theta}(z,t)$					
		Fitting case		Dry case		Wet case	
		Day 1	Day 51	Day 15	Day 30	Day 1	Day 30
Areal	1.5	0.321	0.092	0.232	0.081	0.313	0.130
	32.5	0.171	0.152	0.085	0.087	0.302	0.225
		Volumetric soil water content $\theta_{AREA}(z,t)$					
Set 1	1.5	0.325	0.099	0.235	0.085	0.313	0.140
	32.5	0.163	0.158	0.085	0.085	0.306	0.234
Set 2	1.5	0.324	0.099	0.236	0.086	0.312	0.141
	32.5	0.163	0.158	0.085	0.085	0.305	0.235
Set 3	1.5	0.309	0.115	0.219	0.095	0.324	0.155
	32.5	0.168	0.151	0.085	0.091	0.323	0.204
Set 4	1.5	0.253	0.106	0.203	0.091	Crashed	
	32.5	0.216	0.150	0.085	0.087	(initial $\theta > \theta_s$)	
Set 5	1.5	0.310	0.123	0.220	0.099	0.328	0.169
	32.5	0.170	0.151	0.086	0.100	0.325	0.206
Set 6	1.5	0.310	0.062	0.199	0.049	0.296	0.094
	32.5	0.170	0.136	0.085	0.087	0.302	0.205
Set 7	1.5	0.303	0.083	0.212	0.069	0.295	0.119
	32.5	0.169	0.147	0.085	0.087	0.303	0.211

The best-performing parameter sets were then selected. The performance of these sets is illustrated in more detail by Fig. 12.8a–f, which gives the daily evaporation rates (Fig. 12.8a,c and e) and the soil water content profiles for some days (Fig. 12.8b,d and f) of the three cases examined.

Of the two parameter sets fitted on soil moisture profile data, the set with all six parameters fitted (parameter set 1, Table 12.1) performed slightly better. Since parameter sets 1 and 2 (Table 12.1) hardly differ, large differences should not be expected. It appears that the match between the results of parameter sets 1 and 2 and the areal values is quite satisfactory for the fitting problem as well as for the dry and wet cases.

The fit on evaporation data only (parameter set 3, Table 12.1) calculates the evaporation of the fitting problem only slightly better than the fit on moisture profiles (parameter set 1, Table 12.1). Its performance on the

Table 12.3. As Table 12.2 but for evaporation, i.e. $\overline{\text{FLUX}}\,(o,t)$ or $\text{FLUX}_{\text{AREA}}(o,t)$

Case	Day	Daily evaporation (mm)							
			Set 1	Set 2	Set 3	Set 4	Set 5	Set 6	Set 7
		$\text{FLUX}_{(o,t)}$				$\text{FLUX}_{\text{AREA}}(o,t)$			
Fit-	2	12.9	14.8	14.7	16.1	14.0	16.2	18.3	16.4
ting	3	3.73	2.90	2.91	3.33	3.32	3.34	3.98	3.45
	30	0.37	0.26	0.26	0.32	0.38	0.32	0.49	0.41
Dry	4	0.11	0.05	0.03	−0.1	−0.0	−0.3	0.27	0.07
	6	2.19	2.47	2.47	2.49	2.29	2.49	3.23	2.57
	7	0.54	0.40	0.39	0.42	0.37	0.42	0.61	0.40
	14	0.12	0.09	0.09	0.12	0.09	0.15	0.17	0.09
	16	4.17	4.35	4.35	3.52	4.01	2.77	5.19	4.50
	17	0.91	0.74	0.74	0.60	0.70	0.42	0.93	0.75
	30	0.12	0.10	0.10	0.15	0.11	0.20	0.15	0.10
Wet	2	19.8	17.3	17.0	28.9	c	29.7	21.5	20.6
	3	7.78	6.20	6.02	12.2	r	12.5	7.87	7.97
	4	5.84	4.91	4.77	9.56	a	9.69	6.30	6.54
	6	13.3	13.7	13.5	18.9	s	18.9	15.7	15.9
	7	5.45	5.04	4.92	7.59	h	7.56	6.03	6.24
	14	2.53	2.64	2.59	3.10	e	2.99	3.18	3.27
	16	9.03	10.3	10.1	10.5	d	10.3	11.1	11.1
	17	3.47	3.50	3.46	3.75		3.62	3.97	4.01
	30	1.40	1.51	1.50	1.31	$\theta > \theta_s$	1.21	1.80	1.70

evaporation of both test cases is even less good than that of set 1. Due to the rather high soil water retention curve of parameter sets 3, 4, 5 (Fig. 12.4), the pressure head corresponding to the initial water content of the dry case was lower than the specified pressure head at the upper boundary. Hence, during the first few days infiltration occurred instead of evaporation (Table 12.3; Fig. 12.8c). After that, parameter set 3 underestimated evaporation and resulted in too wet a profile (Fig. 12.8d). The opposite holds for the wet case. Although the $k(\theta)$ curve (Fig. 12.5) does not look strange, the hydraulic conductivity is remarkably high for pressure heads lower than -100 cm (due to the relatively high water contents shown in Fig. 12.4). Hence, large amounts of water could flow to the soil surface and evaporate before upward flow was seriously limited. As a result, the evaporation of the wet test case is overestimated (Fig. 12.8e), and the profile after 30 days is too dry (Fig. 12.8f). In both test cases, parameter set 1 performed much better. In an attempt to improve the fit of the soil water retention curve, water contents near the soil surface were included in the data set on which the parameters were optimized. The resulting parameter set 4 has a very low saturated water content of 0.265. The initial water content of the wet test case exceeded that value for part of the profile. Therefore, the wet test could not be performed for this parameter set. Such a low θ_s obviously limits the water storage in the profile, giving rise to

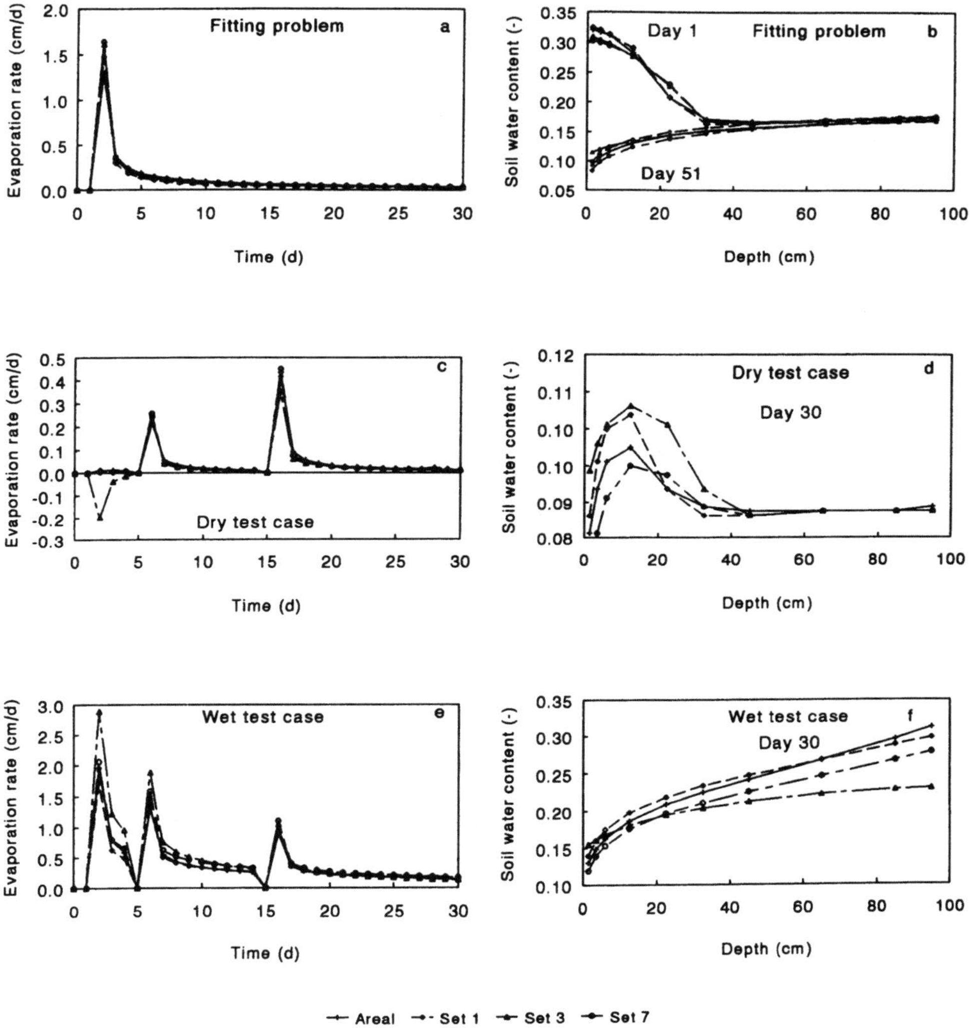

Fig. 12.8a–f. Areal average and simulated daily evaporation and soil moisture profiles. The selected parameter sets are the best-performig ones

inaccurate or useless results for wetter problems than the one it was determined on. We tried to avoid this problem by optimizing the same data subset but with θ_r and θ_s fixed (parameter set 5). Although better than parameter set 4, parameter set 5 was still slightly outperformed by parameter set 3.

Apparently, the combination of shallow water contents (0.5-cm depth) and cumulative fluxes through the upper boundary is a difficult one for the optimization routine. When fitting all six parameters it hardly modifies the parameters α and K_s, but ends up with a useless fit (parameter set 4). When the number of fitting parameters is restricted to four by fixing θ_r and θ_s, it assigns to α a rather extreme value without much change in K_s and l. Although it is not really bad, the resulting fit performs less well than the one based on less distinct data (cumulative fluxes only). Unlike the other cases, the optimizations resulting in parameter sets 4 and 5 are based on two kinds of input variables. In such cases, the influence of the individual observations on the final result of the optimization should not be unbalanced. Hence, much attention has to be paid to the selection of the input data used for optimization, and to the tuning of the weighting factors that determine the relative importance of the individual observations during the optimization process. Tentative tests showed that the optimization resulting in parameter set 4 (the poorest set) could be improved significantly by increasing the weight of the moisture contents.

Of the reference curves (parameter sets 6 and 7) set 7 performed best, especially in the dry case when set 6 strongly overestimated evaporation during the entire simulation period. Note the similarity between the values of the parameters (Table 12.1) from parameter set 1, based on soil moisture profiles, and the reference values (parameter set 7). The simulation results (notably the calculated evaporation) are also somewhat similar. Since the reference curve can only represent the area and the distribution of the scaling factors that go with it, it performs surprisingly well. In general. the results of parameter set 7 are somewhat less smooth than those of set 1, with higher evaporation peaks and lower tail evaporation. This may be due to the fact that the reference parameterization is supposed to represent a column, albeit an imaginary one, while the optimized set is expected to represent an entire area.

Clearly, optimization of areal fluxes does not give satisfactory results, while optimization of average moisture profiles does. The similarity between the best-performing optimized parameterization (parameter set 1) and the reference parameters obtained by scaling (parameter set 7) is striking.

12.5 Validation of the Water Balance of Grass

In the preceding section the derived effective hydraulic parameter sets were validated for artificial dry and wet conditions. In this section the two best-performing parameter sets (sets 1 and 3 of Table 12.1) will be judged by simulating the water balance of a grass crop under natural meteorological conditions. If a parameter set is valid under extreme, artificial conditions, it will probably also be valid under less extreme but natural conditions. On the other hand, parameter sets that deviate under extreme conditions may show acceptable results when simulating moderate situations.

Growing of a grass crop has two effects. It causes a higher soil water loss by transpiration as compared with the bare soil. At the same time, the $\theta(z,t)$ profile will probably be more homogenized, because of water extractions by roots from soil depths that are relatively wet.

The roots were assumed to be homogeneous with depth and root water extraction was calculated according to Feddes et al. (1978). To account for water uptake by roots, the Richards' equation has been extended to include a sink term. The simulations were performed with the SWACROP model for the growing season of the fairly dry year 1982 and the extremely dry year 1976, as these years show the largest variation in pressure head profiles with depth over time. The potential evapotranspiration has been calculated as described by the Penman-Monteith equation, using daily averages of precipitation, net radiation, air temperature, air humidity and wind velocity. Actual transpiration and soil evaporation have been computed according to Belmans et al. (1983).

The bottom boundary condition chosen at 1-m depth is unit hydraulic head gradient, which results in:

$$q = k(\theta), \tag{3}$$

where q is the flux density (LT^{-1}) through the lower boundary. As the initial condition we took the average downward flux to be the mean precipitation surplus (precipitation minus evapotranspiration) during the month preceding the growing season. Given the specified boundary condition at the bottom and the site-specific soil hydraulic properties, the initial, uniform distribution of water content could be computed.

Simulations were again performed for the same 32 homogeneous soil columns of the Hupselse Beek catchment area. On the first day of each month we calculated the mean and standard deviation of the cumulative actual transpiration, T_a, the cumulative flow through the bottom boundary Q_b, and the volumetric water contents at 1.5- and 32.5-cm depth. These quantities were selected because of their relevance to, e.g., crop growth, recharge to the ground water-table and runoff generation. The results were compared with simulations using the soil hydraulic functions derived from scaling (parameter set 7 of Table 12.1), optimization on soil moisture profiles (parameter set 1) and optimization on evaporation data (parameter set 3).

Table 12.4 gives a summary of the SWACROP output. The years 1976 and 1982 showed similar trends for all the parameter sets. The order of best performance is given in Table 12.5. T_a and Q_b are accurately simulated by both parameter set 7 and parameter set 1. The water contents in the top soil are closely predicted by parameter set 1. Parameter set 7 underestimates the water contents in the top soil, although the fluxes at both boundaries are correctly simulated. Parameter set 3 shows less extreme water content values than the mean. This was expected, as the optimization on evaporation data only results in soil hydraulic functions with less steep gradients. Hence, the final conclusion is that optimization on soil moisture profiles yields the parameter set which performs best.

12.6 Conclusions

1. From the performance of parameter set 1 (fit on soil moisture profiles) and the reference parameter set 7 it appears that in the absence of surface runoff a porous medium is able to mimic rather well the average vertical water movement of an area in the simplified setup of this study.

Table 12.4. Summary of SWACROP simulations of the grass covered soil over the period from 1 April to 1 October of the extremely dry year 1976 and the fairly dry year 1982

Cumulative actual transpiration T_a (cm)

	1976					1982				
Daynr	Mean	⟨Std⟩	Refer	Profil	Evap	Mean	⟨Std⟩	Refer	Profil	Evap
120	4.13	0.141	4.13	4.16	4.08	3.81	0.025	3.82	3.82	3.77
150	8.76	0.771	8.61	8.82	9.20	9.48	0.207	9.51	9.54	9.25
180	13.82	1.213	13.60	13.91	14.34	13.81	0.813	13.66	13.90	14.07
210	16.19	1.283	15.94	16.26	17.22	18.89	1.343	18.58	18.95	18.88
240	19.23	1.495	18.95	19.34	19.99	21.13	1.336	20.86	21.22	21.64
270	21.48	1.334	21.31	21.63	22.28	24.65	1.333	24.45	24.81	24.82

Cumulative flow to the groundwater table Q_b (cm)

	1976					1982				
Daynr	Mean	⟨Std⟩	Refer	Profil	Evap	Mean	⟨Std⟩	Refer	Profil	Evap
120	0.85	0.059	0.85	0.84	0.75	0.85	0.054	0.85	0.84	0.78
150	1.24	0.186	1.19	1.19	1.05	1.35	0.145	1.30	1.30	1.21
180	1.44	0.275	1.34	1.34	1.05	1.64	0.291	1.55	1.55	1.38
210	1.56	0.338	1.43	1.43	1.09	1.81	0.291	1.68	1.68	1.50
240	1.64	0.386	1.49	1.49	1.12	1.92	0.334	1.77	1.77	1.56
270	1.70	0.426	1.53	1.53	1.14	2.00	0.370	1.83	1.84	1.60

Volumetric water content θ at 1.5-cm soil depth

	1976					1982				
Daynr	Mean	⟨Std⟩	Refer	Profil	Evap	Mean	⟨Std⟩	Refer	Profil	Evap
120	0.077	0.058	0.053	0.075	0.123	0.186	0.072	0.159	0.187	0.194
150	0.221	0.073	0.192	0.218	0.201	0.108	0.067	0.080	0.106	0.128
180	0.064	0.051	0.048	0.067	0.068	0.262	0.075	0.231	0.259	0.241
210	0.165	0.068	0.140	0.164	0.145	0.059	0.048	0.048	0.066	0.072
240	0.064	0.044	0.053	0.067	0.080	0.206	0.071	0.179	0.205	0.186
270	0.144	0.065	0.121	0.145	0.131	0.172	0.067	0.148	0.172	0.155

Volumetric water content θ at 32.5-cm soil depth

	1976					1982				
Daynr	Mean	⟨Std⟩	Refer	Profil	Evap	Mean	⟨Std⟩	Refer	Profil	Evap
120	0.166	0.072	0.141	0.168	0.168	0.177	0.073	0.151	0.179	0.182
150	0.138	0.063	0.113	0.137	0.158	0.155	0.070	0.129	0.156	0.163
180	0.127	0.060	0.105	0.127	0.121	0.155	0.056	0.137	0.156	0.188
210	0.122	0.058	0.099	0.120	0.126	0.128	0.060	0.108	0.131	0.127
240	0.115	0.058	0.094	0.115	0.105	0.130	0.054	0.104	0.124	0.152
270	0.112	0.057	0.090	0.111	0.117	0.118	0.058	0.098	0.120	0.121

Mean, arithmetic mean of 32 locations in the Hupsel catchment;
⟨Std⟩, standard deviation of the results for the 32 locations;
Refer, reference soil hydraulic functions as derived by scaling;
Profil, optimized soil hydraulic functions using $\theta(z,t)$ data;
Evap, optimized soil hydraulic functions using evaporation data.

Table 12.5. Accuracy of the results of SWACROP simulations (see Table 12.4) with the parameter sets 1, 3 and 7 of Table 12.1

Actual transpiration, T_a	Set 1 = set 7 > set 3
Bottom boundary flow, Q_b	Set 1 = set 7 > set 3
Soil water content, θ ($z = -1.5$ cm)	Set 1 > set 3 > set 7
Soil water content, θ ($z = -32.5$ cm)	Set 1 > set 3 > set 7

2. The performance of set 7 (reference parameters) offers the possibility of using reference curves as "effective" curves representing the area.
3. The optimizations on moisture profile data yielded good results (parameter set 1), while those on soil surface data performed poorly (parameter sets 3 and 4). The combination of cumulative fluxes through the soil surface and shallow water contents resulted in bad fits. Here, the optimization was performed on two different variables. In such cases, more attention should be paid to the selection of the data used for optimization and, probably, if appropriate, on the tuning of the weighting factors that determine the relative importance of the individual observations in the optimization process.
4. A serious drawback for the proper formulation of the problem is the approximate nature of the model. The porous medium does not represent a small volume of undisturbed soil. Instead, the effective parameters describe the properties of a porous medium acting as an analogue of a large area. The resulting differences between model output and physical measurements may lead to local minima in the response surface of the objective function. The Levenberg–Marquardt optimization algorithm uses derivatives of the objective function to determine the search direction. This means that in order to arrive at the global minimum, the initial parameter values should not be too far from their optimal values. If measurement errors are taken into account, a sound initial estimate is even more important.

References

Arya LM, Farrell DA, Blake GR (1975) A field study of soil water depletion patterns in presence of growing soybean roots. I. Determination of hydraulic properties of the soil. Soil Sci Soc Am Proc 39: 424–430

Belmans C, Wesseling JG, Feddes RA (1983) Simulation model of the water balance of a cropped soil: SWATRE. J Hydrol 63: 271–286

Bouma J (1977) Soil survey and the study of water movement in unsaturated soil. Soil Survey Institute, Wageningen, The Netherlands (The soil survey institute is now incorporated in the Staring Centre) Soil Surv Pap 13: 47–52

Bresler E, Dagan G (1983) Unsaturated flow in spatially variable fields. 2. Application of water flow models to various fields. Water Resour Res 19: 421–428

Dane JH, Hruska S (1983) In situ determination of soil hydraulic properties during drainage. Soil Sci Soc Am J 47: 619–624

Dirksen C (1979) Flux-controlled sorptivity measurements to determine soil hydraulic property functions. Soil Sci Soc Am J 43: 827–834

Feddes RA, Kowalik PJ, Zaradny H (1978) Simulation of field water use and crop yield. Simulation Monograph. Pudoc, Wageningen, The Netherlands 189 pp
Hopmans JW (1987) A comparison of various methods to scale soil hydraulic properties. J Hydrol 93: 241–256
Hopmans JW (1988) Treatment of spatially variable groundwater levels in one-dimensional stochastic unsaturated water-flow modelling. Agric Water Manage 15: 19–36
Hopmans JW, Stricker JNM (1987) Soil hydraulic properties in the study area Hupselse Beek as obtained from three different scales of observations: an overview. Publ 78, Dep Hydraulics and Catchment Hydrology, Agricultural University Wageningen, The Netherlands, 112 pp (The department is now incorporated in the Department of Water Resources)
Hopmans JW, Stricker JNM (1989) Stochastic analysis of soil water regime in a watershed. J Hydrol 105: 57–84
Kool JB, Parker JC (1987) Estimating soil hydraulic properties from transient flwo experiments: SFIT user's guide. Soil and Environmental Sciences, Virginia Polytechnic Institute and State University Blackburg, VA 59 pp
Kool JB, Parker JC (1988) Analysis of the inverse problem for transient unsaturated flow. Water Resour Res 24: 817–830
Kool JB, Parker JC, van Genuchten MTh (1985) Determining soil hydraulic properties from one-step outflow experiments by parameter estimation. I. Theory and numerical studies. Soil Sci Soc Am J 49: 1348–1354
Kool JB, Parker JC, van Genuchten MTh (1987) Parameter estimation for unsaturated flow and transport models – a review. J Hydrol 91: 255–293
Mantoglou A, Gelhar LW (1987) Effective hydraulic conductivities of transient unsaturated flow in stratified soils. Water Resour Res 23: 57–68
Marquardt DW (1963) An algorithm for least-squares estimation of nonlinear parameters. J Soc Appl Math 11: 431–441
Miller EE, Miller RD (1955a) Theory of capillary flow. I. Practical implications. Soil Sci Soc Am Proc 19: 267–271
Miller RD, Miller EE (1955b) Theory of capillary flow. I. Experimental information. Soil Sci Soc Am Proc 19: 271–275
Miller EE, Miller RD (1956) Physical theory of capillary flow phenomena. J Appl Phys 27: 324–332
Mishra S, Zhu JL, Parker JC (1990) How effective are effective medium properties? In ModelCARE 90: calibration and reliability in groundwater modelling. Proc Conf, The Hague, Sept 1990, IAHS Publ No 195
Philip JR (1980) Field heterogeneity: some basic issues. Water Resour Res 16: 443–448
Russo D, Bresler E, Shani U, Parker JC (1991) Analysis of infiltration events in relation to determining soil hydraulic properties by inverse problem methodology. Water Resour Res 27: 1361–1373
Sharma ML, Luxmoore RJ (1979) Soil spatial variability and its consequences on simulated water balance. Water Resour Res 15: 1567–1573
Sposito G, Jury W (1985) Inspectional analysis in the theory of water flow through unsaturated soil. Soil Sci Soc Am J 49: 791–798
Stakman WP, Valk GV, van der Harst GG (1969) Determination of soil moisture retention curves. I. Sandbox apparatus. Range pF 0 to 2.7. ICW, Wageningen, The Netherlands (The ICW is now incorporated in the Staring Centre)
Tillotson P, Nielsen DR (1984) Scale factors in soil science. Soil Sci Soc Am J 48: 953–959
Van Dam JC, Stricker JNM, Droogers P (1992) Evaluation of the inverse method for determining soil hydraulic functions from one-step outflow experiments. Soil Sci Soc Am J (in press)
Van Genuchten MTh (1980) A closed-form equation for predicting the hydraulic conductivity of unsaturated soils. Soil Sci Soc Am J 44: 892–898
Warrick AW, Amoozegar-Fard A (1979) Infiltration and drainage calculations using spatially scaled hydraulic properties. Water Resour Res 15: 1116–1120
Warrick AW, Mullen GJ, Nielsen DR (1977a) Scaling field-measured soil hydraulic properties using a similar media concept. Water Resour Res 13: 355–362

Warrick AW, Mullen GJ, Nielsen DR (1977b) Prediction of the soil water flux based upon field-measured soil-water properties. Soil Sci Soc Am J 41: 414–419

Wesseling JG, Kabat P, van den Broek BJ, Feddes RA (1989a) Instruction for input. SWACROP. The Winand Staring Centre for Integrated Land, Soil and Water Research. Wageningen, The Netherlands, 29 pp

Wesseling JG, Kabat P, van ven Broek BJ, Feddes RA (1998b) SWACROP: simulating the dynamics of the unsaturated zone and water limited crop production. Winand Staring Centre, Wageningen, The Netherlands

Yeh TC, Gelhar LW, Gutjahr AL (1985) Stochastic analysis of unsaturated flow in heterogeneous soils. Parts 1, 2, 3. Water Resour Res 21: 447–471

13 Impacts of Vertical Heterogeneity on Simulated Water Flow in Hawaiian Basaltic Saprolite: Relation to Recharge

Timothy R. Green[1], *Richard E. Green*[2], *and Mark E. Miller*[3]

13.1 Introduction

Recharge of aquifers by rainfall is a subject of increasing interest to both scientists and the general public in Hawaii. On the island of Oahu, current freshwater use for agricultural, industrial and urban purposes is about 1.5 million m^3/day. Approximately 90% is obtained from groundwater. The Pearl Harbor aquifer (Fig. 13.1) provides about 48% of all water used on Oahu (C. Lao, pers. comm.). Both the quantity and quality of Oahu's groundwater are important to the social and economic well-being of the populace. Detection of pesticides and nonagricultural synthetic organic chemicals in water from many public-supply wells completed in the Pearl Harbor aquifer has led to costly well-head treatment and closing of wells. An understanding of the groundwater contamination problem is gradually unfolding (Lau and Mink 1987; Oki and Giambelluca 1987), but much remains to be learned. The sources of chemicals are not always clearly identified, and the mechanisms by which chemicals are transported as deep as 300 m to groundwater are not yet well defined.

The timing of major rainfall events, relative to the time of pesticide application to pineapple fields, is an important determinant of pesticide movement to groundwater, especially for volatile fumigants (Wong 1987; Oki and Giambelluca 1989). Reliable estimates of downward water flux below the root zone are required for realistic simulation of water movement in the soil profile. Previous modeling of pesticide movement toward groundwater in Hawaii (Green et al. 1986; Oki 1987; Loague et al. 1989a, b) did not have access to the detailed data on hydraulic properties, which provide input to the physically based, numerical model of the present study. The possibility of lateral flow and rapid vertical drainage from transient perched water-tables is suggested by observations of perched water during drilling for soil samples at some locations on Oahu in 1983 and 1985.

The objective of the study reported here is to assess the location and frequency of zones of positive pressure in the deep soil/saprolite column at

[1] Stanford University, Department of Civil Engineering, Stanford, CA 94305, USA and United States Department of the Interior, Geological Survey, Water Resources Division, 345 Middlefield Rd., MS-496, Menlo Park, CA 94025, USA
[2] University of Hawaii, Department of Agronomy and Soil Science, Honolulu, HI 96822, USA
[3] Daniel B. Stephens and Associates, 6020 Academy NE, Albuquerque, NM 87109, USA

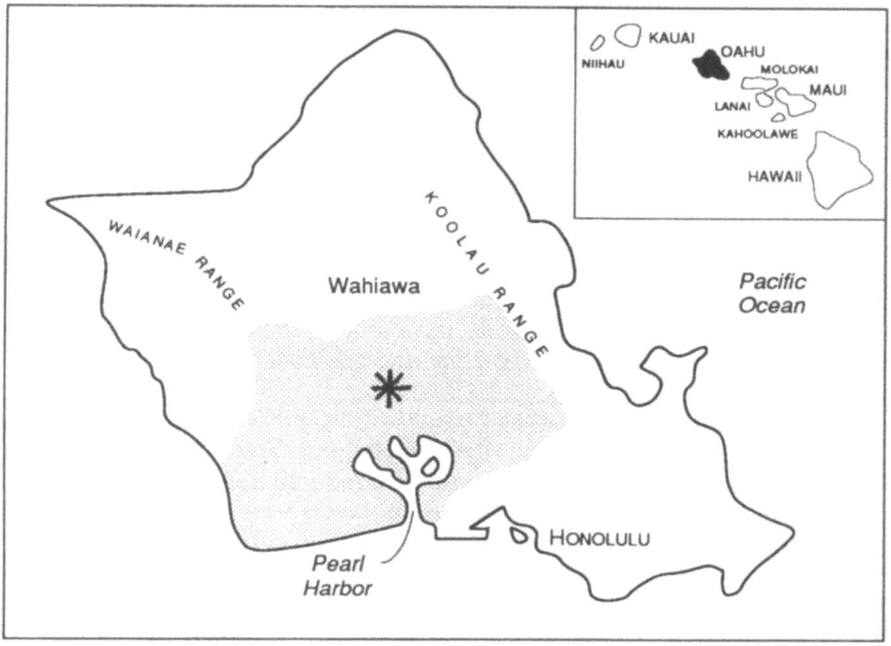

Fig. 13.1. Map showing general setting of recharge area (*shaded zone*) for Pearl Harbor aquifer and study location (*) on the island of Oahu, Hawaii

two locations on Oahu, Hawaii during successive periods of relatively low and high rainfall. The occurrence of perched water tables at depth might indicate lateral flow; such flow could exit at seepage faces in the gulches on either side of the broad ridges that are cropped with pineapple. Seepage has been observed in central Oahu during high rainfall periods (P.C. Ekern, pers. comm.). The existence of lateral flow would mean less deep percolation and, hence, reduced migration of contaminants to the basal aquifer than might be expected from calculations based on surface-infiltrated water. Also, some of the water and its solutes may flow rapidly downward during saturation by traveling through the largest pores.

The present investigation is possible because of the existence of soil-water characteristic data obtained on intact soil cores to a depth of 30.5 m in central Oahu (Miller et al. 1988). A one-dimensional, numerical model (Sect. 13.3.3) uses the above data to evaluate the spatial and temporal features of vertical flow and storage of water in the soil/saprolite sequence.

13.2 Physical Setting of Study Area

The pertinent information on the geology and hydrology of the Pearl Harbor aquifer is given by Lau (1987). The Wahiawa plateau (Fig. 13.1), at an elevation of 180–300 m, consists of lavas from the younger Ko'olau volcano ponded against the eroded lower slopes of the Waianae volcano. The vadose zone consists of four layers: surface soil, subsoil, saprolite (weathered basalt) and basalt (Miller et al. 1988). The textural and structural transitions between the layers are gradual, and surface soil is generally defined by the tillage zone, having a depth of about 0.5 m. The subsoil, a zone of illuviation, may have a thickness of 3–10 m; this zone often has a higher bulk density and lower hydraulic conductivity at saturation than either the surface soil or the underlying saprolite. The subsoil grades into saprolite, which has some textural features of the parent rock, but generally much lower values of saturated hydraulic conductivity; the thickness of this layer is about 20–40 m. Beneath the saprolite lies relatively unweathered basalt, which comprises the remainder of the unsaturated zone and the basal aquifer (Miller et al. 1988).

Hydraulic conductivities of the Oxisol soils of the Wahiawa plateau have been measured in situ at several sites (Bresler and Green 1983; Green et al. 1982), but there are no actual measured conductivities for the saprolite. Estimated conductivities of the various sublayers to a depth of 30.5 m are calculated from water characteristic curves for soil cores (Miller et al. 1988) using the procedure of Marshall (1958). Hydraulic conductivity and rainfall data from sites 4201A and 4213, located about 2 km apart, are used in the simulation study described here. The elevation and annual rainfall for the two site are: 310 m and 2000 mm/yr for 4201A, and 240 m and 1300 mm/yr for 4213. Mineralogies and organic carbon contents are also provided by Miller et al. (1988).

13.3 Methodology

13.3.1 Estimates of $K(h)$ from Water Characteristic Data

Details of soil core sampling and measurement of water retention at suctions of 1–1500 kPa are described by Miller et al. (1988). Relatively undisturbed core samples were collected to a depth of 30.5 m in thin-walled Shelby tubes (inside diameter of 7.6 cm). Boreholes were advanced with a system of hollow-stem augers through which the Shelby tube was lowered to the appropriate sampling depth and pressed into the soil or saprolite. Cores were generally about 0.5 m long. After extrusion from the tube, the core was encased in a closely fitting fiberglass/epoxy shell. Two 5-cm-long segments were then cut from the original core; these segments were used to determine water characteristic curves by tension table and pressure chamber techniques.

Water retention data for each core were used to calculate hydraulic conductivity as a function of water content, $K(\theta)$, or pressure head, $K(h)$, by the method of Marshall (1958). Although such calculated conductivities are expected to depart significantly from direct measurements (e.g., Green and Corey 1971), they should reflect differences among the various layers sampled. In order to provide conductivity data in the Mualem–van Genuchten form (van Genuchten 1980) as input for the water flow model, van Genuchten parameters (α and N) were obtained by least-squares fit (van Genuchten 1978) of the measured data to the analytical water retention equation.

Lacking measured K_{sat} values on the cores, we used the saturated conductivities calculated by the Marshall method, adjusted by a single matching factor, to obtain an estimate of K_{sat} for each core. The matching factor was obtained by relating the Marshall saturated conductivity, K^m_{sat}, to a saturated conductivity, K^i_{sat}, measured in situ at a depth of about 1 m on a similar soil (Green et al. 1982). The use of a common matching factor (K^m_{sat}/K^i_{sat}) for each location requires the assumption that the Marshall conductivity has the same relative error at all depths in the profile. Scaling K^m_{sat} by a constant changes the spatial mean and variance of K_{sat}. However, it changes only the mean of log K_{sat} and does not affect the variance of log K_{sat}. Thus, if K_{sat} is log-normally distributed, our scaling of K^m_{sat} shifts the mean, without altering the spread of the spatial distribution of log K_{sat}.

13.3.2 Rainfall and Evaporation Data

Daily rainfall data for fields 4201A and 4213 were provided by D.S. Oki (pers. comm.). A first-order Markov chain model was used to disaggregate monthly rainfall at each field into synthetic daily rainfall. Daily rainfall (1949–1983) at a nearby station was used to characterize the Markov transitional probabilities and seasonal distribution parametes for the study area (Oki and Giambelluca 1989).

Daily pan evaporation data were calculated from monthly measured values for a site near the study area (Ekern and Chang 1985). Derived daily values are constant within a given month. The same evaporation data were used for both field sites. A constant pan coefficient of 0.3 was used to represent evaporation from a planted pineapple field (Ekern 1965). Therefore, the potential evaporation is less than half of what one might estimate for a bare soil using a pan coefficient of 0.7. The monthly rainfall at location 4201A and the evaporation demand are shown in Fig. 13.2.

13.3.3 Variability Saturated Zone Simulation Model

Water flow through variably saturated porous media is simulated using Richards' equation [Eq. (1)] with capillary pressure head (h) as the dependent variable:

$$[C(h) + s{*}S_s]\frac{\partial h}{\partial z} = \frac{\partial}{\partial z}\left[K_{sat}K_r(h)\frac{\partial H}{\partial z}\right], \tag{1}$$

where θ is the volumetric soil-water content (cm^3/cm^3), $C(h)$ is $\partial\theta/\partial h$ (specific

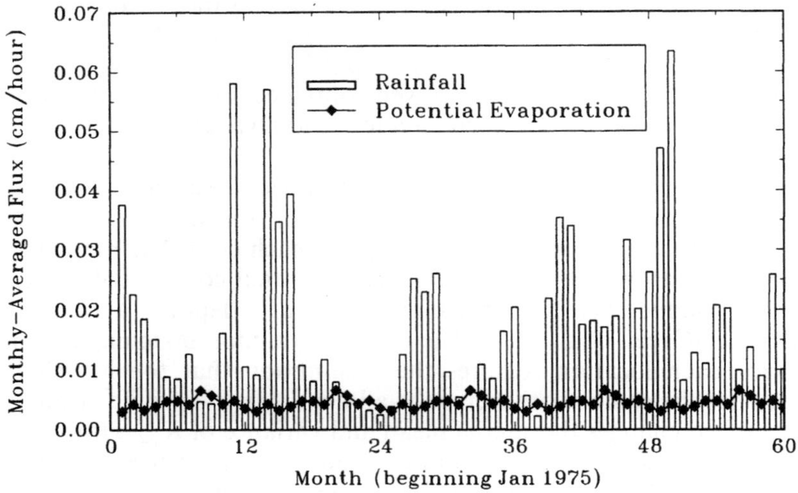

Fig. 13.2. Monthly rainfall and potential evaporation at 4201A (1975–1979). Potential evaporation is 30% of the measured pan evaporation according to Ekern (1965)

moisture capacity; cm^{-1}), s is θ/θ_{sat} (water saturation; dimensionless), S_s is specific storage (important at saturation; cm^{-1}), t is time (h), z is relative elevation (positive up; cm), H is $z+h$ (cm), K_{sat} is saturated hydraulic conductivity (cm/h), and $K_r(h)$ is relative hydraulic conductivity as a function of h (dimensionless). This nonlinear partial differential equation is solved numerically with VS2D (Lappala et al. 1987) which uses a block-centered finite-difference method.

Four types of boundary conditions are implemented in the present study. The first three occur at the land surface and are available in the published version of VS2D: infiltration (ponded and nonponded) and evaporation. Daily rainfall minus the potential evaporation (determined above) is input as a surface flux when the difference is positive. If the soil is not able to transmit the entire influx, the excess water begins to pond on the surface and VS2D changes the boundary condition from nonponded to ponded infiltration. In this case, the maximum depth of ponding is set to 2.0 cm. The excess inflow becomes surface runoff and is removed from the subsurface system. When the potential evaporation is greater than rainfall, water is removed from the simulated column at a rate determined by near-surface hydraulic gradients and conductivities, without exceeding the potential evaporation minus rainfall. We modified VS2D to include a pressure head gradient condition for the bottom boundary. At that boundary, the capillary pressure gradient is set equal to zero, such that gravity drainage exists. The published version of VS2D has been validated by several scenarios with analytical and/or experimental solutions. The new gradient boundary condition was checked with known steady-state solutions.

Table 13.1. Hydraulic properties[a] used in simulations

	4201A						4213					
Layer	z_b	θ_{sat}	θ_r	α	N	K_{sat}	z_b	θ_{sat}	θ_r	α	N	K_{sat}
1	0.4	0.584	0.250	0.0406	1.42	3.10	0.4	0.584	0.250	0.0406	1.42	3.10
2	0.8	0.539	0.255	0.0737	1.30	3.00	0.8	0.539	0.255	0.0737	1.30	3.00
3	2.0	0.584	0.446	0.0710	1.203	1.45	1.6	0.548	0.463	0.1340	1.160	0.16
4	3.5	0.665	0.505	0.1245	1.155	1.53	2.6	0.550	0.463	0.0553	1.210	0.26
5	5.5	0.554	0.350	0.0524	1.180	1.59	4.8	0.559	0.434	0.0936	1.190	1.96
6	6.8	0.566	0.383	0.0941	1.157	2.06	9.2	0.598	0.411	0.0237	1.360	0.76
7	8.4	0.712	0.447	0.0704	1.150	3.37	13.0	0.628	0.433	0.0481	1.135	0.50
8	10.7	0.660	0.433	0.1421	1.180	13.89	14.8	0.623	0.400	0.0490	1.120	2.27
9	13.0	0.605	0.444	0.1470	1.160	5.11	17.2	0.671	0.428	0.1970	1.070	3.01
10	15.0	0.661	0.395	0.0921	1.187	3.52	19.0	0.628	0.418	0.2692	1.080	1.49
11	18.0	0.652	0.396	0.0713	1.140	2.11	20.6	0.654	0.441	0.0154	1.193	0.15
12	24.0	0.617	0.395	0.0918	1.260	0.88	22.1	0.674	0.450	0.0117	1.275	0.99
13	25.0	0.649	0.403	0.0936	1.117	1.26	23.7	0.662	0.389	0.0054	1.277	1.26
14	26.6	0.660	0.451	0.0183	1.313	0.53	25.4	0.625	0.411	0.0337	1.103	0.72
15	28.0	0.656	0.407	0.0307	1.150	0.87	26.7	0.617	0.384	0.1597	1.093	2.12
16	30.0	0.667	0.324	0.0793	1.202	8.19	28.2	0.639	0.425	0.0823	1.133	1.66
17	35.0	0.620	0.294	0.0092	1.143	1.32	29.6	0.658	0.470	0.0901	1.167	2.84
18							32.0	0.663	0.445	0.0306	1.163	1.27

[a] z_b is depth to bottom of each layer (m). θ_{sat} and θ_r are saturated and residual water contents (dimensionless). α (cm^{-1}) and N (dimensionless) are parameters of the van Genuchten water characteristic function. K_{sat} is saturated hydraulic conductivity (cm/h). Soils in layers 1 and 2 were measured by Green et al. (1982) at a nearby site. Layers 3 and below comprise the saprolite sequence (Miller et al. 1988).

Given the governing equation and auxiliary conditions, the problem is solved numerically at discrete points in space and time. VS2D uses a central-difference approximation in space (second-order accurate) and the Backward Euler time march method (first-order accurate).

13.3.4 Hydraulic Properties and Stratigraphy of the Soil/Saprolite

Core data described in Section 13.3.1 are averaged (Section 13.3.5) for each layer, resulting in the hydraulic properties used for our simulations. For modeling purposes, heterogeneity in the sampled material is assumed to be perfectly stratified in distinct, uniform layers. Locally horizontal stratification implicit in our one-dimensional simulations approximates reality, because lava flows here have only mild slopes. Each layer is given hydraulic properties corresponding to an average of up to six replicate core samples. K_{sat} is the geometric mean of the estimated conductivities of the replicates, while the van Genuchten water retention parameters are given the arithmetic mean of the fitted parameters.

Table 13.1 lists the hydraulic conductivity and water retention parameters for each simulated layer at 4201A and 4213. Including the two layers of surface soils, there are 17 modeled layers to a depth of 35 m at 4201A, and 18 layers to a depth of 32 m at 4213. Layers are numbered from the surface down and are shown graphically in the profiles that follow (Fig. 13.6–13.9). Beneath the surface soils, no obvious spatial trend exists. Variability in K_{sat}, for instance, may reflect deposition/cooling of the parent basalts where interbed conductivity is much greater than that of dense basalt layers.

13.3.5 Simulation Scenarios

The spatial grid is designed by trial-and-error to resolve the movement of the wetting front, especially at material interfaces. We reduced the grid spacing until the maximum difference in pressure heads at each node converged within 1 cm. As a result, the minimum grid spacing is 1 cm and gradually increases up to 20 cm away from an interface.

A time series of rainfall data was chosen to include a relatively dry period (1976–1977) and a wet period (1978–1979). Preliminary simulations at 4201A showed that a pulse of infiltration above a steady-state flux of 0.005 cm/h produced a peak drainage rate at 35-m depth within 6 to 9 months. Of course, the system response is nonlinear and can vary significantly with the initial conditions and applied pulse, but the simulated response served as a guideline for choosing the appropriate simulation period. Starting with initial conditions of an assumed steady flow of 0.007 cm/h, an average year (1975) of rainfall is used to bring the system to a more realistic state before entering the dry period (1976–1977).

13.4 Results

Transience of the subsurface flow, water content (especially saturated conditions) and surface runoff generation are highlighted. First, boundary fluxes are observed with time, including the recharge rate at the bottom of the simulated zone for each location. Within each profile, zones of potential lateral flow and/or rapid verical flow are indicated by the temporal statistics of capillary pressure heads and water saturation throughout the profile. Of particular interest is the percentage of simulated times, sampled every 15 days, that the pressure head exceeds some critical value (e.g., $h = 0$) at each node (depth). Finally, we look at snapshots of the profile during extremes in the depth-averaged water storage ("wetness").

13.4.1 Time Series: Infiltration, Drainage and Water Content

Transient behaviors of the soil/saprolite systems' responses to infiltrated rainfall are displayed in time series of simulated water flow (Figs. 13.3 and 13.4) and storage (Fig. 13.5). Although significant dampening exists, as expected in conveying water more than 30 m through the vadose zone, rapid changes in recharge relative to the time scale of simulation are prominent. Evaporative losses (other than those which were subtracted directly from the rainfall) are not shown in Figs. 13.3 and 13.4. On dry days, the evaporation was generally at or near the potential (see Fig. 13.2) except after uncommonly long periods without rainfall.

Figures 13.3 and 13.4 illustrate the time lag between peaks of infiltration and deep drainage, but the maximum water stored in the system occurs at

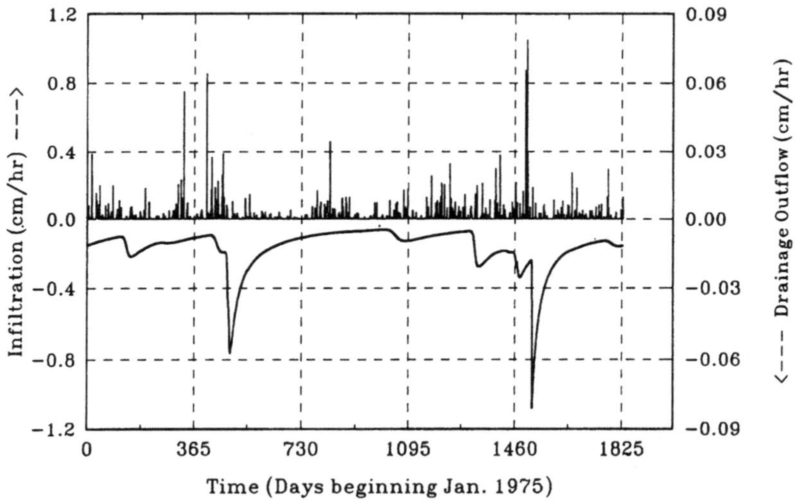

Fig. 13.3. Infiltration and drainage for 4201A (1975–1979)

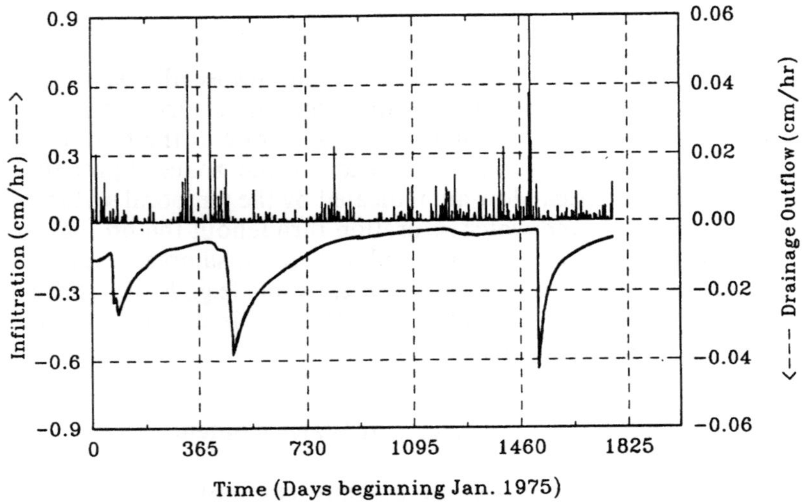

Fig. 13.4. Infiltration and drainage for 4213 (1975–1979)

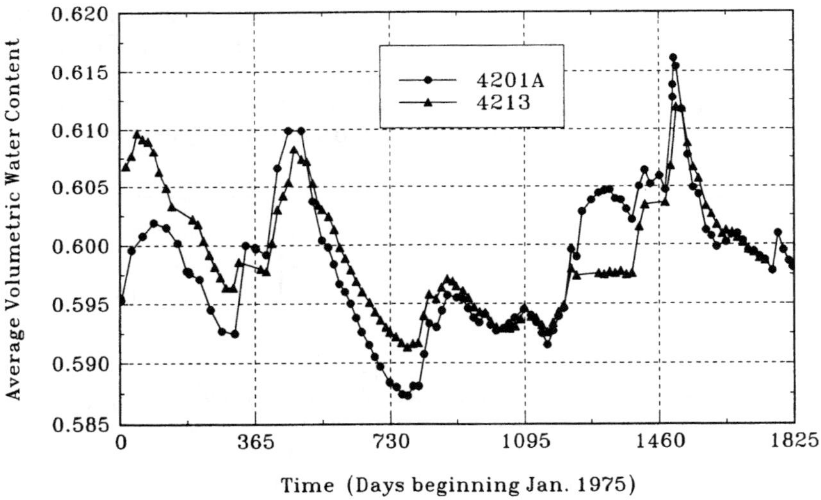

Fig. 13.5. Depth-averaged water contents for 4201A and 4213 (1975–1979)

some intermediate time between these peaks. The system's water storage or "wetness" is indicated by the spatially averaged soil-water content at discrete times. Thus, Fig. 13.5 is used to determine the extreme wet and dry times and to further compare the two sites. Reference levels of the average θ_{sat} for 4201A and 4213 are 0.637 and 0.625, respectively.

13.4.2 Pressure Head Profiles for Extreme Periods

From the time series plots above, we have chosen "snapshots" of the pressure head profiles at relatively dry and wet times at both locations (Figs. 13.6 and 13.7). Note that the two locations are plotted at somewhat different times, given by the extremes in Fig. 13.5.

A third profile of h at $t = 1496$ days is included for 4201A (Fig. 13.6) to show the transient formation of perched water and movement of the wetting front. The near-surface layers saturate and desaturate within 2 weeks at 4201A. When the entire column stores the maximum volume of water ($t = 1499$ days in Fig. 13.5), many sublayers approach saturation, but $h(z) < 0$ for all z; i.e., no "perched water" forms. Conversely, perched water does persist at 4213 (Fig. 13.7).

13.4.3 Profiles of Pressure Exceedence

Finally, the spatial distribution of commonly saturated or near-saturated zones are illustrated by looking at the temporal statistics of h at each node. We discovered that the time histograms of h in most layers were quite skewed. Therefore, the mean and variance give a false positive indication

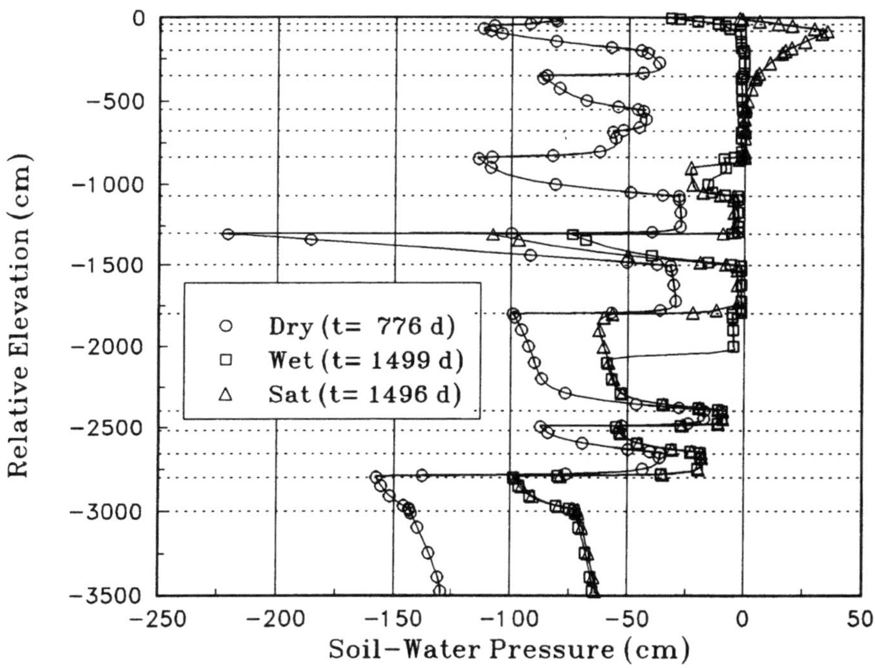

Fig. 13.6. Pressure head profiles for 4201A (1975–1979) during relatively dry and wet periods. *Dotted lines* correspond to the layers in Table 13.1

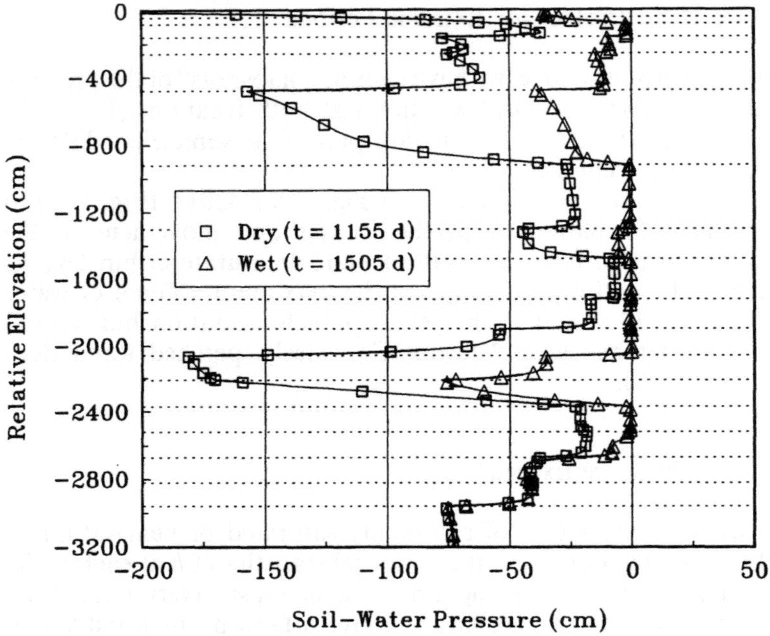

Fig. 13.7. Pressure head profiles for 4213 (1975–1979) during relatively dry and wet periods. *Dotted lines* correspond to the layers in Table 13.1

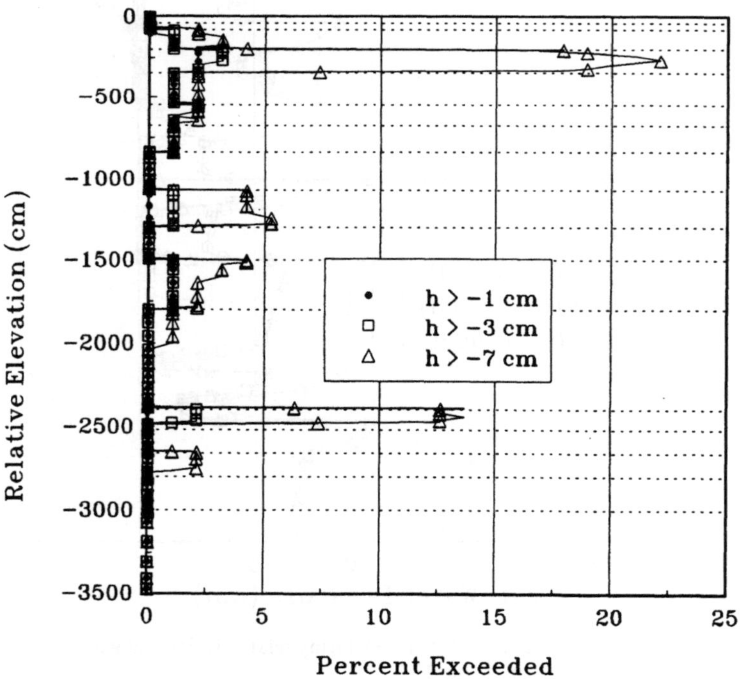

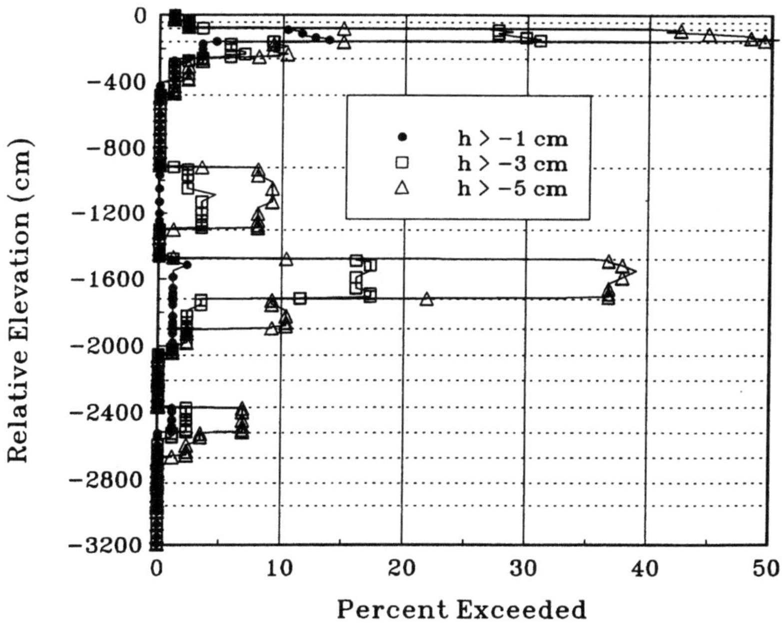

Fig. 13.9. Profile of percentage of observation (simulation times) when *h* exceeds key values for 4213 (1975–1979). *Dotted lines* correspond to the layers in Table 13.1

of the frequency of layer saturation, i.e., $h = 0$ falls well within two standard deviations of the mean h in layers which never saturate. Both temporal statistics and the histograms (neither is shown here) show that capillary pressure fluctuations (e.g., variance in h) are lower in wet zones (mean h near zero) and higher in zones which tend to dewater (mean $h \ll 0$).

Figures 13.8 and 13.9 show the percentage of discrete times for which the indicated values of h are exceeded at each elevation (node). This is more lucid than plotting the first two temporal moments, because no assumptions about the form of the distribution are needed; critical values h are tested for exceedence directly.

Results in Figs. 13.8 and 13.9 for near-surface layers are affected by the sampling frequency. Saturation of the surface soils is initiated by perching of water above the underlying layers, and runoff is generated during peak rainfall days. However, the surface soils drain rapidly at 4201A and simulated periods of saturation are not sampled (see the top of Fig. 13.8). The lower K_{sat} values of the subsoils at 4213 (see Table 13.1) create perched water into the surface soils which are reflected in Fig. 13.9 using the 15-day sampling interval. Dampening of the wetting front with depth makes the temporal sampling density adequate for most of the saprolite sequence.

Fig. 13.8. Profile of percentage of observations (simulation times) when *h* exceeds key values for 4201A (1975–1979). *Dotted lines* correspond to the layers in Table 13.1

13.5 Discussion

Before summarizing the results we will discuss some of the numerical modeling aspects, which limited the scope of our study. Heterogeneity in the soil/saprolite necessitates the use of a fine spatial mesh at material interfaces. Consequently, the maximum time step is controlled by the minimum grid spacing together with the extreme nonlinearities in $K(h)$ and its spatial distribution associated with sharp wetting fronts. The time step was reduced to less than 10^{-6} h during high rainfall days, particularly following a dry period.

The present results display a marked transience in recharge to a depth of 35 m. Simulations indicate that the occurrence of perched water is consistent with field observations in 1983, and that runoff is generated by saturation from below the soil layers, rather than being infiltration limited. Our model also simulates the location and duration of positive pressure heads, which indicates the potential for lateral flow and discharge along the seepage faces of nearby gulches. The lateral dimension, i.e., the width of the plateau between gulches, will influence the amount of perched water (and partially saturated lateral flow) which will reach the seepage faces.

Richards' equation [Eq.(1)] with the Mualem–van Genuchten parameterization of $K(h)$ appears to be appropriate for the simulated conditions. The soil-water retention curves and the resulting $K(h)$ curves provide sufficient changes in conductivity near saturation for the soil/saprolite columns to respond quite rapidly to large infiltration events. Thus, the water-filled pores transmit the pressure wave, resulting in a recharge rate which can rise and fall by an order of magnitude within a few weeks or months. In conclusion, rapid vertical flow is simulated without invoking a two-domain ("macropore") model to bypass the saprolite matrix in impeding layers.

13.6 Conclusions

when to monitor zones of saturation and seepage faces. Finally, it drainage and possible lateral flow to seepage faces. The former increases recharge rates and the potential for groundwater exposure to agricultural chemicals in the Pearl Harbor aquifer,while the latter removes water and solutes from the subsurface in this vicinity. Our one-dimensional simulations indicate that lateral seepage is likely, but we cannot quantify the volume. Vertical drainage in this environment, where percolation rates and water contents change significantly over a period of a few weeks, must be studied in a transient mode. The present study may provide guidance on how and when to monitor zones of saturation and seepage faces. Finally, it complements previous studies in the island that have focused on understanding and preventing the transport of toxic chemicals to the groundwater supply.

Acknowledgments. We are grateful to Delwyn Oki and April Kam of the Water Resources Research Center, University of Hawaii, for providing daily rainfall data and preparing Fig. 13.1, respectively. Partial support has been provided by the American Geophysical Union's 1991 Horton research grant. It is our privilege to contribute this chapter in memory of our friend and colleague Eshel Bresler. He provided exceptional leadership in the application of physical principles and mathematics to many practical problems in soil hydrology. His two sabbaticals in Hawaii were stimulating and memorable times for us.

References

Bresler E, Green RE (1983) Soil parameters and sampling scheme for characterizing soil hydraulic properties of a watershed. WRRC Tech Rep 148, Univ Hawaii, Honolulu

Ekern PC (1965) Evapotranspiration of pineapple in Hawaii. Plant Physiol 40: 736–739

Ekern PC, Chang J-H (1985) Pan evaporation: State of Hawai'i, 1894–1983. Rep 74, State of Hawaii, Department of Land and Natural Resources, Div Water and Land Development, Honolulu

Green RE, Corey JC (1971) Calculation of hydraulic conductivity: a further evaluation of some predictive methods. Soil Sci Soc Am Proc 35: 3–8

Green RE, Ahuja LR, Chong SK, Lau LS (1982) Water conduction in Hawaii oxic soils. WRRC Tech Rep 143, Univ Hawaii, Honolulu

Green RE, Liu CCK, Tamrakar N (1986) Modeling pesticide movement in the unsaturated zone of Hawaii soils under agricultural use. In: Garner WY, Honeycutt RC, Nigg HN (Ed) Evaluation of pesticides in ground water. ACS (Am Chem Soc) Symp Ser 315: 366–383

Lappala EG, Healy RW, Weeks EP (1987) Documentation of computer program VS2D to solve the equations of fluid flow in variably saturated porous media. Water-Resources Investigations Report 83–4009. US Geological Survey, Lakewood, Colorado

Lau LS (1987) Organic chemical contamination of O'ahu groundwater. WRRC Tech Rep 181. Univ Hawaii, Honolulu

Lau LS, Mink JF (1987) Organic contamination of groundwater: a learning experience. J Am Water Works Assoc 79: 37–42

Loague KM, Green RE, Liu CCK, Liang TC (1989a) Simulation of organic chemical movement in Hawaii soils with PRZM 1. Preliminary results for ethylene dibromide. Pac Sci 43: 67–95

Loague K, Giambelluca TW, Green RE, Liu CCK, Liang TC, Oki DS (1989b) Simulation of organic chemical movement in Hawaii soils with PRZM. 2. Predicting deep penetration of DBCP, EDB, and TCP. Pac Sci 43: 362–383

Marshall TJ (1958) A relation between permeability and size distribution of pores. J Soil Sci 9: 1–8

Miller ME, Green RE, Peterson FL, Jones RC, Loague K (1988) Hydrogeologic characteristics of subsoil and saprolite and their relation to contaminant transport, central O'ahu, Hawai'i. WRRC Tech Rep 178, Univ Hawaii, Honolulu

Oki DS (1987) Modeling drainage flux in pineapple cultivation areas of Central Oahu and its impact on EDB transport in the unsaturated zone. MS Thesis, Univ Hawaii, 407 pp

Oki DS, Giambelluca TW (1987) DBCP, EDB, and TCP contamination of ground water in Hawaii. Ground Water 25: 693–702

Oki DS, Giambelluca TW (1989) Groundwater contamination by nematicides: influence of recharge timing under pineapple crop. Water Resour Bull 25: 285–294

van Genuchten MTh (1980) A closed-form equation for predicting the hydraulic conductivity of unsaturated soil. Sci Soc Am J 44: 892–989

van Genuchten R (1978) Calculating the unsaturated hydraulic conductivity with a new, closed-form analytical model. Res Rep 78-WR-08, Water Resour Program, Dept Civil Engineering, Princeton Univ, Princeton

Wong L (1987) Analysis of ethylene dibromide distribution in the soil profile following shank injection for nematode control in pineapple culture. In: Rao PSC, Green RE (eds) Toxic organic chemicals in Hawaii's water resources. HITAHR Res Exten Ser 086 pp 28–40

14 Estimating Infiltration at Waste Sites: Methodology Development[1]

C.T. Kincaid and G.W. Gee[2]

14.1 Introduction

The description of spatial and temporal variability of soil hydraulic properties and their relationship to flow phenomena are current research topics of great interest to many soil scientists and hydrologists. How variations in hydraulic properties affect the movement of water into and through cover soils is an issue of importance to regulatory agencies charged with licensing waste-site repositories.

Several methods have been used to describe soil variability. Warrick and Nielsen (1980) provided an excellent summary of soil variability and the use of Monte-Carlo methods (i.e., repetitive simulations using deterministic values of hydraulic properties) to predict flow and transport in soils. Their review demonstrated that for certain soils there are large variations in saturated and unsaturated hydraulic conductivity, ranging from a coefficient of variation (CV) of 190% for saturated conductivity to a CV of over 400% for unsaturated conductivity. Using Monte-Carlo methods, Warrick and Nielsen demonstrated that transport of solute into spatially variable soils generally went deeper and faster than in uniform soils.

Another method for describing soil variability has been detailed by Dagan and Bresler (1979, 1983) and Bresler and Dagan (1979, 1983, 1988). This method recognizes the stochastic nature of flow in surface soils. Field variability is assumed to take place in the horizontal plane, while uniformity is assumed in the vertical plane. The saturated hydraulic conductivity is assumed to be a random variable, and the suction and water content are related to it by simple analytical (Brooks–Corey-type) relationships (Brooks and Corey 1964). The single stochastic hydraulic parameter, saturated hydraulic conductivity, is related to a dimensionless scaling parameter which is assumed to be log-normally distributed. Infiltration is visualized as occurring as vertical flow in a homogeneous column. The concept of field-scale variability is combined with the concept of one-dimensional infiltration to achieve a probabilistic simulation of the ensemble of soil columns that comprise the field or cover soil; i.e., the field is composed of many individual

---

[1] This chapter is based on work done jointly with Dr. Eshel Bresler during his sabbatical leave at Pacific Northwest Laboratory during the summer and fall of 1989

[2] Hydrology Section, Geosciences Department, Pacific Northwest Laboratory, Box 999, Richland, Washington 99352, USA

Table 14.1. Statistical characterization of radon barrier soil properties

	Bulk density (g/cm^3)	Porosity (cm^3/cm^3)	Saturated hydraulic conductivity, K_s(cm/h)	Natural logarithm of K_s	Relative saturation
Mean	1.822	0.33	0.003774	−7.070000	0.848444
Variance	0.0151	0.00221	4.255×10^{-5}	3.497330	0.004859
Standard deviation	0.123	0.047	0.006523	1.870115	0.069705
Correlation coefficient, r			−0.119108	0.144020	−0.107230
Correlation, r^2			0.014187	0.020742	0.011498

columns of unifrom soil, each representing a portion of the distribution of saturated hydraulic conductivities.

This probabilistic methodology was applied to analyze infiltration through compacted clay layers placed over uranium mill tailings. Bulk density, porosity, saturated hydraulic conductivity (K_s), and relative saturation data were obtained from measurements made on 18 samples taken from various depths within the clay layers of three covers. Clay is used for a radon barrier to prevent escape of radon gas. In the following analysis we assume that a 1-m-thick clay layer is used. Table 14.1, in which these data are summarized, shows that radon barriers exhibit a relatively large variability of K_s (i.e., ≈ 3 orders of magnitude), from 3×10^{-5} to 2×10^{-2} cm/h (i.e., 10^{-8} to 10^{-5} cm/s) and relatively small variability in relative saturation (ϕ). The arithmetic average of porosity or saturated water content is 0.33 with a standard deviation of ± 0.047. The arithmetic average of relative saturation (i.e., field water content divided by saturated water content) is 0.848 with a standard deviation of ± 0.0697 (see Table 14.1). Statistical analysis suggests that there is no correlation between either saturated hydraulic conductivity or saturation and depth in the soil. Values of correlation, r^2, (i.e., between soil depth and three parameters: saturated hydraulic conductivity, natural logarithm of conductivity, and saturation) were 0.0142, 0.0207, and 0.0115, respectively. These small correlation values indicate that the soil variables given in table 14.1 may be treated as random variables distributed spatially over the barrier, independent of the soil depth.

14.2 Methodology and Assumptions

Theoretical development of the closed-from analytical approach is detailed in Dagan and Bresler (1979, 1983) and Bresler and Dagan (1983, 1988). Although some equation development and assumptions required for applying this technique may be found in these references, they are presented here in terms of the application to cover and barrier systems. Further equation development and a complete list of equations to calculate probabilities and expected values are found in Smyth et al. (1990).

The objectives of this modeling effort are to estimate the infiltration rate of water through a hypothetical radon barrier, and to demonstrate whether or not the water flow rate through such a barrier and into a tailings pile is low enough to ensure compliance with standards for groundwater protection. A unit hydraulic gradient (i.e., gravitational) flow has been assumed. Additional simplifying assumptions and their justification include the following:

1. We assume that the thickness of the radon barrier (e.g., 1 m) is small compared to the areal dimensions of the barrier (100–1000 m). It is further assumed that changes in properties with depth in the barrier are small so that each vertical profile can be considered homogeneous, (i.e., there are no distinct horizons in the clay radon barrier). Parameter variability in the barrier is assumed to be statistically homogeneous or stationary in the areal plane.
2. The flow in each profile is assumed to be vertical. Gravitational flow implies that the flux, q, which is generated by the recharge, R, may be either equal to K_s or to R when R is constant (Dagan and Bresler 1979).
3. The unsaturated hydraulic conductivity function $K(\Theta)$ is given by the Brooks and Corey (1964) relationship:

$$K(\Theta) = K_s[(\Theta - \Theta_r)/(\Theta_s - \Theta_r)]^{1/\beta} = K_s\Phi^{1/\beta}, \tag{1}$$

where K_s is the saturated hydraulic conductivity; Θ is the water content at the point of interest; Θ_s and Θ_r are the saturated and residual water contents, respectively; β is the constant soil parameter; and Φ is the relative saturation.
4. K_s is considered to be a stochastic parameter and is assumed to be related to a dimensionless scaling parameter δ (Peck et al. 1977, Warrick et al. 1977), i.e.,

$$K_s = K_s^* \delta^2, \tag{2}$$

where K_s^*, sometimes referred to as the deterministic average, is defined by:

$$K_s^* = [E(K_s^{1/2})]^2 = \left[\int_0^\infty K_s^{1/2} f(K_s)\, \mathrm{d}K_s\right]^2, \tag{3}$$

where E stands for the expectation operator.
5. All other soil properties, e.g., Θ_s, Θ_r, and β are assumed to be deterministic (Bresler et al. 1978; see also Dagan and Bresler 1979; Russo and Bresler 1982).
6. The probability density function (PDF) of δ is log-normal, i.e.,

$$Y = \log\delta : N(m_Y, \sigma_Y^2), \tag{4}$$

where $N(m_Y, \sigma_Y)$ is the normal distribution function, and m_Y and σ_Y^2 are the arithmetic mean and variance of Y, (i.e., $\log\delta$), respectively. Rewriting

Eq. (2) in terms of Y yields:

$$K_s = K_s^* \exp(2Y).$$

7. Y is a random variable of x and y (i.e., a coordinate function), and is determined from measurements taken at a few locations in the barrier. The necessary parameters for the stochastic analysis are K_s^*, m_Y and σ_Y.
8. Ergodicity is assumed to apply to the analysis of the cover system. Ergodicity, when combined with stationarity, implies that ensemble and space averages for a given radon barrier realization over the x, y plane are equivalent (Dagan 1989).
9. In addition to their dependence on the variability of K_s or Y, the uncertainty in V or q and Θ or ϕ is also associated with the uncertainty in the spatial distribution of recharge rates. For the case considered here this additional uncertainty is also assumed to be random and is confined to the surface recharge rate, R. The surface recharge PDF, $f(R)$, is assumed to be a uniform (rectangular) distribution:

$$f(R) = 0 \text{ for } R < (\bar{R} - d_R) \text{ and for } R > (\bar{R} + d_R)$$

$$f(R) = 1/(2d_R) \text{ for } (\bar{R} - d_R) < R < (\bar{R} + d_R). \tag{5}$$

The PDF for surface recharge is constant in a recharge interval of width $2d_R$ around the average surface recharge, $\bar{R}$. A maximum possible value of d_R (i.e., $d_R \equiv \bar{R}$, forcing R to remain positive), is considered in this study.
10. Water flow in each profile is assumed not to influence other profiles within the overall barrier.
11. The flow of water into the clay random barrier layer is generated by recharge at the barrier–atmosphere surface which results from natural precipitation. For an arid site an annual precipitation of ≈ 15 cm (6 in.) is assumed; neglecting evapotranspiration (ET) and runoff this corresponds to an annual average recharge rate of 1.7×10^{-3} cm/h (4.7×10^{-7} cm/s). Accounting for 80 to 90% of that being ET and runoff, an even lower average yearly recharge might be between 1.7×10^{-4} and 3.4×10^{-4} cm/h (4.7×10^{-8} and 9.4×10^{-8} cm/s). To account for the range of possible recharge, recharge assumed for the stochastic analytical model was between 3.6×10^{-6} and 3.6 cm/h (10^{-9} and 10^{-3} cm/s).

As a result of these assumptions V, q, Θ, and ϕ depend on the deterministic parameters K_s^*, β, and Θ_s and on the random variables R and Y.

14.2.1 Estimation of the Parameters

The values of K_s^*, m_Y, and σ_Y were estimated from the interpreted values given in Table 14.1. Assuming that K_s is log-normally distributed with $\mu = \log(K_s)$, then [substituting $f(\mu)d\mu$ for $f(K_s)dK_s$, and using Eq. (A8) from Smyth et al. (1990) with $\alpha = 0.5$, $\beta' = -\infty$ and $\delta' = \infty$] the righthand side of Eq. (3) becomes:

$$K_s^* = \{E[\exp(0.5\mu)]\}^2 = \left[\int_{-\infty}^{\infty} \exp(0.5)\mu) f(\mu) d\mu\right]^2$$

$$= [\exp(0.5\mathrm{m}_\mu + 0.125\sigma_\mu^2)]^2.$$

Substituting the estimated values of the mean and variance of $\mu = \log(K_s)$ from Table 14.1, an estimated value of 0.00204 cm/h (5.7×10^{-7} cm/s) is obtained for K_s^*.

To estimate the value of m_Y, the expectations of the log of both sides of Eq. (2) yields [after rearrangement and substitution for $E(\mu)$ its estimated value from Table 14.1, and for $\log K_s^*$ its calculated value]:

$$m_Y = E(Y) = 0.5[E(\mu) - \log(K_s^*)] = -0.43735.$$

To estimate σ_Y, rewrite Eq. (2) in terms of Y in Eq. (4), and [using Eq. (A8) from Smyth et al. (1990) with $\alpha = 2$, $\beta' = -\infty$, $\delta' = \infty$] calculate the expectations to obtain:

$$E[K_s] = K_s^* E[\exp(2Y)] = K_s^* \exp(2\mathrm{m}_Y + 2\sigma_Y^2).$$

Substituting for $E(K_s)$ its estimated value from Table 14.1 (0.003774 cm/h) and calculated values of -0.43735 for m_Y and 0.00204 for K_s^*, the value of 0.863 is obtained for σ_Y. With these estimated values of K_s^*, m_Y, and σ_Y, the probabilities of V, q, Θ and Φ can be calculated.

14.2.2 Calculations for Steady-State Flow

Explicit closed-form solutions for the parameters of interest in the heterogeneous randon barrier are given by Dagan and Bresler (1979), and are summarized in Bresler (1987). For deterministic recharge, clarifying details are given in Smyth et al. (1990) on the cumulative probabilities and expected values of velocity, flux, saturation, travel time and ponding.

For *stochastic* recharge, the cumulative probability (i.e., the probability that the random output variable of interest will be equal to or less than a prescribed value) of the output variable V is:

$$P(V) = \int_{R - d_R}^{\bar{R} + d_R} \int_{-\infty}^{Y(R,V)} f(R) f(Y) dR dY,$$

where $f(R)$ is given by Eq. (5) and $f(Y)$ is the normal PDF and is given by:

$$f(Y) = \frac{1}{\sqrt{2\pi\sigma_Y}} \exp(Y - m_Y)^2/(2\sigma_Y^2)]. \tag{6}$$

The expansion of this and other integrals of the steady-state formulation can be found in Smyth et al. (1990).

14.2.3 Calculaions for Nonsteady Flow

A nonsteady analysis requires consideration of the movement of wetting fronts through the barrier with time. An approximation for water flow through an areally extensive soil layer suggested by Dagan and Bresler (1983) assumes piston-type flow under unit hydraulic gradient for everywhere except at the wetting front. The front itself is assumed to be sharp (i.e., it occurs over an infinitesimally small vertical distance) and to have the appearance of a square wave as shown in Fig. 14.1. This approximation for calculation of flow variable averages of various soils gives essentially the same results as a more sophisticated numerical approach (Bresler and Dagan 1983).

Conceptually, water flow through the barrier is visualized as exhibiting one of three typical profiles at any given moment in time. For a nonsteady analysis with an initial water content, Θ_L, water will initially drain from the barrier at a corresponding rate, q_L, equal to the unsaturated hydraulic conductivity at the initial water content [i.e., $K(\Theta_L)$], dictated by unit hydraulic gradient theory. After a period of time, the wetting front will break through and flow will be established at one of two rates, the saturated hydraulic conductivity, K_s (i.e., if the surface recharge rate exceeds K_s), or the surface recharge rate, R (i.e., if the surface recharge rate is less than or equal to K_s). It is necessary to determine the saturated hydraulic conductivity that defines the breakpoint between $q = q_L$, $q = K_s$, and $q = R$ water flux contributions to the overall flux.

Expected Value Limits Under Nonsteady Flow. From integration of the continuity equation [see Eqs. (28) and (29) of Dagan and Bresler 1983], one obtains:

$$L(t) \approx (K_s - K_L)t/(\Theta_s - \Theta_L); \quad R \geqslant K_s \tag{7}$$

and

$$L(t) \approx (R - K_L)t/[(\Theta_s - \Theta_r)(R/K_s)^\beta + \Theta_r - \Theta_L]; \quad R < K_s, \tag{8}$$

where $L(t)$ is the location of the equivalent front of the square wave. Solving

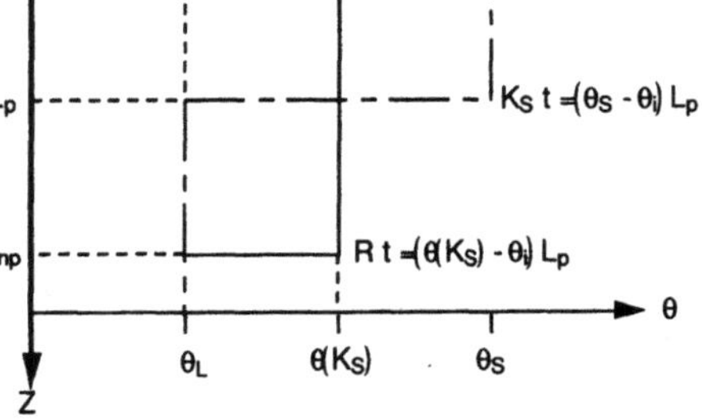

Fig. 14.1. Schematic presentation of ponded and nonponded profiles

Eq. (7) for saturated hydraulic conductivity, i.e.,

$$K_s^a = \frac{z}{t}(\Theta_s - \Theta_L)/(1 - \phi_L^{1/\beta})$$

defines the breakpoint hydraulic conductivity (i.e., the conductivity value below which the initial condition governs) as a function of thickness, z, and time, t, for saturated breakthrough, i.e., the contribution to specific flux from a ponded region of the barrier. For recharge rate $R > K_s^a$ there is a contribution from the ponded region. On the other hand, if $R < K_s^a$ there is no contribution from a ponded area, and one solves the nonlinear Eq. (8) for K_s^b. Equation (8) is rewritten to facilitate a Newton–Raphson solution as follows:

$$(K_s^b)^{-\beta} + K_s^b\{\phi_L^{1/\beta}t/[R^\beta(\Theta_s - \Theta_r)L]\}$$
$$-\{[Rt + (\Theta_L - \Theta_r)L]/[R^\beta(\Theta_s - \Theta_r)L]\} = 0.$$

The solution, K_s^b, defines the breakpoint saturated hydraulic conductivity for unsaturated breakthrough (i.e., contributions to specific flux from unsaturated regions at saturations greater than the initial saturation, $\Theta_L \leqslant \Theta < \Theta_s$).

Calculating Expected Values of Specific Flux Under Nonsteady Flow. Two intervals corresponding to saturated and unsaturated soil water flow were required to perform the integrations in the steady-state flow analysis. Now. for the first case of nonsteady flow, which includes ponding, three intervals must be defined: unsaturated flow at initial saturations, saturated or ponded flow after breakthrough when $R > K_s$, and unsaturated flow after breakthrough when $R < K_s$. Figure 14.1 shows the contributions from the field ponded area where L_p is greater than the barrier thickness (i.e., $L_p > z$), and $R > K_s^a$; that portion of the barrier has saturated hydraulic conductivity bounded by $K_s^a < K_s < R$. When $L_p < z$ and $K_s < K_s^a$, the contribution to specific flux is q_L, corresponding to the initial saturation, ϕ_L. Finally when $L_p > z$ and $K_s > R$ the contribution to specific flux is equal to the recharge R. The expected value of specific flux is written in integral form:

$$E(q) = \int_0^\infty qf(q)dq = \int_0^{K_s^a} q_L f(K_s)dK_s + \int_{K_s^a}^{R} K_s f(K_s)dK_s + R\int_R^\infty f(K_s)dK_s.$$

The first term on the righthand side represents flux from initial condition regions (i.e., flux from the portion of the barrier delivering water from below the wetting front, $L_p < z$, $q = q_L$). The second term represents flux from ponded areas where $q = K_s$, and the third term represents flux from unsaturated portions of the field where $R < K_s$ and therefore $q = R$.

In the second case, when there is no contribution from ponding, only unsaturated conditions exist. This occurs wherever $R < K_s^a$. At early times in the nonsteady flow analysis, breakthrough of the saturated front has not occurred through the barrier and contributions to overall flux come from only two areas: flux associated with the initial condition [i.e., $\phi = \phi_L$ and $q = q_L = K(\phi_L)$], and flux associated with unsaturated profile breakthrough, i.e., $q = R$. The expected value of specific flux is written as:

$$E(q) = \int_0^\infty q f(q) dq = \int_0^{K_s^b} q_L f(K_s) dK_s + R \int_{K_s^b}^\infty f(K_s) dK_s.$$

The first term on the righthand side represents flow from the initial condition region, while the second term represents flow from unsaturated flow breakthrough.

This brief description of the integration intervals for analysis of the nonsteady flow cases provides an understanding of the way in which these analyses differ from the steady-state analyses. Greater detail on specific equations necessary for encoding the stochastic analytical approach can be found in Smyth et al. (1990).

14.3 Results

Before presenting the results for the hypothetical barrier, one overall assumption must be stressed. All these results are based on a limited data set and were calculated using *estimated* values of $m_Y = -0.43735$, $\sigma_Y = 0.86$ and $K_s^* = 0.00204$ cm/h (see Section 14.2.1) to represent a hypothetical barrier. Equations of probability and expected values for saturation and other parameters are calculated as for velocity and given in Smyth et al. (1990).

14.3.1 Estimating Surface Recharge Flux by Calibration with Measured Values of Saturation

Calibration against measured data is required to obtain a reasonable value for the recharge at the barrier surface boundary. Unfortunately, the only measured values of any soil variable at the radon barrier sites considered are those of saturation, Φ, in Table 14.1. If more precise information on the surface recharge was available (e.g., from water balance studies employing lysimeters or the use of tracers), the model results could be more rigorously tested or validated. However, the available information on Φ (Table 14.1) can be used to calibrate the model with respect to the reasonable range of surface recharge. The expected value of saturation, $E\{\Phi\}$, as a function of recharge for a wide range of $1/\beta$ values [including the highly probable value of $1/\beta = 7.2$ (Bresler et al. 1978)] is shown in Fig. 14.2. Examination of information plotted in the figure indicates that for the range of estimated average water content, $\Phi \pm \sigma_\Phi$ (Table 14.1), an average recharge value of

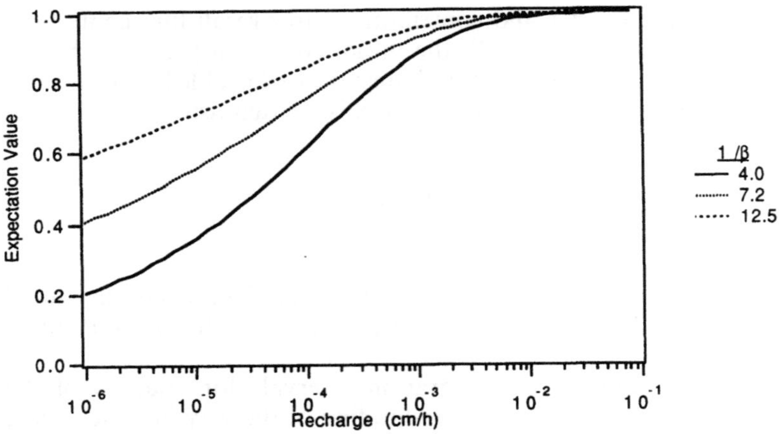

Fig. 14.2. Expected value of Φ $E\{\Phi\}$, as a function of recharge for three values of $1/\beta$

3×10^{-4} cm/h is found with ranges of between 4×10^{-5} and 2×10^{-3} cm/h. This range of values can be used as the highly probable range of values for stochastic prediction purposes of flow problems over the barrier. However, if desired, a more liberal or more conservative (i.e., a higher or lower) recharge value may be assumed. It should be noted that these "calibrated" values are in line with the previously assumed range which was based on 10–20% of the annual precipitation (see assumption 11).

14.3.2 Cumulative Probability of Saturation

The relationship between the probability of saturation, $P\{\Phi\}$, and surface recharge, R, is given in Fig. 14.3. In this figure the cumulative probability of Φ, $P\{\Phi\}$, (or the fraction of the barrier that is less than or equally saturated) for a wide range of surface recharge rates (3.6×10^{-6} to 3.6 cm/h) is presented. Examination of Fig. 14.3 reveals an approximately 20% probability of observing saturations less than or equal to 0.5 at a surface recharge rate of 3.6×10^{-5} cm/h. For the higher recharge of 3.6×10^{-3} cm/h, there is a 20% probability of observing saturations at or below 0.99. As a corollary, the probability of observing pore water at saturations between 0.99 and 1.0 is approximately 80%; thus, the probability of ponding at recharge of 3.6×10^{-3} cm/h is 80%. Examining the 70% probability line, one finds saturations at or below approximately 0.7 at surface recharge of 3.6×10^{-5} cm/h and saturations at or below approximately 0.99 at recharge of 3.6×10^{-4} cm/h. Again, the corollary is the approximately 30% probability of observing pore water at saturations between 0.99 and 1.0 for recharge of 3.6×10^{-4} cm/h. Thus, the probability of ponding at this recharge is 30%. At the lowest surface recharge examined, 3.6×10^{-6} cm/h, there is a probability of nearly 100% of observing saturations at or below

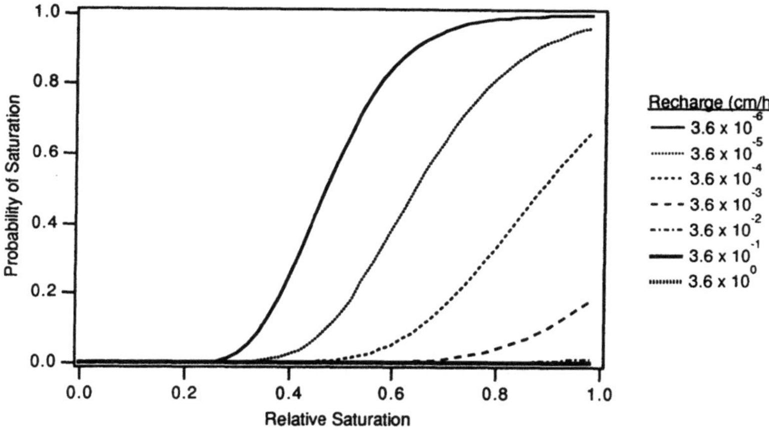

Fig. 14.3. The probability of saturation, $P\{\Phi\}$, where $1/\beta = 7.2$ is shown for seven recharge rates

approximately 0.9; therefore, there is little or no probability of observing water at saturations between 0.9 and 1.0. Thus, ponding has negligible probability at surface recharge of 3.6×10^{-6} cm/h (10^{-9} cm/s).

Further study has shown that for values of $R < 0.036$ cm/h, the $P\{\Phi\}$ becomes sensitive to the $1/\beta$ values. For values of recharge greater than or equal to 0.036 cm/h there is a virtual certainty for saturation. The probability of ponding as a function of the recharge rate may also be observed at the extreme right of Fig. 14.3 where $\Phi = 1.0$. This value is not affected by the value of $1/\beta$.

The data in Fig. 14.3 have an important practical implication because of the high dependence of radon gas diffusivity on the degree of saturation. Analysis of Fig. 14.3, $R = 3.6 \times 10^{-4}$ cm/h, indicates that there is a 50% probability of observing saturations at or below approximately 0.87 for the $1/\beta$ value of 7.2. This implies a 50% probability of observing air phase saturations at or above 0.13. Smallar recharge values exhibit an even greater likelihood of observing significant air phase saturations. Hence, in the practical surface recharge range between 10^{-5} and 10^{-3} cm/h, the probability of having lower saturation values (and increased chances for a higher radon gas diffusivity) is relatively high. It should be noted that recharge conditions, which increase the probability of saturation and act favorably to decrease radon gas diffusivity, will act to increase the flux of water through the barrier.

14.3.3 Probability of Ponding and of Fluxes for $q < R$

When examining infiltration and its evaluation under stochastic conditions, it is interesting to calculate the probability that ponding will occur, i.e.,

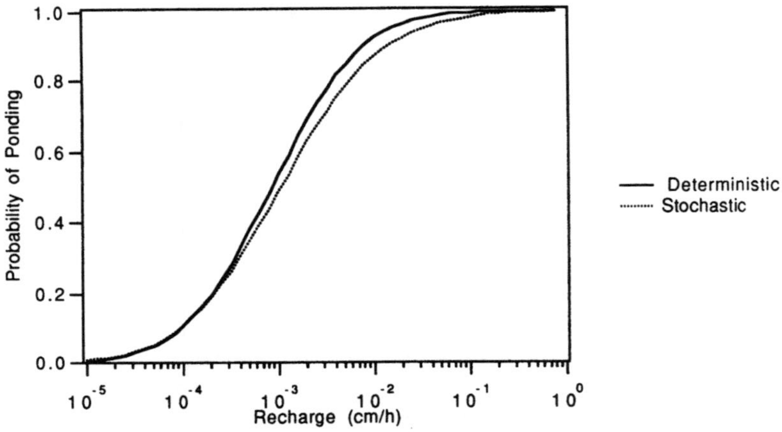

Fig. 14.4. Probability of ponding for deterministic ($d_R = 0$) and stochastic ($d_R = E\{R\}$) recharge

$P\{\text{pond}\}$. The results for the full range of possible surface recharge between 10^{-5} and 1 cm/h are given in Fig. 14.4. The probability of observing any value of ponding depends on both the magnitude of the recharge and whether the recharge is deterministic or stochastic. The maximum possible value of deterministic R (i.e., $d_R = 0$), was used for illustration purposes. In practice, there is a probability of 10% that the steady flux through the barrier will be below 1×10^{-4} cm/h and a probability of 50% that it will be below 1×10^{-3} cm/h if the recharge is deterministic. Note that by the ergodic hypothesis, the $P\{\text{pond}\}$ represents the fraction of the barrier surface area that is ponded, i.e., the physical fraction of the barrier that is under water ponded conditions and may give rise to surface runoff.

Under deterministic surface recharge conditions (i.e., $d_R = 0$), the probability of ponding is equivalent to the probability of the flux, $P\{q\}$, when $q < R$. In this case, $P\{\text{pond}\}$ on the ordinate of Fig. 14.4 is replaced by $P\{q\}$ and recharge on the abscissa is replaced by q. Note that there is a discontinuity in the probability of flux, $P\{q\}$, at $q = R$; $P\{q\} \equiv 1.0$ for all $q \geqslant R$. The probability of observing a flux, q, less than the applied recharge rate, R, is given by the curve in Fig. 14.4. However, the probability of observing flux equal to the recharge rate is always one, and, furthermore, the flux can never exceed the recharge rate in the deterministic model.

It is important to recall that the deterministic ($d_R = 0$) and stochastic ($d_R = \bar{R}$) cases represent the maximum range of barrier response to stochastic recharge. The deterministic recharge represents a degenerative case of the stochastic analysis, where the range of recharge values experienced is assumed to be infinitesimally small, i.e., $d_R = 0$. Physically, this range of values represents a site free from vegetation and with a uniform application of water. The stochastic recharge shown in all cases corresponds to a maximum range of recharge, i.e., $d_R = \bar{R}$. Physically, the recharge range

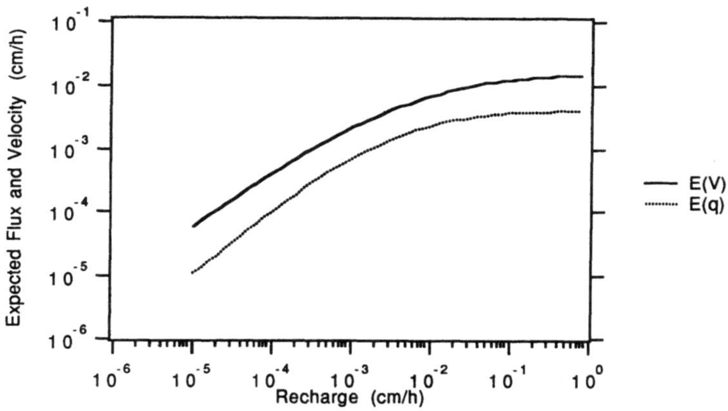

Fig. 14.5. Expected values of flux and velocity, $E\{q\}$ and $E\{V\}$ respectively, as a function of recharge ($1/\beta = 7.2$)

represents the greatest possible variability in water application caused by processes such as plant canopy interception, diversion, and stem flow. Therefore, the stochastic recharge curve presents the range of recharge variability a site could experience.

14.3.4 Expected Values of Flux

Expected values (i.e., means) of flux are an important consideration in making decisions based on evaluation of the infiltration processes. For illustration purposes the results of $E\{V\}$ and $E\{q\}$ for $1/\beta = 7.2$ are presented together in Fig. 14.5. Obviously, the expected flux is equivalent to the product of the expected velocity and water content (i.e., $E\{q\} = E\{V\Theta\}$), and $E\{q\}$ will always be smaller than $E\{V\}$ because maximum water content is equal to total porosity, which is always less than one.

14.3.5 Expected Values of Saturation and Flux at $z = 1$ m for Nonsteady Infiltration

The expected values of saturation at $z = 1$ m as a function of time and for an initial saturation of 0.1 are presented in Fig. 14.6 for seven different recharge values. Independent of the value of $1/\beta$, the system always comes to a steady unsaturated condition (i.e., expected value less than 1) in the barrier as long as the recharge is less than or equal to 3.6×10^{-3} cm/h. The expected value of saturation at these recharges is related to the unsaturated nonponded area. Clearly, at lower surface recharge rates a greater portion of the K_s distribution remains unsaturated. Figure 14.6 also reveals that less time is required for successively greater surface recharge rates to overcome the initial condition. Breakthrough takes over 100 years for the

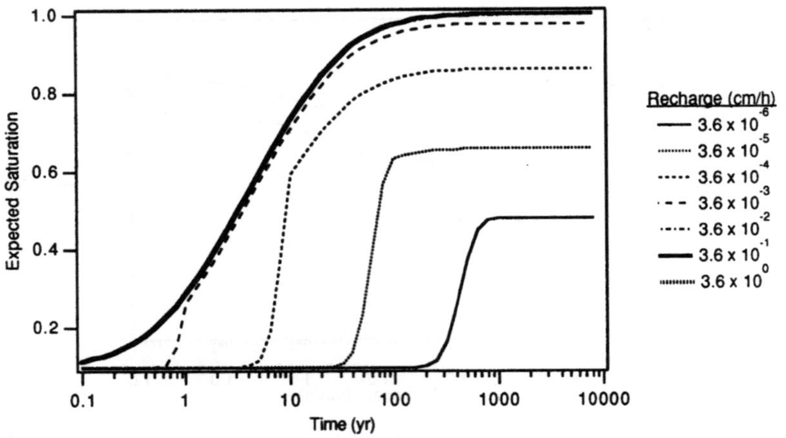

Fig. 14.6. Expected value of saturation as a function of time where $1/\beta = 7.2$ and initial saturation = 0.10 is shown for seven recharge rates

$R = 3.6 \times 10^{-6}$ cm/h case, while breakthrough occurs in less than 1 year for $R = 3.6 \times 10^{-3}$ cm/h. At a recharge of 3.6×10^{-4} cm/h the initial saturation value lasts for approximately 5 years; it then increases with time at a decreasing rate until it reaches a steady-state value (which is identical to those of Fig. 14.2) after about 500 years. At this surface recharge rate, the effect of $1/\beta$ on $E\{\Phi\}$ over time is seen in Fig. 14.7. The abrupt change from $L_p < z = 1\,\text{m} < L_{np}$ to $z = 1\,\text{m} < L_p$ is most pronounced in the case of the highest $1/\beta$ value. When the recharge rate equals or exceeds 3.6×10^{-2} cm/h, the shape of the curve becomes independent of the recharge. The expected value of saturation increases with time due to an increasing contribution

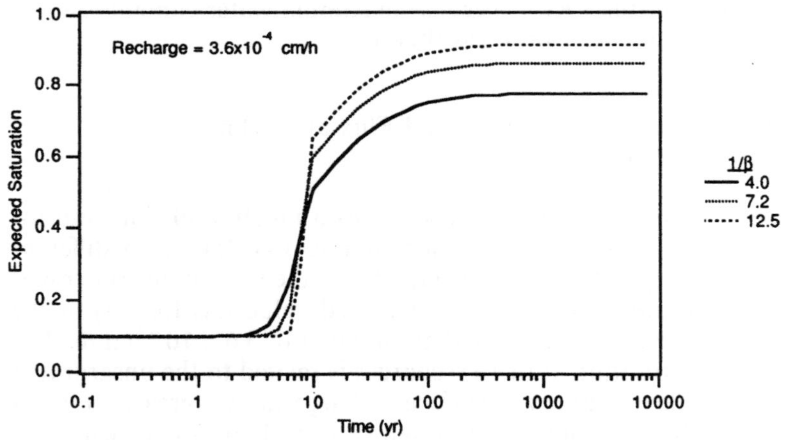

Fig. 14.7. Expected value of saturation as a function of time for three values of $1/\beta$ and recharge = 3.6×10^{-4} cm/h

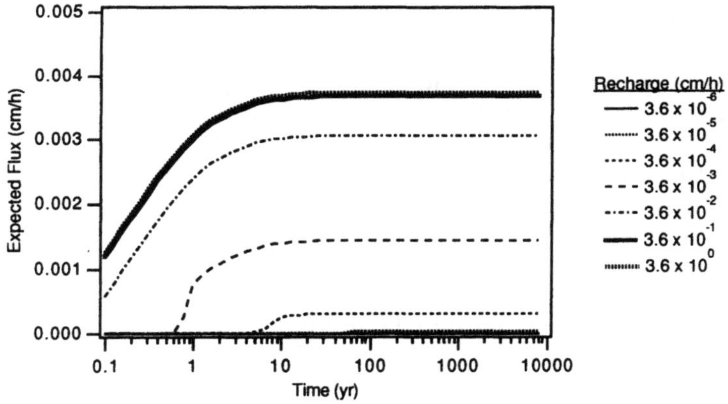

Fig. 14.8. Expected value of flux, $E\{q\}$, as a function of time where $1/\beta = 7.2$ is shown for seven values of recharge rate

from the saturated portion of the barrier, until all the barrier becomes saturated at about 500 years.

The effect of $1/\beta$ on the expected value of flux (Fig. 14.8) is negligibly small because the effect of beta is due only to the initial water content or saturation. The initial saturation value was assumed to be 0.10 in this and all previous calculations. Note that the timing of breakthrough for various surface recharge rates is, as expected, coincident for both expected values of saturation and flux. Figure 14.8 also reveals that at successively higher surface recharge rates, more water becomes runoff as a greater portion of the saturated hydraulic conductivity distribution acts to control flux.

To see the effect of initial saturation, Φ_L, on the expected values of saturation and flux, sensitivity calculations using Φ_L of 0.01 and 0.50 were also performed. Initial saturation values greater than or equal to 0.50 can result in expected values of saturation less than the initial condition; this is a situtation that may require an analysis of hysteresis, which is neglected in this analysis. The expected saturation is impacted by the initial condition in its early response but there is no influence on the long-term steady-state response which is achieved in all cases after approximately 500 years. There is no affect on expected flux for $1/\beta$ of 7.2 and initial saturations at or below 0.50. This is because the initial surface recharge rate, corresponding to such low saturations, is at or below 3.6×10^{-6} cm/h, the lowest recharge value analyzed. This suggested that the stochastic analytical approach can be used to investigate higher initial saturations, provided that attention is restricted to higher surface recharge values.

14.4 Summary

This analysis emphasizes the evaluation of uncertainty embodied in estimates of infiltration through areally extensive and relatively thin covers. The stochastic analytical approach to the evaluation of water infiltration is based on the distribution of two stochastic parameters which directly influence cover response: saturated hydraulic conductivity and surface recharge rate. Both steady-state and unsteady-state infiltration responses were investigated using the stochastic approach.

Recharge rates were shown, using model calibration with measured saturation values, to range between 10 and 20% of annual precipitation (4×10^{-5} to 2×10^{-3} cm/h, or 1.1×10^{-8} to 5.6×10^{-7} cm/s). Analyses were performed using both deterministic and stochastic (varying) recharge rates.

In general, the analysis indicated that the goals of using a clay barrier both to prevent radon gas diffusion and to inhibit infiltration of water into a waste disposal unit may be mutually exclusive.

A sensitivity analysis of the results with respect to the initial saturation level of the clay barrier showed that the expected value of saturation occurring at an early time depends strongly on initial saturation; however, long-term results (i.e., 500 years for saturation, 50 years for flux) are not influenced. Expected values of flux were found to be insensitive to initial saturation values ranging from 0.01 to 0.50. Simulations of higher initial saturations must be restricted to the study of higher surface recharge rates. Higher values of initial saturation provided evidence of hysteretic soil behaviour, a physical process which was not included in the model.

Unsteady analyses did provide insight into the timing of barrier response to events following construction; however, flux estimates became asymptotic to steady-state values in roughly 10 to 50 years. Saturation estimates became asymptotic in roughly 100 to 500 years. Differences in timing reflect the relative significance of pores (i.e., within the hydraulic conductivity distribution) to estimates of flux and saturation. The longer time required for low conductivity pores to exhibit breakthrough does not strongly influence flux but does influence saturation. Steady-state analyses of the stochastic analytical approach should be sufficient to provide estimates of flux needed by subsequent elements of an overall performance assessment.

Acknowledgments. This work was supported by the US Nuclear Regulatory Commission (US NRC) under US Department of Engergy Contract DE-AC06-76RLO 1830, NRC FIN L10072. We thank T.J. Nicholson of the US NRC for his enthusiastic support of this effort. We also express thanks to two of our colleagues at the Pacific Northwest Laboratory: C.R. Cole for his thoughtful review and J.L. Stroh for his efforts to encode the equations and generate the graphics.

References

Bresler E (1987) Modeling of flow, transport, and crop yield in spatially variable fields. Adv Soil Sci 7: 1–51

Bresler E, Dagan G (1979) Solute dispersion in unsaturated heterogeneous soil at field scale. 2. Applications. Soil Sci Soc Am J 43: 467–472

Bresler E, Dagan G (1983) Unsaturated flow in spatially variable fields. II. Application of water flow models to various fields. Water Resour Res 19: 421–428

Bresler E, Dagan G (1988) Variability of yield of an irrigated crop and its causes. 1. Statement of the problem and methodology. Water Resour Res 24: 381–388

Bresler E, Russo D, Miller RD (1978) Quick estimate of unsaturated hydraulic conductivity function. Soil Sci Soc Am J 42: 170–172

Brooks RH, Corey AT (1964) Hydraulic properties of porous media. Hydrology Paper No. 3, Colorado State University, Fort Collins, CO

Dagan G (1989) Flow and transport in porous formations. Springer Berlin Heidelberg New York

Dagan G, Bresler E (1979) Solute dispersion in unsaturated heterogeneous soil at field scale. 1. Theory. Soil Sci Soc Am J 43: 461–467

Dagan G, Bresler E (1983) Unsaturated flow in spatially variable fields. 1. Derivation of models of infiltration and redistribution. Water Resour Res 19: 413–420

Peck AJ, Luxmoore RJ, Stolzy JL (1977) Effects of spatial variability of soil hydraulic properties in water budget modeling. Water Resour Res 13: 348–354

Russo D, Bresler E (1982) A univeriate versus a multivariate parameter distribution in a stochastic conceptual analysis of unsaturated flow. Water Resour Res 18: 483–488

Smyth JD, Bresler E, Gee GW, Kincaid CT (1990) Development of an infiltration evaluation methodology for low-level waste shallow land burial sites. NUREG/CR-5523, US Nuclear Regulatory Commission, Washington, DC

Warrick AW, Nielsen DR (1980) Spatial variability of soil physical properties in the field. In: Hillel D (ed) Applications of soil physics. Academic Press, New York, pp 319–344

Warrick AW, Mullen GJ, Nielsen DR (1977) Scaling field measured soil hydraulic properties using a similar media concept. Water Resour Res 13: 355–362

15 Irrigation Scheduling Considering Soil Variability and Climatic Uncertainty: Simulation and Field Studies

D. Or[1] *and R.J. Hanks*[2]

15.1 Introduction

The effects of uncertainty, stemming from spatial variability of soil properties, uncertain climate, and nonuniform water application, on irrigation management and the resulting crop yields have received much attention in recent years. Zaslavsky and Buras (1967) studied crop yield response to nonuniform application of irrigation water. Warrick and Gardner (1983) and Russo (1986) have studied the effects of spatially variable soil-water conditions on crop yield. Rhenals and Bras (1981) considered the effects of uncertain potential evapotranspiration (ETp) on irrigation scheduling. English (1981) used historical yield and climate data to inter future irrigation practices based upon a farmer's attitude toward risk. Bresler and Dagan (1988) developed a methodology for incorporating various sources of uncertainty into crop yield analyses.

To offset the effects of spatially variable soil properties such as variable water holding capacity, a farmer can use excessive amounts of irrigation water to reduce the extent of water stress. Such an approach can be sustained only where irrigation water is abundant and its price is sufficiently low. However, dwindling water resources and mounting public concern over drainage-caused groundwater pollution are likely to avert widespread use of such excessive irrigation practices. Consequently, when the adverse effects of excessive drainage are considered, deficit irrigation over some portions of the field and reduced crop yields are inevitable.

The role of improved irrigation management schemes becomes increasingly important with the dwindling of water supplies. Improved irrigation decisions are inhibited for highly variable soils and uncertain climate. Thus, risk is involved in irrigation decisions because of incomplete information. Acquiring more information may help reduce such risks but the finances available for information gathering are limited. The expected improvement in performance resulting from perfect information (Avriel and Williams 1970) in a typical irrigated field is likely to be small because crop prices are low and the margins of profit are narrow. Thus, practical irrigation management schemes are likely to rely on only a few field measurements (Russo 1986;

[1] Department of Civil Engineering, University of California, Berkeley, California, USA
[2] Department of Plants, Soils, and Biometeorology, Utah State University, Logan, Utah, USA

Or and Hanks 1992a). Furthermore, incorporating uncertainty considerations, which can substantially alter decision making, in irrigation management schemes (English 1981) may result in only a marginal improvement in performance (Rhenals and Bras 1981).

The effects of a modified irrigation scheme are not always clear due to (1) the complexity of irrigated fields, and (2) possible "flatness" of response surfaces to irrigation alternatives (Roger and Fiering 1986). It is therefore prudent to field-test any proposed irrigation managemenet scheme, to check whether the benefits from the new scheme justify the additional cost and effort.

In this chapter, we present a simple approach for incorporating information on the spatial variability of soil properties and on ETp uncertainty into an irrigation scheduling scheme. We assume that spatial information on soil properties pertinent to soil water and yield is available and can be transformed through a classification (grouping) process into a more convenient form by defining a "certainty equivalent" field. Uncertain ETp can be accounted for by using a stochastic model based on historical data to obtain ETp predictions necessary for planning irrigation schedules. The specific objective of this study is to develop and field-test a scheme for scheduling a limited amount of seasonal irrigation water to a field having highly variable soil properties, under conditions of uncertain crop requirements as governed by ETp.

15.2 Theoretical Considerations

15.2.1 Crop Yield Model

The core of many economic evaluations of irrigation scheduling programs is the assumption that crop yield is a function of relative evapotranspiration ($ETr = ETa/ETp$, where ETa is the actual ET), or of relative transpiration (Yaron and Bresler 1983). ETa is dependent on atmospheric conditions and on the soil-water level; ETp is assumed to be known. A seasonal crop production function may be expressed in a form similar to that in Russo (1986):

$$\frac{Y}{Yp} = \sum_{i=1}^{n} \frac{T_i}{Tp_i}, \tag{1}$$

where n is the number of time periods (days or weeks) in a growing season; Y is the actual seasonal yield; Yp is the potential (maximum) yield; Tp_i is potential transpiration for a nonstressed crop during period i estimated by $Kc_i ETp_i$, and Kc_i is a crop coefficient determined experimentally (Kincaid and Heermann 1974). T_i is actual transpiration for period i estimated by:

$$T_i = Kc_i SWS_i ETp_i, \tag{2}$$

were SWS_i is a soil-water stress factor (Hanks 1983) given by:

$$SWS_i = \begin{bmatrix} 1.0 & \text{if } \dfrac{WC_i - WP_i}{FC_i - WP_i} > TR \\ \dfrac{(WC_i - WP_i)}{TR(FC_i - WP_i)} & \text{otherwise} \end{bmatrix}, \tag{3}$$

where WC_i is the soil water depth, WP_i is the wilting point, FC_i is the field capacity, and TR is the soil-water deficit threshold from which any further decrease in soil water result in a reduction in transpiration with $0 < TR < 1.0$. A value of $TR = 0.5$ was chosen for corn (Hanks 1983). FC_i and WP_i are time dependent because the rooting depth changes with time and as available soil water increases. If soil-surface evaporation is negligibly small, T_i is equal to ETa_i, and Kc_i is then a transpiration coefficient (Hanks 1985). In this simplified formulation, water quality is not considered.

15.2.2 Soil-Water Balance

Soil-water balance, assuming no precipitation and no runoff, is given by:

$$WC_{i+1} = WC_i + IRR_i - Kc_i SWS_i ETP_i - DR_i + (RZD_{i+1} - RZD_i)W_{ini}, \tag{4}$$

where WC_i and WC_{i+1} are soil-water storage depths at the beginning of time periods i and $i+1$, respectively; IRR_i is the applied irrigation during period i; DR_i is drainage; RZD_i is the root zone depth in period i; and W_{ini} is the initial volumetric soil-water content in the profile. DR_i is greater than zero only when IRR_i is sufficient to exceed the soil-water holding capacity. Stored soil water that becomes available as roots reach deeper soil layers is given in the last term on the righthand side of Eq. (4).

15.2.3 Stochastic ETp Model

Let ET_i be the *normalized and standardized* ETP_i given by:

$$ET_i = \frac{ETp_i - \overline{ETp_i}}{\sigma_{ET_i}}, \tag{5}$$

where $\overline{ETp_i}$ is mean ETp for day i, and σ_{ETp_i} is the standard deviation of ETp_i (estimated from climatic records of the study site). A stochastic model may then be fitted to the stationary time series; an example is the widely used generalized univariate autoregressive moving-average model [ARMA (p, q), Box and Jenkins 1976] given by:

$$ET_i = \phi_1 ET_{i-1} + \ldots + \phi_p ET_{i-p} + u_i - \theta_1 u_{i-1} - \ldots - \theta_q u_{i-q}, \tag{6}$$

where ϕ's are the autoregression (AR) parameters, p is the order of the AR

model, θ's are the moving-average (MA) parameters, q is the MA model order, and u is a zero mean, uncorrelated and normally distributed random term. Procedures for selecting a particulars model order and estimating its parameters are given elsewhere (e.g., Box and Jenkins 1976; Salas et al. 1980). In several cases (and in the present study), a simple model such as an $AR(1)$ given by:

$$ET_{i+1} = \varphi_1 ET_i + u_i \tag{7}$$

was sufficient for modeling real ET time series (Rhenals and Bras 1981; Aboitiz et al. 1986), where ϕ_1 is a fitted autoregression parameter, and u_i is a zero mean and normally distributed term with a variance equal to $\sigma^2_{ETi}(1-\phi^2_1)$, where σ^2_{ETi} is ET variance obtained from historical records (daily variance).

The main purpose of the ETp model is to generate ETp-sequences having similar statistical properties to the historical time series and which can be used as predictions of ETp, or in a state-space method (Or and Hanks 1992a) as part of a model for providing soil-water estimates and predictions.

15.2.4 Mathematical Programming (Optimization)

Many factors, such as ETp and rainfall, that affect irrigation decisions are unpredictable. Therefore, some form of feedback control is needed to account for these uncertain events as they occur, while at the same time considering the entire season to ensure crop maturity with limited water. One such control method is *open loop feedback control* (OLFC), usually employed within the framework of dynamic programming (Bertsekas 1987; Bras and Seo 1987). The main idea is to obtain an irrigation schedule for the entire season at each time period (stage) based on the information available, and then to exercise only the immediate control recommended by the solution (i.e., irrigation decision for present period), with further periodic resolution of the problem as more information becomes available (Smith 1973; Bras and Seo 1987).

Although stochastic dynamic programming would have been a natural choice for solving the stochastic multistage decision process of irrigation scheduling (Rhenals and Bras 1981), we decided to use linear programming (LP) for the following reasons:

1. The irrigation scheduling problem under the previously stated assumptions is linear.
2. The availability of easy-to-implement commercial software (e.g., LINGO from LINDO Systems Inc.) creates an incentive to structure the problem as an LP (Matanga and Marino 1979; Loucks et al. 1981).
3. Spatial variability is included in the scheduling scheme in a linear form.

Nevertheless, the resulting "staircase" LP scheme can be converted into a dynamic programming form if desired.

15.2.4.1 Irrigation Scheduling in a Homogeneous Field

We first consider and LP formulation of irrigation scheduling in a field with homogeneous soil and this may be represented by a single soil profile. The main processes that affect crop yield [e.g., Eqs. (1)–(5)] require only minor modification for their incorporation into a linear program. A constraint of the form:

$$\sum_{i=1}^{n} IRR_i \leqslant Q_{IRR} \tag{8}$$

indicates a limited available seasonal amount of irrigation, where Q_{IRR} is seasonal available irrigation water (which may also be subject to uncertainty, but was considered as known in this study). Constraints on specific IRR_i can be imposed, e.g., IRR_1 and $IRR_n = 0$, which means no irrigation during the first and last time periods.

The soil-water stress factor in Eq. (3) is a piecewise linear function of actual soil water which can be expressed as two simultaneous constraints:

$$SWS_i = \frac{1}{TR} * \frac{(WC_i - WP_i)}{(FC_i - WP_i)} \tag{9}$$

$$SWS_i \leqslant 1.0.$$

WC_i is bounded between FC_i and WP_i, given by the following time-dependent constraints:

$$WC_i \leqslant FC_i$$

$$WC_i \geqslant WP_i, \tag{10}$$

for each period i, FC_i and WP_i are the products of rooting depth (RDZ_i) and volumetric water contents at FC and WP, respectively, where FC and WP are constants which determine the soil-water holding capacity. Other data required by the LP include: a sequence of ETp_i predictions generated by the stochastic model, Kc_i, and RZD_i sequences.

The objective function for selecting an optimal schedule by the LP may be to maximize yield over the field, considering the entire season, and this is given by:

$$\max \quad \frac{\sum_{i=1}^{n} kc_i SWS_i ETp_i}{\sum_{i=1}^{n} kc_i ETp_i} \quad . \tag{11}$$

Note that Eq. (11) is different from the sum of ratios in Eq. (1), because it considers the entire season. The objective function in Eq. (11) can be expanded to maximize net benefits considering other costs such as the price of irrigation water, or environmental costs as a penalty on drainage. The modified objective function acquires the following form:

$$\max \left[P_Y Y_{\max} \left(\frac{\sum_{i=1}^{n} kc_i SWS_i ETp_i}{\sum_{i=1}^{n} kc_i ETp_i} \right) - P_w \sum_{i=1}^{n} IRR_i - P_{DR} \sum_{i=1}^{n} DR_i \right], \tag{12}$$

where P_Y, P_w, and P_{DR} are unit prices of yield, irrigation water, and drainage penalty, respectively.

15.2.4.2 Irrigation Scheduling in a Heterogeneous Field

In a heterogeneous field, we assume that spatial information regarding pertinent soil properties is available. Often, the statistical information requires transformation into another accessible form for decision making, for example, defining soil types by grouping and creating parcels of similar properties. Such a classification may be affected by the type of available information, irrigation method, crop, and the objectives of the irrigator. For example, the amount of water entering the soil under sprinkler irrigation in a flat field is strongly affected by soil texture and saturated hydraulic conductivity. Thus, sampling those properties and obtaining estimates at unmeasured locations by interpolation (e.g., kriging) enables one to identify parcels having similar properties with respect to the irrigation system.

Similarly, one can group soil properties with respect to their effect on crop yield, resulting in parcels of soil classes (**m**) which may occupy different areas in the field. Representing each of the **m** soil classes as a single soil column, and weighting yields realized on soil class by the relative area of each class provides an estimate of the expected yield over the entire field. Such a hypothetical field may be viewed as the "certainty equivalent" of the original random field.

Different soil-water holding capacities (FC-WP) were selected to identify different soil types [Eq. (10)] in the present formulation. A separate water balance [Eq. (5)] was solved for each soil class j. The following variables: ET, IRR, Kc, and RDZ, and their appropriate constraints, were shared by all soil classes. The resulting water content WC_i^j, drainage DR_i^j, and soil-water stress SWS_i^j, factors determine the relative yield realized on each soil class. The combined objective function in a net benefit form, including the expected yield and drainage over the field, is given by:

$$\max \left[P_Y Y_{\max} \sum_{j=1}^{m} w^j \left(\frac{\sum_{i=1}^{n} kc_i SWS_i^j ETp_i}{\sum_{i=1}^{n} kc_i ETp_i} \right) - P_w \sum_{i=1}^{n} IRR_i \right.$$

$$\left. - P_{DR} \sum_{j=1}^{m} w^j \sum_{i=1}^{n} DR_i^j \right], \tag{13}$$

where w^j is the relative area of soil class j (i.e., soil class weight).

15.3 Materials and Methods

15.3.1 Simulation Study

The main purpose of the simulation study was to test the effects of different amounts of irrigation water on yield crop of grown in a heterogeneous field. Soil-water content at *FC* was shosen as the single spatially variable soil property for the simulation study. *FC* is directly related to soil-water holding capacity and thus plays a major role in the water-balance simulation model PLANTGRO (Hanks 1974), which was selected to simulate crop growth in an irrigated field. The turning bands method (TBM, Journel and Huijbregts 1978) was used to generate a multivariate normal (MVN) random field of *FC* on a 30 × 20 grid (representing 600 soil profiles) using an exponential covariance model having a correlation range of 5 m. The estimated correlation range in the generated field was about 5.7 m. The assumed mean and standard deviation (SD) were 0.25 (m^3/m^3) and 0.0375 (m^3/m^3), respectively, and the generated values were 0.256 (m^3/m^3) and 0.0332 (m^3/m^3), respectively.

For simulation purposes the field was regarded as a collection of noninteracting soil profiles, each with its own *FC* (Dagan and Bresler 1979). Initial soil water stored in each profile was set at *FC*. Real *ETp* data of 1988 were used as input to the model. A seasonal irrigation amount of 200 mm was uniformly applied (simulating a sprinkler-irrigated field) in four amounts of 40, 50, 60 and 50 mm on the 4th, 7th, 10th and 13th weeks, respectively. A deterministic simulation of an entire growing season was performed at 600 locations.

15.3.2 Field Study

A field experiment was conducted at the Utah State University experimental farm in Logan, Utah, to test the irrigation scheduing scheme in a field having highly variable soil-water conditions. The experimental setup is described in detail by Or, (1990) and Or and Hanks (1992a); thus only a brief description is given here.

Results from an early study on the spatial variability of various soil properties (soil-water content at saturation and *FC*, bulk density, saturated hydraulic conductivity, and retention data) indicated that the field area was essentially homogeneous with respect to the sampled properties. Therefore, soil-water spatial variations were induced by an irrigation system with spatially varying water application.

The field was divided into six 30 × 15 m plots, each of which spanned 20 corn (Zea mays) rows (0.75 m apart). All plots received equal seasonal irrigation amounts of 150 mm; the value was chosen such that water would be limiting and deficit irrigation would be inevitable over portions of the field. Three replicate plots were designated to receive an optimized irrigation schedule (OPT). The other three replicate plots (NPT) were irrigated according to a fixed schedule on 30 July (30 mm), 15 August (60 mm), and 7 September (50 mm). An irrigation system that delivered spatially variable amounts of water by using different flow-rate drippers was installed on each plot. Each system consisted of 600 drippers with seven different flow rates (0, 2, 4, 6, 8, 10 and 12 l/h) mounted on 20 lines (polyethylene tubing) with 30 drippers at 1-m intervals on each line. The flow-rate sequences on the lines determined the spatial pattern of water application and this was identical on all plots (Fig. 15.1). The spatial pattern was based on a TBM-generated random field of flow rates having a correlation range of 5.27 m, a mean flow rate of 5.6 l/h, and an SD of 3.18 l/h. The correlation range was chosen to exceed the range of influence of a typical corn plant root zone, and to facilitate the use of small experimental plot (15 × 30 m). Irrigation amounts refer to amounts applied with a mean flow rate of 5.6 l/h. Actual amounts at different locations were dependent upon local flow rates simulating different soil-water conditions. For example, the seasonal amount in a location having a flow rate of 12 l/h would be 321 mm (12/5.6 × 150), while it would be only 53.6 mm in locations receiving 2 l/h.

Measurements of soil-water content were made with a neutron probe (model 503 from Campbell Pacific Nuclear Inc.). Seven access tubes were installed in each plot to a depth of 2.4 m adjacent to drippers with flow rates of 10 and 12 l/h, where drainage was expected, and to a depth of 1.8 m elsewhere. Access tubes were located within clusters of drippers with similar flow rates (referred to as *standard locations*). The placement of access tubes relative to the irrigation system was similar for all six plots.

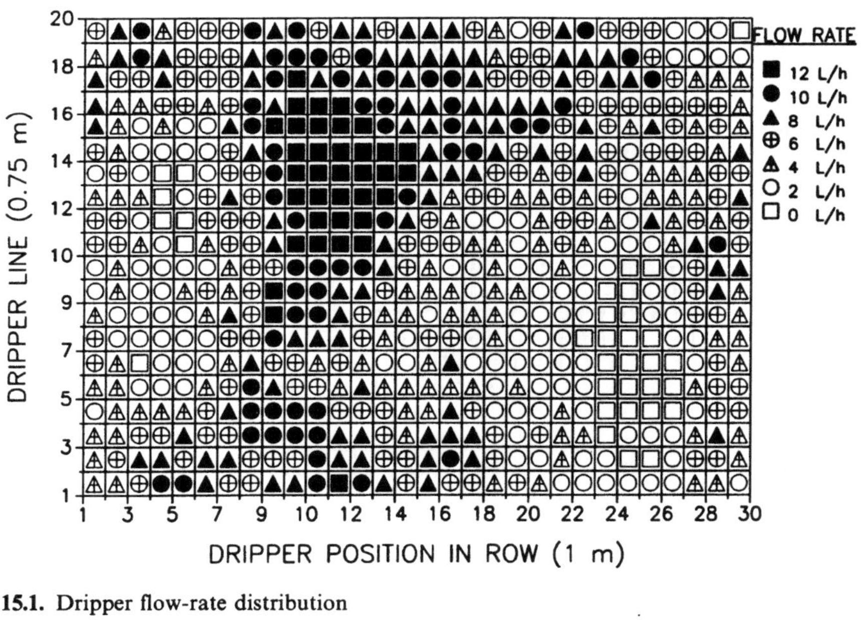

Fig. 15.1. Dripper flow-rate distribution

15.3.3 Stochastic *ETp* Model for the Study Site

A stochastic model for generating synthetic *ETp* sequences was estimated from a 6-years (1980–1985, and 1988) *ETp* time series based on observations made at North Farm, Logan, Utah. The Penman equation (Allen 1986) was used to compute daily *ETp* for 144 days each year, representing a typical growing season (for which a complete data set was available). Daily average *ETp* values and corresponding standard deviations (SD) were obtained from the 6-year data set and are shown in Fig. 15.2a. A clear seasonal trend exists in the *ETp* time series which had to be removed prior to fitting a stochastic model. A Fourier series as fitted to both the daily *ETp* and its SD using procedures described by Salas et al. (1980). Only two harmonics were needed to explain 81% of the variations in the average daily *ETp*, indicating a

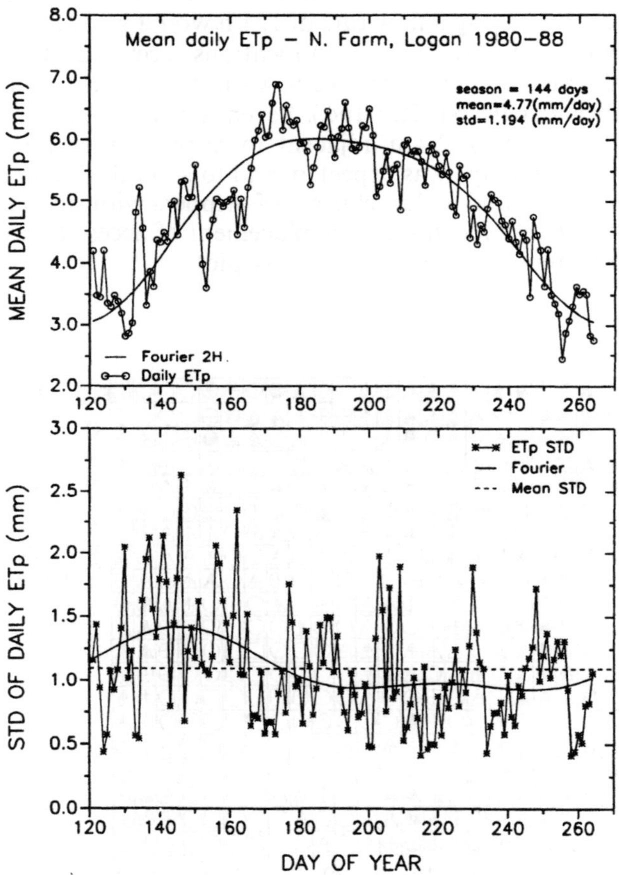

Fig. 15.2. *Top* mean daily *ETp* and *bottom* its standard deviation at North Farm, Logan, Utah, based on 6 years of records. Seasonal trends were fitted with two-harmonics Fourier series

Table 15.1. Estimated model parameters for standardized mean daily ETp series

Parameters	ARMA(p,q) model order				
	AR(1)	AR(2)	MA(1)	MA(2)	ARMA(1, 1)
ϕ_1	0.515	0.537			0.471
ϕ_2		−0.029			
θ_1			−0.423	−0.550	−0.063
θ_2				−0.253	
Residuals variance σ_ε^2	0.138	0.138	0.147	0.139	0.138
Porte Manteau test-$Q_{L=36}$[a]	28.9	31.0	47.8	33.1	29.3
$\chi^2_{95\%}(L\text{-}p\text{-}q)$	49.8	48.6	49.8	48.6	48.6
AIC	−282.9	−280.6	−273.8	−279.5	−281.0

[a] L is the maximal lag in computing the residuals autocorrelation. AIC, Akaika information criterion.

strong seasonal trend as is evident from the data. A two-harmonic Fourier series was also fitted to the SD of mean daily ETp (Fig. 15.2b) but this explained only 14.8% of the variations in SD and suggests a lack of a clear seasonal trend. However, note that higher variations (SD) in daily ETp values would probably be expected during the transition season in the spring, as depicted in Fig. 15.2b.

An approach similar to a classical time-series analysis was adopted for fitting a stochastic model to the residuals of the mean daily ETp series, i.e., after removing the deterministic trend (Salas et al. 1980). Several possible ARMA (p,q) stochastic models were tested, based on criteria of best fit and parsimony. Parameters estimated by the maximum likelihood method, statistics, and tests of the residuals are given in Table 15.1. The AR(1) model was selected as the best model fof the daily ETp data based on the Akaika information criterion (AIC).

Measurements of daily ETp, based on data from a weather station, were obtained during the field experiment growing season (1989), and were used as input to the linear program to update the optimized irrigation schedule.

15.3.4 Irrigation Scheduling Scheme

A computer program for optimizing irrigation scheduling on OPT plots was developed with the objective of maximizing yield over the entire plot. A flexible LP computer code (LINGO fron LINDO Systems Inc.) was used to generate and solve the model. The different dripper flow rates resulted in seven basic soil-water regimes (simulating soil types) governed by applied irrigations at each location (interactions were not considered). The only necessary modification of the model presented in Section 15.2.4.2 was to substitute local soil properties (FC and WP), which represent the spatial variability, by local applied irrigation as determined by dripper flow rate (denoted as $FLOW^j$). A soil-water balance equation [Eq. (4)] for each regime (soil class) was modified such that mean applied irrigation was adjusted by the local flow rate (e.g., $IRR_i^j = IRR_i\ FLOW^j/5.6$, where IRR_i is a average irrigation during period i). The weight factors ($w^j, j = 1, 2, \ldots 7$) needed for the computation of expected yield over the plot are the relative areas occupied by each flow rate (i.e., ratio of the number of drippers of a given flow rate to the total number of drippers; see Table 15.2).

Other inputs required by the program are given in Table 15.3 and include: crop coefficients for corn, obtained by Kincaid and Heermann's (1974) method; and weekly rooting depth, estimated from the PLANTGRO simulation model (Hanks 1974). Due to the relatively short season and the drip irrigation method, a maximum rooting depth of 1350 mm was assumed. A sequence of weekly ETp predictions for the entire season, generated by

Table 15.2. Number of drippers and the relative area occupied by each flow rate

Flow rate (l/h)	No. of drippers	Relative area (w^j)
0	33	0.055
2	117	0.195
4	110	0.183
6	144	0.240
8	101	0.168
10	59	0.098
12	36	0.060

Table 15.3. Weekly inputs into the irrigation scheduling LP, and projected irrigation schedule based on the first-week solution

Week	ETp (generated)	Irrigation (projected) (mm)	Root zone depth	Crop coefficients
1	40.2	—	270	0.20
2	40.1	20.0	410	0.23
3	40.7	10.0	550	0.35
4	43.6	15.0	690	0.51
5	41.4	20.0	830	0.69
6	40.4	23.0	970	0.84
7	41.8	27.0	1110	0.94
8	40.5	21.0	1250	0.98
9	37.9	20.0	1350	0.99
10	37.0	—	1350	0.97
11	34.0	—	1350	0.92
12	32.0	—	1350	0.85

a stochastic model based on historical *ETp* records of the study site, provided the schedule with a long-term "look ahead". A schedule obtained in the first week of the season and based on predicted *ETp* only is also given in Table 15.3.

Weekly *ETp* was estimated from daily climatic data collected on-site using the Penman combination equation and was used to update the generated *ETp* predictions. Each week, an updated schedule for the entire season was computed based on the newly availale *ETp* and soil-water information, and such schedules determined the timing and amount of the next irrigation.

15.4 Results and Discussion

15.4.1 Simulation Study – The Effect of Seasonal Irrigation Amount on Crop Yield

The simulation study examined the effects of different seasonal irrigation amounts on crop yields using the same fixed-dated irrigation schedule for each amount (Table 15.1). Simulated crop relative yields are depicted in Fig. 15.3 as a function of *FC* at a location. Different curves resulted from using three seasonal amounts of irrigation water: 150, 200, and 250 mm. The probability density function (PDF) of relative yield vs. the relative area in the field for which relative yield was less than a certain value is described in Fig. 15.4. Yield production over larger portions of the field increased with the increase in irrigation amounts, and is manifested by larger skewness

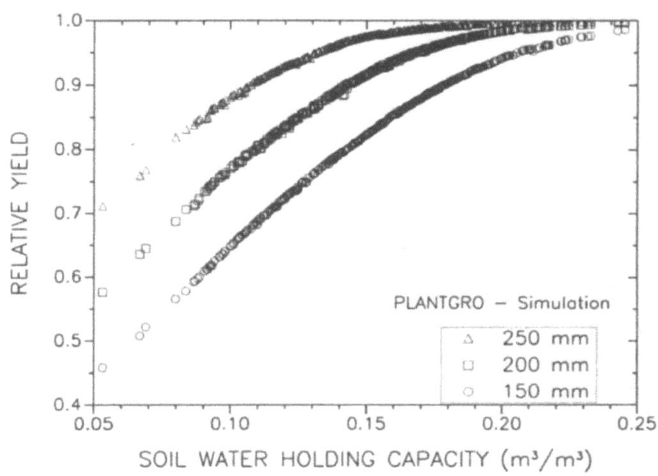

Fig. 15.3. Simulated crop relative yield vs. volumetric water holding capacity at the generated random FC field with three seasonal irrigation amounts

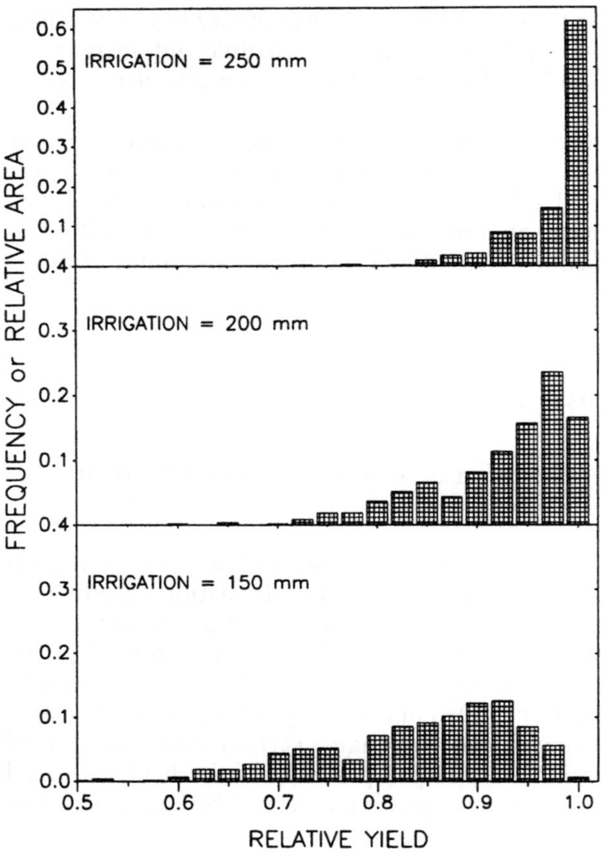

Fig. 15.4. Relative crop yield distribution with three seasonal irrigation amounts

of the relative yield's PDF. An increase of 50 mm in the seasonal irrigation water from 150 to 200 mm (with a similar irrigation schedule) resulted in an increase in average relative yield over the entire field from 83 to 91%; an additional increase of 50 mm increased the relative yield to 96.5% (Table 15.4). The increase in yield was coupled with a decrease in the yield's coefficient of variation (CV) resulting in a more uniform yield over the field.

The marginal contribution of additional irrigation water decreased as irrigation increased. It was assumed that over-irrigation causes no adverse effects on crop yields; thus, maximum yield over the entire field could be obtained with sufficient irrigation amounts.

The results of the simulation study help to explain the tendency of some farmers to over-irrigate (where irrigation water is abundant and the price is low) rather than under-irrigate their fields. Such practice is supported by other findings (Or and Hanks 1992c), showing that yield depression due to over-irrigation of a vegetative crop (Zea mays) is much less than the

Table 15.4. Simulated relative yield for FC random field as affected by the seasonal amount of irrigation water

Case	Day of year					Relative yield		
	175	200	220	240		Mean	SD	CV(%)
	Irrigation (mm)				Total			
I	40	70	70	70	250	0.965	0.039	4.1
II	40	50	60	50	200	0.912	0.071	7.7
III	30	40	50	30	150	0.830	0.095	11.5

reduction resulting from water stress due to under-irrigation. However, this may not be true for crops with special sensitivity to over-irrigation, e.g., cotton for cotton fiber production.

15.4.2 Field Experiment

15.4.2.1 Water Balance and Crop Yield

Experimental results pertaining to the spatial structures of crop yield and of soil-water conditions have been reported elsewhere (Or 1990; Or and Hanks 1992a,b). Average biweekly water balances for the three replications are shown in Tables 15.5 and 15.6 for the standard locations on OPT and NPT, respectively. We neglected deep percolation since soil-water content for the 1.5- to 1.8-m depth remained nearly constant below *FC* for most access tubes. However, deep percolation may have occurred at several locations, mainly in the NPT plots, where the estimated ETa/ETp ratios were greater than 1.0 (italic entries in Tables 15.5 and 15.6). Seasonal irrigation amounts and relative yields (Yr) at each of the standard locations are also given. The relative yeild was based on a six-plot average potential (maximum) yield (Yp) of 7 kg/m^2, where *Yp* for each plot was determined as the average to 5% of the yield records. Tables 15.5 and 15.6 indicate that spatially variable water application rates resulted in large differences in various soil-water balance components.

Mean crop fresh yield on the optimized schedule, OPT plots, was 13% higher than on NPT plots (Table 15.7). A detailed irrigation schedule for each treatment is provided in Table 15.8. Above-ground fresh yield is reported because the ratio of fresh to dry biomass was constant at 0.188 for all samples. Yield differences between paired plots, OPT vs. NPT (plots 1–2, 3–4, and 5–6; Table 15.7), resulted in a significant t value at the $P = 0.05$ level. This indicates that the optimized irrigation schedule improved the yields of the OPT plots compared with the NPT plots. The observed mean difference of about 13% in crop above-ground biomass is substantial considering the short growing season and the low seasonal irrigation amount.

Table 15.5. Irrigation applied (Irr.), potential *ET* (*ETp*), actual *ET* (*ETa*), *ETa*/*ET*max, *ETa*/*ETp*, and relative yield (Yr) for all flow rates at the standard locations in the OPT plots

Days of year									
from:	191	201	214	230	244	259	272	Sum (mm)	Yr
to:	201	214	230	244	259	272	280	539	
ETp (mm)	78	107	106	85	76	60	27		
12 l/h									
Irr. (mm)	0	64	103[b]	88	81	10[c]	0	346	
ETa (mm)	5	64	94	88	84	29	25	389	0.78
ETa/*ET*max[a]	—	1.00	1.00	1.00	1.00	0.67	1.00	—	
ETa/*ETp*	—	0.60	0.89	1.02	*1.10*	0.48	0.93	—	
10 l/h									
Irr. (mm)	0	53	88[b]	73	68	10[c]	0	292	
ETa (mm)	5	56	90	75	75	41	18	360	0.84
ETa/*ET*max	—	0.88	0.96	0.85	0.89	0.95	0.72	—	
ETa/*ETp*	—	0.52	0.85	0.88	0.99	0.68	0.67	—	
8 l/h									
Irr. (mm)	0	43	74[b]	59	54	10[c]	0	240	
ETa (mm)	5	46	82	72	59	43[d]	12	319	0.78
ETa/*ET*max	—	0.72	0.87	0.82	0.70	1.00	0.48	—	
ETa/*ETp*	—	0.43	0.77	0.84	0.78	0.71	0.44	—	
6 l/h									
Irr. (mm)	0	32	59[b]	44	41	10[c]	0	186	
ETa (mm)	5	41	66	56	56	38	15	277	0.65
ETa/*ET*max	—	0.64	0.70	0.64	0.67	0.88	0.60	—	
ETa/*ETp*	—	0.38	0.62	0.65	0.74	0.63	0.56	—	
4 l/h									
Irr. (mm)	0	21	37[b]	29	27	10[c]	0	124	
ETa (mm)	5	28	49	45	46	29	13	215	0.64
ETa/*ET*max	—	0.44	0.52	0.51	0.56	0.67	0.52	—	
ETa/*ETp*	—	0.26	0.46	0.53	0.62	0.48	0.48	—	
2 l/h									
Irr. (mm)	0	11	30[b]	15	14	10[c]	0	80	
ETa (mm)	5	17	44	36	35	27	5	169	0.54
ETa/*ET*max	—	0.27	0.40	0.40	0.42	0.63	0.20	—	
ETa/*ETp*	—	0.16	0.42	0.41	0.46	0.45	0.19	—	
0 l/h									
Irr. (mm)	0	0	15[b]	0	0	10[c]	0	25	
ETa (mm)	5	14	25	13	22	23	5	107	0.439
ETa/*ET*max	—	0.22	0.27	0.15	0.28	0.53	0.20	—	
ETa/*ETp*	—	0.13	0.24	0.15	0.29	0.38	0.19	—	

[a] *ETa* at a flow rate of 12 l/h is assumed as *ET*max on *OPT*.
[b] Sprinkler irrigation (15 mm) to enhance nodal root support.
[c] Rain (10 mm).
[d] Assumed as *ET*max for the time period.

15.4.2.2 ETp Predictions

During most of the 1989 growing season, daily *ETp* was considerably higher than the expected 6-year average, as depicted in Fig. 15.5. *ETp* estimates based on daily measurements and predictions produced by the AR(1) were

Table 15.6. Irrigation applied (Irr.), potential *ET* (*ETp*), actual *ET* (*ETa*), *ETa*/*ET*max, *ETa*/*ETp*, and relative yield (Yr) for all flow rates at the standard locations in the NPT plots

Days of year	from: 191	201	211	230	244	259	272	Sum (mm)	Yr
	to: 201	211	230	244	259	272	280	539	
ETp (mm)	78	90	123	85	76	60	27		
12 l/h									
Irr. (mm)	0	0	101[b]	129	107	10[c]	0	347	
ETa (mm)	5	6	102	103	99	37	12	364	0.71
ETa/*ET*max[a]	—	1.00	1.00	1.00	1.00	1.00	0.71	—	
ETa/*ETp*	—	0.07	0.83	*1.20*	*1.30*	0.62	0.44	—	
10 l/h									
Irr. (mm)	0	0	96[b]	107	89	10[c]	0	302	
ETa (mm)	5	6	93	92	84	37	11	328	0.64
ETa/*ET*max	—	1.00	0.91	0.89	0.85	1.00	0.65	—	
ETa/*ETp*	—	0.07	0.76	1.08	*1.11*	0.62	0.41	—	
8 l/h									
Irr. (mm)	0	0	72[b]	86	71	10[c]	0	239	
ETa (mm)	5	6	90	78	72	35	17[d]	303	0.71
ETa/*ET*max	—	1.00	0.88	0.76	0.73	0.95	1.00	—	
ETa/*ETp*	—	0.07	0.73	0.91	0.95	0.58	0.63	—	
6 l/h									
Irr. (mm)	0	0	58[b]	64	54	10[c]	0	186	
ETa (mm)	5	6	72	57	56	34	10	240	0.61
ETa/*ET*max	—	1.00	0.71	0.55	0.56	0.92	0.59	—	
ETa/*ETp*	—	0.07	0.59	0.67	0.73	0.57	0.37	—	
4 l/h									
Irr. (mm)	0	0	44[b]	43	36	10[c]	0	133	
ETa (mm)	5	6	65	38	46	28	13	201	0.60
ETa/*ET*max	—	1.00	0.64	0.37	0.46	0.76	0.76	—	
ETa/*ETp*	—	0.07	0.53	0.44	0.60	0.47	0.48	—	
2 l/h									
Irr. (mm)	0	0	29[b]	21	18	10[c]	0	78	
ETa (mm)	5	6	40	32	34	23	7	147	0.45
ETa/*ET*max	—	1.00	0.39	0.31	0.34	0.62	0.41	—	
ETa/*ETp*	—	0.07	0.33	0.37	0.45	0.38	0.26	—	
0 l/h									
Irr. (mm)	0	0	15[b]	01	0	10[c]	0	25	
ETa (mm)	5	6	25	23	15	20	6	100	0.32
ETa/*ET*max	—	1.00	0.25	0.22	0.15	0.54	0.35	—	
ETa/*ETp*	—	0.07	0.20	0.27	0.20	0.33	0.22	—	

[a] *ETa* at a flow rate of 12 l/h is assumed as *ET*max on NPT.
[b] Sprinkler irrigation (15 mm) to enhance nodal root support.
[c] Rain (10 mm).
[d] Assumed as *ET*max for the time period.

obtained by filtering algorithm (Or and Hanks 1992a) known as the Kalman filter (KF) and are shown in Fig. 15.5, with a confidence region of ± 1 SD. Although *ETp* was generally higher than normal, the availability of a daily measurement renders *ETp* predictions quite reliable as can be seen in Fig. 15.5.

Table 15.7. Fresh crop biomass (kg/m^2) in the OPT and NPT plots

OPT			NPT			Difference
Plot	Mean	SD[a]	Plot	Mean	SD[a]	
1	4.67	1.04	2	4.07	1.16	0.60
3	5.32	1.12	4	4.64	1.08	0.68
6	4.64	1.02	5	4.26	0.82	0.38
Mean	4.88			4.32		0.55[c]
SD[b]	0.38			0.29		

[a] Standard deviation within plot.
[b] Standard deviation between plots.
[c] Significant at the $P = 0.05$ level.

Table 15.8. Irrigation schedule and observed *ETp* in OPT and NPT plots

Period days of year	Irrigation (mm)			*ETp* (mm)
	OPT		NPT	
191–197	—		—	53.0
198–204	20		—	54.9
205–211	10		—	60.0
212–218	19		40	54.4
219–225	22		—	47.9
226–231	18	(15[a])	60	43.2
232–239	23		—	41.2
240–246	19		—	44.9
247–253	19		50	32.8
253–260	—		—	37.9
261–267	—	(10[b])	—	29.8
268–274	—		—	29.5
274–281	—		—	19.2
Total	150		150	548

[a] Uniform sprinkler irrigation to enhance nodal root support.
[b] Total rainfall during 3 days (260–262).

15.4.2.3 The Irrigation Schedule

The resulting "optimal" irrigation schedule (Table 15.8) was quite similar to the projected schedule obtained in the first week of the season (Table 15.3). This suggests that although realized *ETp* was considerably higher than expected, its effect on the schedule was minor due to the constraining effect of a small seasonal irrigation amount which was much less than *ETp*.

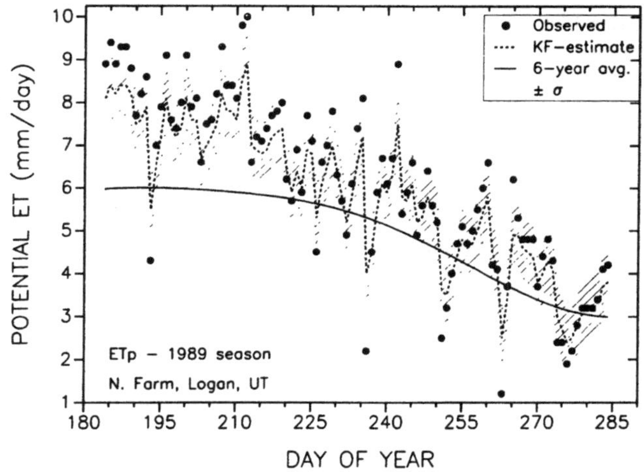

Fig. 15.5. Observed and predicted daily *ETp* by the Kalman filter (± an estimation SD) at North Farm, Logan, Utah, during the 1989 growing season (6-year average daily *ETp* is also shown)

The relative field areas receiving a given flow rate (i.e., the weights in the objective function) were known with certainty (Table 15.2), unlike situations in practice where the relative area of a soil type is seldom known with certainty. Nevertheless, even with the detailed information and the highly controlled experimental conditions, there was a large degree of "fuzziness" in crop yields that was probably a result of interactions between neighboring drippers such interactions were ignored by the scheduling scheme.

A weekly average of stored soil water at each of the seven possible flow rates was supplied as an input to a slightly modified version of the irrigation scheduling program (denoted LP2). The LP2 program was solved considering the updated stored soil water as initial conditions, while other constraints, such as remaining seasonal irrigation water and weekly *ETp* update, were also considered. The resulting schedule by LP2 is given in Table 15.9, and was very similar to the schedule obtained by the regular scheme which did not consider soil-water measurements (denoted LP1). The actual irrigation schedule applied was determined as a weighted average of LP1 and LP2.

The adequacy of the water balance model embodied in the irrigation scheduling program was tested against average stored soil water to a depth similar to that assumed by the LP measured on OPT plots at each flow rate (Table 15.9). One-week-ahead predictions by LP1 and LP2 were compared with measured soil-water storage. As expected, the agreement between measured and predicted storage by LP2 was better than by LP1, mainly because LP2 was updated weekly. In most cases, however, both programs produced acceptable storage predictions and similar optimal schedules. We concluded that LP2 is predictable whenever soil-water measurements are available for updating the irrigation scheduling scheme.

Table 15.9. Projected and applied irrigation and measured and predicted soil-water storage by two LP schemes in OPT plots

	Scheme	Weeks from begining of season									
		2	3	4	5	6	7	8	9	10	11
		Irrigation (mm)									
	LP1	20	10	19	17	22	20	19	—	—	—
	LP2	20	10	18	20	24	18	19	—	—	—
	Applied	20	10	19	18	23	19	19	—	—	—
Flow rate (l/h)		Soil water stored in the root zone (mm)									
	LP1		104	123	134	151	176	189	194	179	168
0	LP2		—	134	155	162	205	223	136	214	196
	Meas.[a]		112	139	148	187	221	242	222	235	232
	LP1		111	132	148	169	196	212	220	207	190
2	LP2		—	140	164	183	222	136	248	231	210
	Meas.		115	142	165	201	226	249	224	231	226
	LP1		118	141	161	186	215	235	245	233	212
4	LP2		—	148	175	197	236	250	263	249	225
	Meas.		120	148	176	212	234	260	236	243	236
	LP1		127	151	175	205	236	260	273	263	237
6	LP2		—	158	190	218	249	268	281	272	244
	Meas.		126	158	193	221	245	274	254	259	246
	LP1		129	156	186	219	252	279	294	285	255
8	LP2		—	162	195	228	261	283	298	292	260
	Meas.		129	157	198	229	255	289	273	278	261
	LP1		129	160	195	228	261	293	313	308	274
10	LP2		—	162	195	228	261	294	317	315	279
	Meas.		124	154	204	233	257	303	286	283	269
	LP1		129	162	195	228	261	294	317	317	281
12	LP2		—	162	195	228	261	294	317	317	281
	Meas.		134	163	206	242	277	312	312	313	307

[a] Measured soil-water storage.

15.5 Summary and Conclusions

Spatial variations in soil properties have important effects on crop yields and on irrigation scheduling. Over-irrigating heterogeneous fields masks the effects of soil variability by increasing vegetative crop yields at locations susceptible to water stress, while resulting in only minor yield depression at over-irrigated locations. However, such an irrigation policy may result in larger volumes of drainage and may enhance the leaching of nutrients and chemicals to surface and underground water resources. When considering these adverse effects, deficit irrigation and yield reduction are inevitable.

A scheme for scheduling a limited amount of irrigation water and which consider spatially variable soils and uncertain ETp was developed and field-tested. The irrigation scheduling problem was cast in a LP optimization scheme which considers the entire season, while updating the schedule as more information becomes available. A weihting scheme, which considers the relative contribution from each soil type (simulated by flow rate) to the total crop yield, was used in the objective function. A field experiment was conducted with soil-water variability induced by irrigation nonuniformity as a surrogate for soil spatial variability. The optimized irrigation schedule resulted in a 13% increase in crop yield over a fixed-dated schedule using similar amounts of irrigation water.

Further research is needed before irrigation scheduling schemes, similar to the one presented here, can be widely used. A major remaining is to assess the extent of spatial variability of relevant soil properties in an irrigated field, so that guidelines for converting such information into a manageable form can be developed. One possibility proposed in this study is to group alike properties into parcels of similar "soil types", or similar units with respect to irrigation scheduling. Another important problem is the "price" or the penalty on excessive drainage; larger penalties would force deficit irrigation and lower expected yields.

References

Aboitiz M, Labadie J, Heermann D (1986) Stochastic soil moisture estimation and forecasting for irrigated fields. Water Resour Res 22(2): 180–190

Allen RG (1986) A Penman for all seasons. Am Soc Civ Eng Proc J Irrig. Drainage Div 112(4): 348–368

Avriel M, Williams AC (1970) The value of information and stochastic programming. Oper Res 18(5): 947–954

Bertsekas DP (1987) Dynamic programming: deterministic and stochastic models. Prentice-Hall, Englewood Cliffs, NJ

Box GEP, Jenkins GM (1976) Time series analysis: forecasting and control. Holden-Day, San Francisco, CA

Bras RL, Dong-Jun Seo (1987) Irrigation control in the presence of salinity: extended linear quadratic approach. Water Resour Res 23(7): 1153–1161

Bras RL, Rodriguez-Iturbe I (1985) Random functions and hydrology. Addison-Wesley, Reading, MA

Bresler E, Dagan G (1988) Variability of yield of an irrigated crop and its causes. 1. Statement of the problem and methodology. Water Resour Res 24(3): 381–387

Bresler E, Dasberg S, Russo D, Dagan G (1981) Spatial variability of crop yield as a stochastic soil process. Soil Sci Soc Am J 45: 600–605

Dagan G, Bresler E (1979) Solute dispersion in unsaturated heterogeneous soil at field scale. I. Theory. Soil Sci Soc Am J 43: 461–467

English MJ (1981) The uncertainty of crop models in irrigation optimization. Trans ASAE 24(4): 917–921,928

Hanks RJ (1974) Model for predicting plant growth as influenced by evapotranspiration and soil water. Argon J 66: 660–665

Hanks RJ (1983) Yield and water-use relationships: an overview. In: Taylor HM, Jordan WR, Sinclair TR (eds) Limitations to efficient water use in crop production. Am Soc Agron Madison, WI, pp 393–411

Hanks, RJ (1985) Crop coefficients for transpiration. Proceedings of the national conference on advances in evapotranspiration. Chicago, IL. ASAE, St. Joseph. MI, pp 431–438

Jensen ME, Wright JL (1987) The role of evapotranspiration models in irrigation scheduling. Trans ASAE 21(1): 82–87
Journel AG, Huijbregts ChJ (1987) Mining geostatistics. Academic Press, London
Kincaid DC, Heermann DF (1974) Scheduling irrigations using a programmable calculator. Rep. ARS-NC-12, USDA, ARS, North Central Region
Loucks DP, Stedinger JR, Haith DA (1981) Water resource systems planning and analysis. Prentice-Hall, Englewood Cliffs, NJ
Matanga GB, Marino MA (1979) Irrigation planning. 1. Cropping pattern. Water Resour Res 15(3): 672–678
Or D (1990) Irrigation management considering soil variability and climatic uncertainty. PhD diss. Utah State Univ, Logan, Utah
Or D, Hanks RJ (1992a) Spatial and temporal soil water estimation considering soil variability and evapotranspiration uncertainity. Water Resour Res 28(3): 803–814
Or D, Hanks RJ (1992b) Soil water spatial variability induced by irrigation non-uniformity. Soil Sci Soc Am J 56: 226–233
Or D, Hanks RJ (1992c) Measurement of crop production functions using point source sprinkler irrigation. Irrig Sci (in press) 13: 55–64
Rhenals AE, Bras RL (1981) The irrigation scheduling problem and evapotranspiration uncertainty. Water Resour Res 17(5): 1328–1338
Rogers PP, Fiering MB (1986) Use of system analysis in water management. Water Resour Res 22(9): 146S–158S
Russo D (1986) A stochastic approach to the crop yield-irrigation relationships in heterogeneous soils. I. Analysis of the field spatial variability. Soil Sci Soc Am J 50: 736–745
Salas JD, Delleur W, Yevjevich V, Lane WL (1980) Apllied modeling of hydrologic time series. Water Resource Publications, Fort Collins, CO
Smith VS (1973) System analysis and irrigation planning. J Irrig Drain Am Soc Civ Eng 99(IRI): 89–107
Warrick AW, Gardner WR (1983) Crop yield as affected by spatial variations of soil and irrigation. Water Resour Res 19: 181–186
Yaron D, Bresler E (1983) Economic analysis of on-farm irrigation using response functions of crops. In: Hillel D (ed) Advances in irrigation vol 2. Academic Press, Orlando, FL, pp 223–255
Zaslavsky D, Buras N (1967) Crop yield response to nonuniform application of irrigation water. Trans ASAE 10: 196–200

16 Use of Geostatistics in the Description of Salt-Affected Lands

S.R. Yates, R. Zhang, P.J. Shouse, and M.Th. van Genuchten[1]

16.1 Introduction

Geostatistical methods are increasingly popular tools in the anaylysis of a variety of agricultural problems. The methods are typically used to determine various spatially related quantities which, in turn, characterize the variability of one or more parameters in space and/or time. Simple, ordinary and universal kriging methods produce linear estimators which are useful for obtaining estimates of a spatially distributed property over a region, especially at locations for which no data are available. For the most part, the final result of an analysis is a map showing the spatial distribution of the property of interest. Many examples appear in the literature (Burgess and Webster 1980a,b; Webster and Burgess 1980; Vieira et al. 1981; Vauclin et al. 1983; Warrick et al. 1986; Yates et al.1986a; Yates and Warrick 1987; ASCE 1990a,b). At other times, descriptors such as variograms and correlation scales are the ultimate goal of a geostatistical investigation.

Geostatistics can also be useful when two or more spatially variable properties are of interest. Cokriging is an extended technique of kriging and incorporates both spatial and intervariable correlation. Cokriging can estimate all or some variables using the same coefficient matrix. Several applications of cokriging for describing soil-water phenomena have been reported during the past 10 years. For example, Vauclin et al. (1983) used kriging and cokriging to predict available water content, pF2.5 values, and sand content. McBratney and Webster (1983) applied cokriging to interpret the topsoil silt content, using subsoil silt or subsoil sand as joint variables. Carr and Myers (1984) studied the application of cokriging to the analysis of satellite data. Yates and Warrick (1987) and Mulla (1988) estimated soil-water content using a cokriging procedure in which the bare soil surface temperature and the sand content were used as auxiliary functions. Using soil-map delineations, Stein et al. (1988) pointed out that the use of cokriging resulted in an average increase in precision of about 10% in 30-year average moisture deficit maps. Zhang et al. (1991) improved the estimation of soil texture by using cokriging with spectral properties.

The key step in cokriging is to formulate cross-variograms, or cross-covariances, of two random variables (or more). One disadvantage to standard approaches for modeling cross-variograms is that only common

[1] USDA-ARS, US Salinity Laboratory, 4500 Glenwood Drive, Riverside, CA 92501, USA

locations can be used. Estimating the cross-variograms requires a large number of locations where data are collected for both variables, a condition that is frequently not satisfied in practice. Cokriging cannot be used to improve the estimation process when this type of information is not available. To alleviate this difficulty, Clark et al. (1989) presented a variation of cokriging, using pseudocross-variograms, and this does not require measurement of both variables at the same locations.

Although primarily implemented for estimation problems, geostatistical methods are also useful for obtaining information that improves sampling efficiency or aids in decision making. A few examples include the determination of the minimum number of samples necessary to produce a certain level of confidence or accuracy in the estimate (Burgess et al. 1981; Vieira et al. 1981), the number of samples and their placement in space such that the estimation of the variogram is optimized (Russo 1984a), and methods which deal with the effects of scale (Burrough 1983a,b). Russo (1984b) described a method for using geostatistics to aid in managing the salinity of a heterogeneous field. Zirschky (1985), Zirschky et al. (1985) and Zirschky and Harris (1986) investigated the use of geostatistics for determining reclamation strategies for the cleanup of hazardous waste sites.

Disjunctive kriging (DK) provides a method for obtaining the conditional probability that a measured indicator variable is above some prescribed tolerance level (see Fig. 16.1). With respect to hazardous waste, the tolerance level could be a threshold level where leakage of a contained material near an impoundment is deemed toxic to humans. In agricultural settings, the tolerance level could be the amount of salts in a field, beyond which plant growth or produce quality suffers. Combining a known tolerance level with the ability to estimate the conditional probability offers a means for making management decisions.

Procedures that allow DK to be used for management decision making generally begin by defining a "management unit". The management unit is the smallest unit of area that can be managed separately from the surrounding units. To support management decision making, information is needed on units for which the indicator variable has become larger than the allowed tolerance level. As opposed to constructing the probability density functions (PDFs) for each unit by intensive sampling, the DK method provides an estimate of the probability distribution using traditional sampling methods.

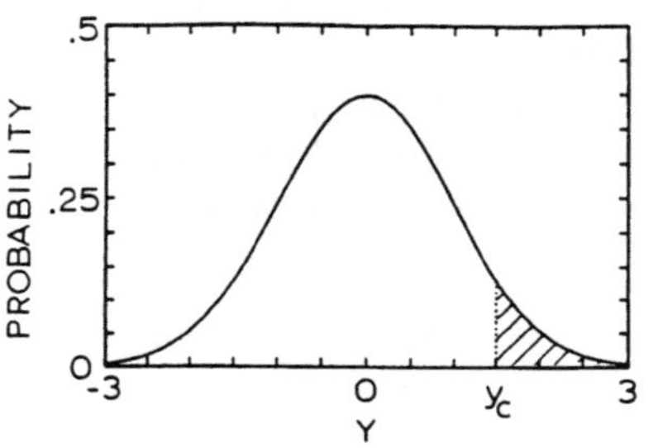

Fig. 16.1. Hypothetical probability density function with respect to the transformed variable. *Shaded area* is the probability of exceeding y_c

A fully functional management decision-making model requires values for the tolerance or cutoff level, the critical probabilty level and estimates of the conditional probability. With these three pieces of information it becomes possible to make management decisions in a quantitative manner. Also, since the tolerance or cutoff level and the critical probability level that causes corrective action are generally known *a priori*, only the conditional probability must be obtained.

The objective of this chapter is to apply geostatistical techniques to characterize the spatial distribution of soil chemicals. Two different approaches will be utilized: pseudocross-variograms and cokriging to improve estimations of the spatial distribution of soil chemicals over relatively large areas of land, and disjunctive kriging to provide a means for managing agricultural fields to minimize the harmful effects due to excess soil chemicals.

16.2 Geostatistical Analyses

16.2.1 Pseudo Cokriging

Let $Z_j(\mathbf{x})$, $Z_k(\mathbf{x})$ be two random functions of interest defined in one-, two- or three-dimensional space. Clark et al. (1989) and Myers (1991) defined the pseudocross-variogram of Z_j and Z_k as:

$$g_{\hat{jk}}(\mathbf{h}) = 0.5E[Z_j(\mathbf{x}) - Z_k(\mathbf{x}+\mathbf{h})]^2, \tag{1}$$

where E is the expected value of the argument. Equation (1) assumes that the pseudocross-variogram depends only on the separation vector h. The pseudocross-variogram is not necessarily symmetric, may not be zero at $\mathbf{h} = 0$, and is not a variogram or a covariance. When $j = k$, and the function satisfies the intrinsic hypothesis, Eq. (1) becomes the variogram. Since, in general, the two random functions Z_j and Z_k do not have the same means, Eq. (1) is not the same as half the variance of the difference, even if both random functions separately satisfy the intrinsic hypothesis. The discrepancy is more serious when the random functions do not have constant means. A more general definition of the pseudocross-variogram is:

$$g_{jk}(\mathbf{h}) = g_{kj}(-\mathbf{h}) = 0.5\,\mathrm{Var}[Z_j(\mathbf{x}) - Z_k(\mathbf{x}+\mathbf{h})], \tag{2}$$

with "Var" indicating the variance of the argument. Since the translation of a second-order stationary random function is also second-order stationary (Journel and Huijbregts 1978), Eq. (2) is the variance of a random function and depends only on h. However, the equation is neither the covariance nor the variogram of a random function unless $j = k$. If $Z(x) = Y(x) + m(x)$, where $Y(x)$ is second-order stationary and $m(x)$ is the mean of $Z(x)$, and if the pseudocross-variograms g_{jk} and g_{kj} are symmetric, Eq. (2) may be rewritten as (Myers 1991):

$$g_{jk}(\mathbf{h}) = 0.5(\sigma_j - 2\sigma_{jk} + \sigma_k) + \gamma_{jk}(\mathbf{h}), \tag{3}$$

where σ_j and σ_k are the variances of variables j and k, respectively, σ_{jk} is the covariance between variables j and k, and $\gamma_{jk}(h)$ is the cross-variogram. Equation (3) demonstrates that the pseudocross-variogram differs from the cross-variogram by a positive constant.

The estimator for $Z_1(x_0)$, using data $Z(x_i) = [Z_1(x_i), \ldots, Z_m(x_i)]$, $i = 1, \ldots, n$, is given by (Myers 1991):

$$Z_1^*(\mathbf{x}_0) = \sum Z(\mathbf{x}_i)\Gamma_i, \tag{4}$$

where Γ is a column vector of weights. If each component of $Z(x_i)$ satisfies the intrinsic hypothesis, then Eq. (4) is unbiased if:

$$\sum \Gamma_i = (1, 0, \ldots, 0)^T, \tag{5}$$

where "T" indicates a transpose. In that case, the variance of the error can be written in the form:

$$\mathrm{Var}[Z_1^*(\mathbf{x}_0) - Z_1(\mathbf{x}_0)] = \sum\sum \Gamma_i^T E(W_i^T W_j)\Gamma_j, \tag{6}$$

where

$$W_i = [Z_1(\mathbf{x}_i) - Z_1(\mathbf{x}_0), \ldots, Z_m(\mathbf{x}_i) - Z_1(\mathbf{x}_0)]. \tag{7}$$

Using Eq. (5), the righthand side of Eq. (6) can be written as:

$$\begin{aligned}\sum\sum \Gamma_i^T E(W_i^T W_j)\Gamma_j = 2\sum [g_{11}^{\wedge}(\mathbf{x}_0 - \mathbf{x}_j), \ldots, g_{1m}^{\wedge}(\mathbf{x}_0 - \mathbf{x}_j)]\Gamma_j \\ - \sum\sum \Gamma_i^T G^{\wedge}(\mathbf{x}_i - \mathbf{x}_j)\Gamma_j,\end{aligned} \tag{8}$$

where

$$G^{\wedge}(\mathbf{x} - \mathbf{y}) = \begin{bmatrix} g_{11}^{\wedge}(\mathbf{x} - \mathbf{y}) & \cdots & g_{1m}^{\wedge}(\mathbf{x} - \mathbf{y}) \\ \cdots & \cdots & \cdots \\ \cdots & \cdots & \cdots \\ \cdots & \cdots & \cdots \\ g_{m1}^{\wedge}(\mathbf{x} - \mathbf{y}) & \cdots & g_{mm}^{\wedge}(\mathbf{x} - \mathbf{y}) \end{bmatrix}. \tag{9}$$

To minimize the variance of the error of an unbiased estimate, m Lagrange multipliers $\mu_1, \ldots, \mu_m$ are introduced. The cokriging equations are then:

$$\sum G^{\wedge}(\mathbf{x}_i - \mathbf{x}_j)\Gamma_j + (\mu_1, \ldots, \mu_m)^T = [g_{11}^{\wedge}(\mathbf{x}_0 - \mathbf{x}_i), \ldots, g_{1m}^{\wedge}(\mathbf{x}_0 - \mathbf{x}_i)]^T, \tag{10}$$

for $i = 1, \ldots, n$. Unlike the matrix variogram function used in Myers (1982), $G^{\wedge}(h)$ does not need to be a symmetric matrix. For the special case where $g_{jk}^{\wedge}(h) = g_{jk}(h) + a_{jk}$ with $a_{jj} = 0$, the above equations are the same as the usual system of cokriging equations in terms of variograms and cross-variograms.

The sample pseudocross-variograms for variable 1 (Z_1) and variable 2 (Z_2) are computed by means of:

$$g_{12}^*(\mathbf{h}) = \frac{1}{2N}\sum_{i=1}^{N} [Z_1(\mathbf{x}_i) - Z_2(\mathbf{x}_i + \mathbf{h})]^2 \tag{11}$$

and

$$g^*_{21}(\mathbf{h}) = \frac{1}{2N}\sum_{i=1}^{N}[Z_2(\mathbf{x}_i) - Z_1(\mathbf{x}_i + \mathbf{h})]^2 \tag{12}$$

where N is the number of sample pairs with lag h. If $g^*_{12}(h)$ and $g^*_{21}(h)$ are symmetrical, i.e., $g_{12}(h) = g_{21}(h)$, we may obtain a cross-variogram from the pseudocross-variogram by utilizing the relationship between $g_{12}(h)$ and $\gamma_{12}(h)$ [Eq. (3)]. The nugget of the cross-variogram may be determined by cross-validation. The cross-validation procedure involves taking the pseudocross-variogram with a smaller nugget value as a cross-variogram, and using the cokriging technique to produce estimates corresponding to measured locations. Based on the statistical analysis of the estimates and measurements, an appropriate nugget of the cross-variogram can be chosen. The choice of a nugget of the cross-variogram should result in a near zero value of the mean error (i.e., $R_\mu = \Sigma^n_{n=1}[Z(\mathbf{x}_i) - Z^*(\mathbf{x}_i)]/(\sigma_k n) \approx 0$), and a value near unity for the reduced kriging variance (i.e., $R_{\sigma^2} = \Sigma^n_{i=1}[Z(\mathbf{x}_i) - Z^*(\mathbf{x}_i)]^2/(\sigma^2_k n)$).

An effective comparison of results estimated with kriging and cokriging requires some criterion for the comparison. The mean squared error (MSE), i.e.,

$$\mathrm{MSE} = \frac{1}{n}\sum_{i=1}^{n}[Z(\mathbf{x}_i) - Z^*(\mathbf{x}_i)]^2 \tag{13}$$

may be used for this purpose. In Eq. (13), $Z(x_i)$ and $Z^*(x_i)$ are the measured and estimated values at x_i, respectively. The relative improvement, or relative reduction of MSE, can be defined by:

$$100(\mathrm{MSE}_k - \mathrm{MSE}_{ck})/\mathrm{MSE}_k, \tag{14}$$

where MSE_k and MSE_{ck} are the mean squared error for kriging and cokriging, respectively. The relative reduction of kriging variance is defined in the same way by replacing MSE_k and MSE_{ck} in Eq. (14) with the kriging variances for kriging and cokriging, respectiely.

16.2.2 Disjunctive Kriging

Since the DK method has been reviewed in detail in several recent publications (Matheron 1976; Journal and Huijbregts 1978; Yates 1986; Yates et al. 1986b,c; Webster 1991), only a brief description will be provided here.

The overall objective of kriging is to obtain a moving average of a property which is distributed in space and/or time. Ordinary kriging (OK) requires the determination of constant weighting coefficients. For DK, unknown functions must be determined and these may or may not be linear. When these functions are linear, and the random function is multivariate normal, the DK method is the same as the OK method. Because of this, OK can be considered a special case of the DK method.

The DK method uses the autocorrelation function to determine the weighting coefficients for a series of Hermite polynomials. Second-order stationarity conditions are required to provide asssurance that the variance exists. For second-order stationarity conditions, the relationship between the autocorrelation function, $\rho(h)$, the covariance function, $C(h)$, and the variogram, $\gamma(h)$, is:

$$\rho(\mathbf{h}) = \frac{C(\mathbf{h})}{C(0)} = \frac{\gamma(\infty) - \gamma(\mathbf{h})}{\gamma(\infty)}, \tag{15}$$

where $\gamma(\infty)$ is the sill value of the variogram.

To obtain the DK estimator, the original data must be transformed into a new variable, $Y(x)$, with a standard normal distribution where pairs of sample values are bivariate normal. The function, $\phi[Y(x)]$, which describes this transformation is:

$$\phi[Y(\mathbf{x})] = Z(\mathbf{x}) = \sum_{k=0}^{\infty} C_k H_k[Y(\mathbf{x})], \tag{16}$$

where the values for $Y(x)$ can be found by taking the inverse of Eq. (16), that is $Y(x) = \phi^{-1}[Z(x)]$, and $H_k[Y(x)]$ is a Hermite polynomial of order k and argument $Y(x)$. The C_k's are the Hermitian coefficients which can be found using the properties of orthogonality. Numerical values for C_k can be obtained using:

$$C_k = \frac{1}{\sqrt{2\pi k!}} \sum_{i=1}^{n} \omega_i \phi(v_i) H_k(v_i) e^{v_i^2/2}, \tag{17}$$

where v_i and ω_i are abscissas and weight factors, respectively, for Hermite integration (Abramowitz and Stegun 1970).

An unbiased estimator with minimum estimation variance is sought and results in the following system of equations:

$$Z^*_{DK}(\mathbf{x}_0) = \sum_{k=0}^{K} C_k H^*_k[Y(\mathbf{x}_0)] \tag{18}$$

$$H^*_k[Y(\mathbf{x}_0)] = \sum_{i=1}^{n} b_{ik} H_k[Y(\mathbf{x}_i)], \tag{19}$$

where the series in Eq. (18) has been truncated to K terms and b_{ik} are the kriging weights which depend on the number of samples used in the estimation process and on the order of the Hermite polynomial. $H^*_k[Y(x_0)]$ is the estimated value of the kth Hermite polynomial at the estimation site, x_0. The sum of these estimates, multiplied by the coefficient C_k makes up the DK estimate at the estimation site. To obtain an estimated value for the Hermite polynomial, the DK weights, b_{ik}, must be found by solving the linear kriging equation for each k:

$$\sum_{i=1}^{n} b_{ik} (\rho_{ij})^k = (\rho_{0j})^k; \quad j = 1, 2, 3 \ldots, n. \tag{20}$$

When $k = 0$, Eq. 20) gives the unbiasedness condition, i.e., that the sum of the weights equals unity. The DK variance is:

$$\sigma^2_{DK} = \sum_{k=1}^{K} k!\, C_k^2 \left[1 - \sum_{i=1}^{n} b_{ik}(\rho_{0j})^k \right]. \tag{21}$$

The DK method may be used to determine the conditional probability that the value at the estimation site is greater than an arbitrary critical value, y_c. This is shown diagrammatically in Fig. 16.1 where the conditional probability (of exceedence) is represented by the shaded area under the curve and to the right of the critical value. y_c. This conditional probability is a useful means for determining the risk of various management alternatives. The conditional probability can be obtained by defining an indicator variable which is equal to unity if $Y(x_i) \geqslant y_c$, but is zero otherwise.

When the indicator variable is defined in this manner, the conditional probability can be written in terms of the conditional expectation to give:

$$P^*_{DK} = 1 - G(y_c) + g(y_c) \sum_{k=1}^{K} \left(\frac{1}{k!} \right) H_{k-1}(y_c) H_k^* [Y(\mathbf{x}_0)] \tag{22}$$

as the estimator of the conditional probability, where $G(y_c)$ and $g(y_c)$ are the standard normal cumulative and PDFs, respectively. The estimated conditional probability density function, $\mathrm{Pdf}^*_{DK}(x_0)$, can be found by taking the derivative of Eq. (22) with respect to y_c.

16.3 Examples

16.3.1 Estimation of NO_3 Using Cokriging with EC and Ca as Auxiliary Variables

The estimation procedure using pseudocross-variograms was applied to a large data set of soil chemical properties described by Warrick et al. (1991). The data were collected in May 1989 near Marana in Arizona. The experiment involved 130 measurement points over a 1000 × 1000 m area. One hundred data points were obtained from a 100 × 100 m grid, and an additional 30 locations were optimally distributed. This optimization was accomplised by adjusting the positions of the 30 samples to fit a predefined distribution of sample separation (Warrick and Myers 1987). Figure 16.2 (top) shows the locations of the 130 points. The variables electrical conductivity (EC), NO_3 and Ca were used for the analysis. It was assumed that the soil chemical properties were log-normally distributed. EC was scaled to approximately the same magnitude as NO_3 by multiplying by 100. A log-transformation was performed for the variables. In the following discussion, all results were based on the log-transformed data.

NO_3 has relatively high correlation coefficients with EC and Ca ($r^2 = 0.689$ and 0.75, respectively). Since EC and Ca are much easier to

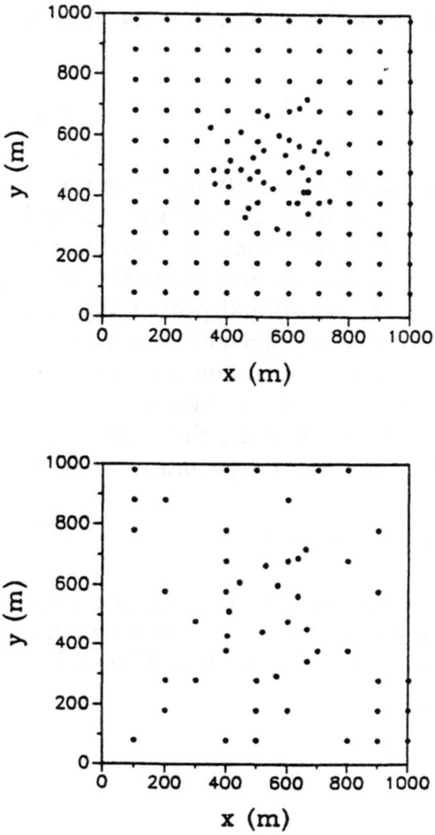

Fig. 16.2. Locations of the *top* 130 soil samples and *bottom* 50 soil samples at the Marana Farm

measure than NO_3, it is cost effective to use information of EC and/or Ca to improve the estimation of NO_3 and to use relatively few measurements of NO_3. We assumed that 130 EC, 130 Ca, and 50 NO_3 data points were available, and used these data to compute pseudocross-variograms and variograms. The locations of the 50 points, chosen randomly from the original 130 points, are shown in Fig. 16.2 (bottom). Plotted in Fig. 16.3 are the sample variograms (circles) for EC (top), Ca (middle), and NO_3 (bottom), along with the fitted variogram (solid line). The fitted variograms are linear models of the forms:

$$\gamma_1(\mathbf{h}) = 0.08 + 1.8(10)^{-4}\mathbf{h} \tag{23}$$

$$\gamma_2(\mathbf{h}) = 0.11 + 2.4(10)^{-4}\mathbf{h} \tag{24}$$

$$\gamma_3(\mathbf{h}) = 0.22 + 6.0(10)^{-4}\mathbf{h}, \tag{25}$$

for EC, Ca and NO_3, respectively. The sample pseudocross-variograms, $g^*_{13}(h)$ and $g^*_{31}(h)$, for variable 1 (EC) and variable 3 (NO_3), respectively, are plotted in Fig. 16.4 (top). The sample pseudo cross-variograms are quite symmetrical as illustrated by the fact that the two sets of results fall on

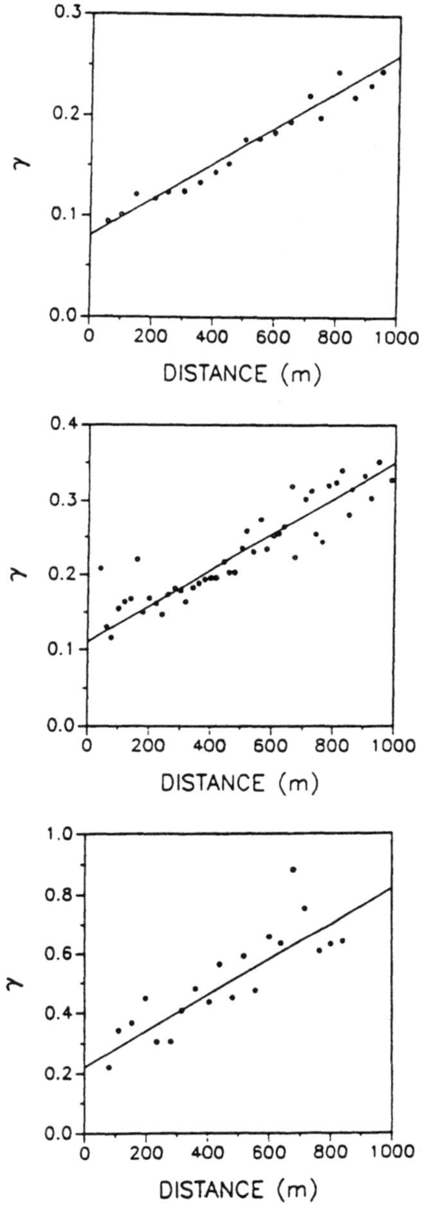

Fig. 16.3. Variograms of *top* EC using 130 data points, *middle* NO_3 using 50 data points, and *bottom* Ca using 130 data points

nearly the same line in Fig. 16.4 (top). The fitted pseudocross-variogram is a linear model:

$$g_{13}(\mathbf{h}) = 0.14 + 3.8(10)^{-4}\mathbf{h}. \qquad (26)$$

The nugget of the cross-variogram of EC and NO_3 was determined by cross-validation, while its slope was taken from Eq. (26). A cross-variogram of

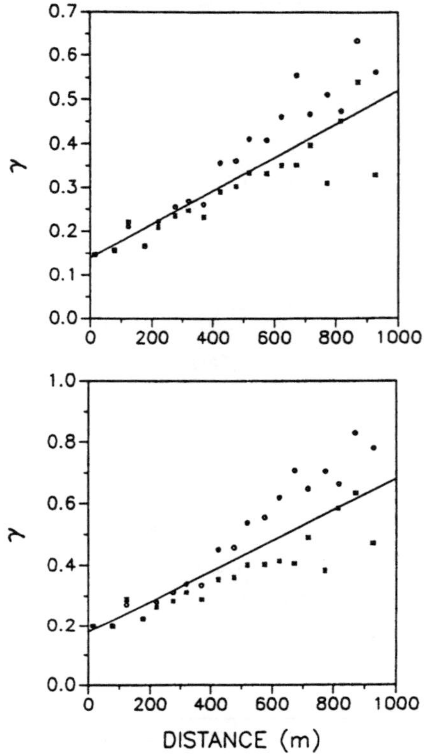

Fig. 16.4. Pseudocross-variograms of *top* NO_3 and EC using 50 NO_3 and 130 EC data, and *bottom* NO_3 and Ca using 50 NO_3 and 130 Ca data. *Solid symbols* in *top* and *bottom* represent g^*_{13} and g^*_{23}, respectively, whereas the *open circles* represent g^*_{31} and g^*_{32}, respectively

the form:

$$\gamma_{13}(\mathbf{h}) = 0.05 + 3.8(10)^{-4}\mathbf{h} \qquad (27)$$

was chosen. The sample pseudocross-variograms, $g^*_{23}(h)$ and $g^*_{32}(h)$, for variable 2(Ca) and variable 3 (NO_3), respectively, are plotted in Fig. 16.4 (bottom). The fitted pseudocross-variogram is also a linear model:

$$g_{23}(\mathbf{h}) = 0.18 + 5.0(10)^{-4}\mathbf{h}. \qquad (28)$$

Based on Eq. (28) and cross-validation, the cross-variogram was determined to be of the form:

$$\gamma_{23}(\mathbf{h}) = 0.05 + 5.0(10)^{-4}\mathbf{h}. \qquad (29)$$

The cross-variogram of EC and Ca, $\gamma_{12}(\mathbf{h})$, was computed by:

$$\gamma_{12}(\mathbf{h}) = 0.5[\gamma^+_{12}(\mathbf{h}) - \gamma_1(\mathbf{h}) - \gamma_2(\mathbf{h})], \qquad (30)$$

where $\gamma^+_{12}(h)$ is the variogram of the sum of EC and Ca at each of the 130 data points; this variogram is a linear function of h:

$$\gamma^+_{12}(\mathbf{h}) = 0.37 + 8.2(10)^{-4}\mathbf{h}. \qquad (31)$$

The results of cross-validation for the variograms and cross-variograms of EC, Ca and NO_3 [Eqs. (23) to (31)] are summarized in Table 16.1. The statistics were calculated from four sets of 50 estimates of NO_3 at the measured locations based on data options: (A) 50 NO_3 (kriging), (B) 50 NO_3 and 130 EC (cokriging), (C) 50 NO_3 and 130 Ca (cokriging), and (D) 50 NO_3 and 130 EC and 130 Ca (cokriging). Table 16.1 indicates that supplemental information of EC and Ca improves the estimation of NO_3, resulting in a smaller MSE (49% reduction), a lower mean kriging variance (29% reduction), and a higher correlation between the estimates and the actual measurements (0.402 vs. 0.766).

Based on the variograms and cross-variograms, kriging and cokriging were carried out to predict 130 NO_3 concentrations at the measured locations. As shown in Table 16.2, the MSE values were 0.267, 0.150, 0.151 and 0.149 for data options (A), (B), (C) and (D), respectively. Cokriging reduced the MSE by 44%, and reduced the kriging variance by 29% as

Table 16.1. Summary statistics of cross-validation for the cross-variograms and variograms

	Data option[a]			
	A	*B*	*C*	*D*
Mean error	−0.058	−0.050	−0.048	−0.047
MSE	0.437	0.230	0.231	0.222
Mean kriging variance	0.329	0.238	0.245	0.233
Reduced kriging variance	1.287	0.951	0.938	0.944
Correlation of estimates and error	0.099	−0.167	−0.245	−0.181
Correlation of estimates and measurements	0.402	0.755	0.763	0.766

[a] The data option indicates the different data sets used in the cross-validation procedure: (A) 50 NO_3 (kriging), (B) 50 NO_3 and 130 EC (cokriging), (C) 50 NO_3 and 130 Ca (cokriging), and (D) NO_3, 130 EC and 130 Ca (cokriging).

Table 16.2. Comparison of four sets of 130 NO_3 estimates, using 50 NO_3, 130 EC and 130 Ca values

	Data option[a]			
	A	*B*	*C*	*D*
MSE	0.267	0.150	0.151	0.149
Reduction of MSE by cokriging		44%	43%	44%
Mean kriging variance	0.199	0.145	0.149	0.142
Reduction of kriging variance by cokriging		27%	25%	29%
Correlation of estimates and measurements	0.672	0.827	0.826	0.828

[a] The data option indicates the different data sets used in the cross-validation procedure: (A) 50 NO_3 (kriging), (B) 50 NO_3 and 130 EC (cokriging), (C) 50 NO_3 and 130 Ca (cokriging), and (D) 50 NO_3, 130 EC and 130 Ca (cokriging).

compared with kriging. In this case, using both EC and Ca did not improve the estimation of NO_3 much more than using EC or Ca only.

Using the same approach with only 20 NO_3 data points, as shown in Fig. 16.5 (top), but again with all 130 EC and Ca measurements, we obtained pseudocross-variograms

$$g_{13}(\mathbf{h}) = 0.08 + 2.9(10)^{-4}\mathbf{h} \tag{32}$$

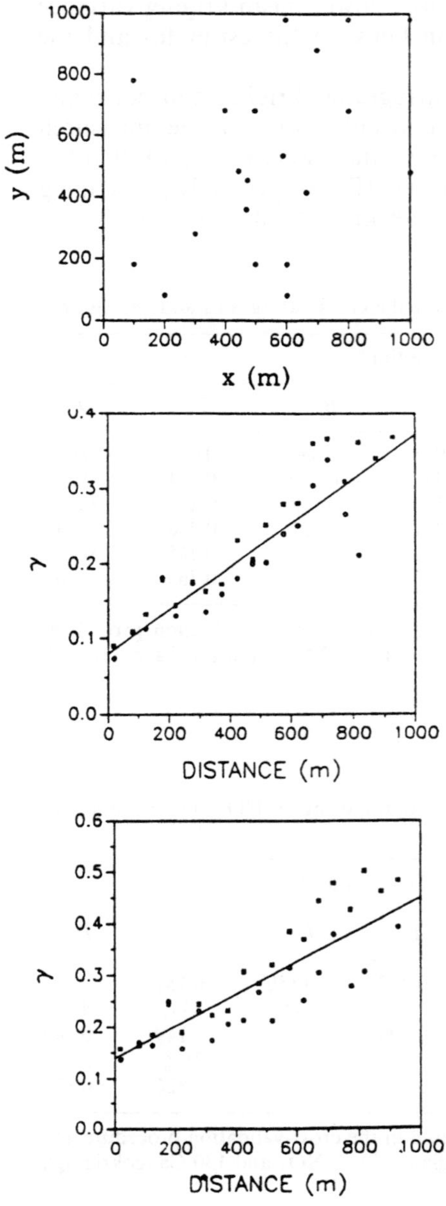

Fig. 16.5. *Top* location of the 20 soil samples; *middle* pseudocross-variogram of NO_3 and EC using 20 NO_3 and 130 EC data; and *bottom* pseudocross-variogram of NO_3 and Ca using 20 NO_3 and 130 Ca data. *Solid symbols* in *middle* and *bottom* represent g_{13}^* and g_{23}^*, respectively, whereas the *open circles* represent g_{31}^* and g_{32}^*, respectively

$$g_{23}(\mathbf{h}) = 0.14 + 3.1(10)^{-4}\mathbf{h} \tag{33}$$

for EC and NO_3, and Ca and NO_3, respectively. The two variograms are shown in Fig. 16.5 middle and bottom, respectively, together with the sample pseudocross-variograms. The cross-variogram models in this case were:

$$\gamma_{13}(\mathbf{h}) = 0.07 + 2.9(10)^{-4}\mathbf{h} \tag{34}$$

$$\gamma_{23}(\mathbf{h}) = 0.1 + 3.1(10)^{-4}\mathbf{h} \tag{35}$$

based on the pseudocross-variograms [Eqs. (32) and (33)] and cross-validation.

Using the 20 NO_3 data and the 130 EC and Ca points, and the variograms and cross-variograms, cokriging and kriging were used to obtain 130 NO_3 estimates for points where measured data were available. The MSE values were 0.407, 0.225, 0.248 and 0.263 for kriging, cokriging with EC, cokriging with Ca, and cokriging with both EC and Ca, respectively. Again, cokriging was found to reduce the MSE by 45%. Other kriging and cokriging results, summarized in Table 16.3, indicate that cokriging improves the estimation of NO_3 significantly.

16.3.2 Managing Soil EC in Agricultural Fields

To illustrate how the DK procedure can be used to manage the EC in agricultural fields, the data described by Al-Sanabani (1982) were used. Al-Sanabani obtained 101 random samples of the electrical conductivity from a 1×1 m grid system superimposed on a Typic Haplargid field soil located in southwest Arisona. The EC was measured by air drying samples of 1–2 kg and extracting a saturated paste.

The distribution of the data was tested for normality and log-normality using the Kolomogorov-Smirnov (KS) test (Rao et al. 1979; Sokal and Rohlf 1981). The test statistic was calculated and compared with the critical value at the 0.1 probability level. When the sample size is greater than

Table 16.3. Comparison of four sets of 130 NO_3 estimates, using 20 NO_3, 130 EC and 130 Ca values

	Data option[a]			
	A	*B*	*C*	*D*
MSE	0.47	0.226	0.247	0.263
Reduction of MSE by cokriging		44%	39%	35%
Mean kriging variance	0.311	0.213	0.265	0.194
Reduction of kriging variance by cokriging		32%	15%	38%
Correlation of estimates and measurements	0.421	0.735	0.718	0.676

[a] The data option indicates the different data sets used in the cross-validation procedure: (A) 20 NO_3 (kriging, (B) 20 NO_3 and 130 EC (cokriging), (C) 20 NO_3 and 130 Ca (cokriging), and (D) 20 NO_3, 130 EC and 130 Ca (cokriging).

Table 16.4. Calculated C_k's for actual and log-transformed data using numerical integration and the properties of the normal and log-normal distributions

Log-normal distribution (i.e., EC data)			Normal distribution [i.e., ln(EC)]	
k	Numerical integration	Log-normal distribution	Numerical integration	Normal distribution
	C_k	C_k	C_k	C_k
0	5.8447	5.795	1.3996	1.4004
1	4.9464	4.893	0.8385	0.8487
2	1.9881	2.066	0.0177	0
3	0.3572	0.581	−0.0125	0
4	−0.0766	0.123	0.0043	0
5	−0.0770	0.0206	−0.0039	0
6	−0.0160	0.0029	−0.0014	0
7	0.0042	3.34E-4	2.88E-4	0
8	0.0021	3.53E-5	1.18E-4	0
9	−1.28E-4	2.44E-6	−2.47E-5	0
Mean	5.845	5.795	1.400	1.400
Var.	34.453	34.926	0.710	0.720
Actual mean		5.846		1.400
Actual variance		35.123		0.720

30, the critical value at the 0.1 probability level can be obtained from $K_{crit} = 0.805(n)^{-1/2}$ (Rao et al. 1979; Rohlf and Sokal 1981). The KS test statistics calculated from the data sets were 0.197 and 0.065 for the original (EC) and natural logarithmic transformed (lnEC) data, respectively, and the critical value was 0.080. The results indicate that the original data are log-normally distributed at the 0.1 probability level.

The first step in the DK process is to find the coefficients which define the transform function given by Eq. (16). For normal and log-normal distributions, the C_k's can be calculated directly from the probability distribution; the results are presented in Table 16.4. Also included in Table 16.4 are values of C_k obtained using numerical integration. These values do not assume any particular shape for the distribution of the data. To obtain values for C_k using Eq. (17), the transform function, $\phi[Y(x)]$, must be known or approximated. This was accomplished by ordering the data sets from smallest to largest in numerical value and calculating for each datum, $Z(x_i)$, a corresponding transfromed datum, $Y(x_i)$, by inverting the probability function $P^{-1}[(i-\frac{1}{2})/n]$, where i is the number of data having a value less than or equal to $Z(x_i)$, n is the total number of data, P is the probability function, and the argument $(i-\frac{1}{2})n$ is an approximation for the probability of datum $Z(x_i)$. Once the pairs $Y(x_i)$, $Z(x_i)$ have been obtained, the relationship $Z(x) = \phi[Y(x)]$ can be determined by a variety of methods.

Plots of EC versus the probability are shown in Fig. 16.6. The figure demosnstrates the fit between the function $\phi[Y(x)]$ and the original data for two cases. In Fig. 16.6A, the C_k's corresponding to the theoretical

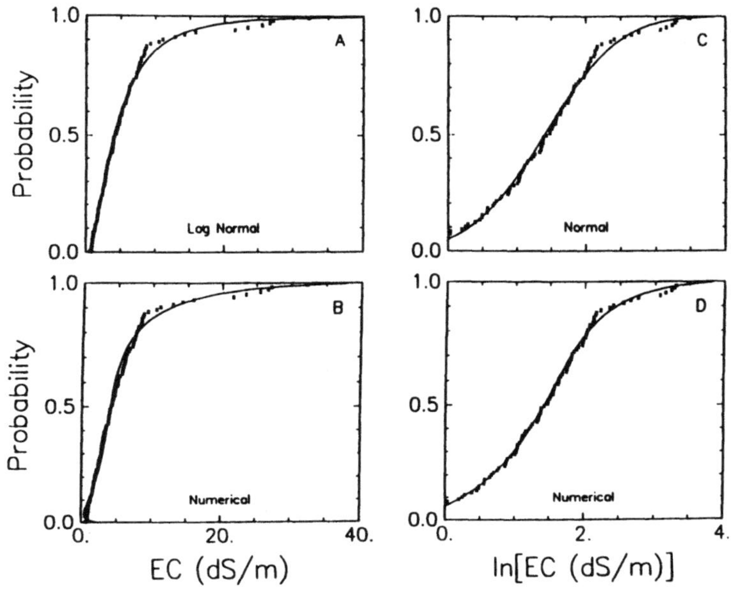

Fig. 16.6A–D. Comparison of the fit between the transform function, ϕ, and the observed data. In **A** and **B** the EC data were used and in **C** and **D** the ln(EC) data were used. *Numerical* indicates that the C_k's were obtained using numerical integration. *Log Normal* and *Normal* indicates that the C_k's were obtained using the properties of these distributions

log-normal distribution were used to generate the $\phi[Y(x)]$ function using Eq. (16). In Fig. 16.6B, the C_k's were calculated from the actual EC data using numerical integration. In both cases ten C_k's were used in each example. The results indicate that the theoretical log-normal distribution fits the data very well in the lower EC range (i.e., 0.5) < EC (dS/m) < 8) but the agreement is not as good for values of EC above 8 dS/m. There is an improvement in the fit between the model and the data when the C_k's are calculated from the data; the improvement gets progressively better when K increases. For the examples described here, the ten-term representation was judged to be adequate. This value was chosen, in part, because a substantial increase in the number of terms in the series causes large increases in computation time.

If the data are transformed by replacing each datum with its natural logarithm, another set of coefficients C_k are obtained. For the Gaussian distribution, the C_k's are $C_0 = \mu_{\ln}$, $C_1 = \sigma_{\ln}$, and C_k is zero for $k > 2$. For this data set, the mean ($\mu_{\ln}$) and standard deviation ($\sigma_{\ln}$) were found to be 1.40 and 0.849, respectively. When numerical integration is used to determine the C_k's, slighty different values are obtained compared with the theoretical values given above. These values are reported in Table 16.4 and the distributions are shown in Figs. 16.6C and D, respectively.

Prior to using the DK method, Eq. (16) must be inverted so that each value for $Y(x_i)$ will correspond to the correct value for $Z(x_i)$. This is necessary since the calculated C_k coefficients are only an approximation and the series in Eq. (16) will be truncated. Such a process will provide assurance that the DK method will produce an exact interpolation at the sample locations, a standard feature of kriging methods.

16.3.2.1 The Sample Correlation Function

All kriging methods require a function which describes the spatial correlation structure. For the DK method, the appropriate function is the autocorrelation. To facilitate a comparison with ordinary kriging, the correlation function was written in terms of the variogram using Eq. (15). The sample variogram for the EC data is shown in Fig. 16.7A. The model used to characterize the variogram for the EC data is:

$$\begin{aligned}
\gamma(\mathbf{h}) &= 0 && \mathbf{h} = 0\\
\gamma(\mathbf{h}) &= 20 + 25\left[\frac{3}{2}\frac{\mathbf{h}}{90} - \frac{1}{2}\frac{\mathbf{h}}{90}\right] && 0 < \mathbf{h} < 90\\
\gamma(\mathbf{h}) &= 45 && \mathbf{h} > 90
\end{aligned} \tag{36}$$

and has a reduced mean of 0.011 and reduced variance of 1.01. A problem with using the reduced mean and variance to determine the adequacy of the variogram models is that there is no independent means for determining how close the reduced mean and variance should be to zero and unity, respectively.

The sample variogram was calculated using the log-transformed data, fitted by a spherical model and validated by the jackknifing procedure. The

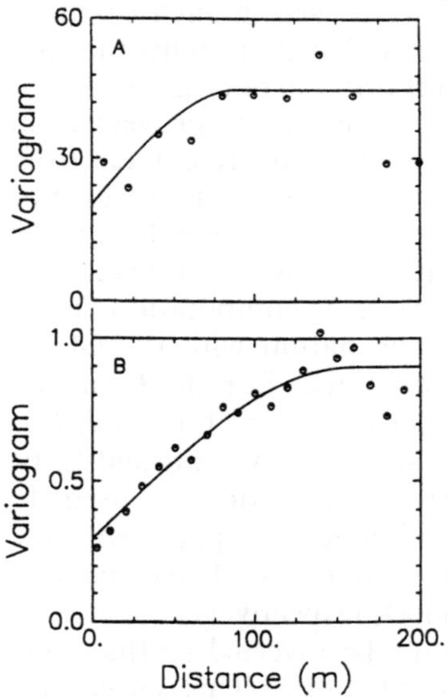

Fig. 16.7A, B. Sample and model variogram for **A** the EC data and **B** ln (EC) data

resulting variogram, shown in Fig. 16.7B, is:

$$\gamma(\mathbf{h}) = 0 \qquad \mathbf{h} = 0.0$$
$$\gamma(\mathbf{h}) = 0.3 + 0.6\left[\frac{3}{2}\frac{\mathbf{h}}{160} - \frac{1}{2}\frac{\mathbf{h}}{160}\right] \qquad 0 < \mathbf{h} < 160$$
$$\gamma(\mathbf{h}) = 0.9 \qquad \mathbf{h} > 160 \qquad (37)$$

with a reduced mean and variance of 0.020 and 1.19, respectively.

16.3.2.2 Estimation

Figure 16.8 shows a contour diagram of 240 EC estimates generated by disjunctive (solid line) and ordinary (dotted line) block kriging using a block size of 24 × 24 m. The block was divided into 2.4 × 2.4 m subblocks which were used to obtain a discrete approximation for the average variogram (Journel and Huijbregts 1978). Contour maps were also generated by using disjunctive and ordinary punctual kriging; they produced similar results. The distribution of EC in this field shows two prominent features, a low EC trough which runs from the coordinate (75, 0) to (175, 300) and a zone of high EC in the lower right quadrant. Kriging in terms of the ln(EC) produces a map with a similar pattern (Yates et al. 1986c). For brevity, only the map of EC estimates will be presented since this will be compared with the conditional probability. Disjunctive kriging also produces an estimation variance which gives an indication of where there is inadequate sampling. A comparison between the estimation variance using disjunctive and ordinary kriging is given by Yates et al. (1986c).

Contour maps such as those shown in Fig. 16.8 provide a convienent but qualitative means for showing the irregularities in the spatial distribution of the EC. As observed by Yates et al. (1986c) the difference between ordinary and disjunctive kriging is not very large. The difficulty in using contour

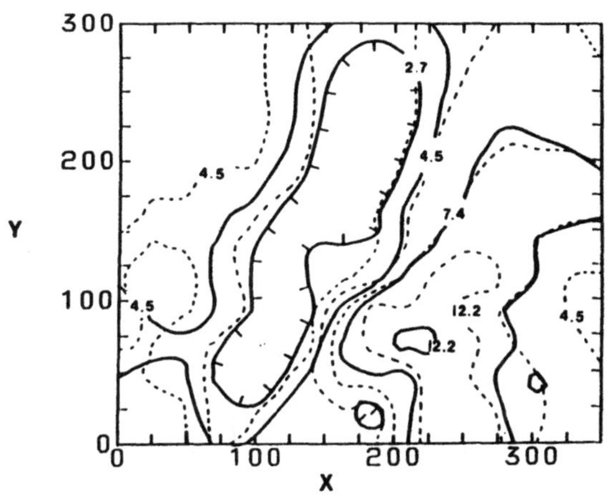

Fig. 16.8. Contour diagram of the EC. *Soild* and *dashed* lines indicate the results from block disjunctive and ordinary kriging, respectively

maps of the estimated value for management purposes is that one must still determine the likelihood that a certain area is greater than the allowed cutoff level. In locations on the map where the EC is much greater than allowed, one would be relatively confident in deciding, without resorting to additional analysis, whether corrective action is necessary. In other areas, however, which may include large areas of interest, the confidence would be much less. Contour maps of the kriging variance can provide additional information but still do not give quantitative measures of the likelihood that the EC is greater than the tolerance level.

16.3.2.3 Conditional Probability

Additional information is needed to provide a quantitative method for locating areas in need of corrective action. The DK estimator can be used to provide this information by calculating the conditional probability that the property of interest is greater than a specified cutoff value. Figure 16.9 shows an example of this by presenting contours of the conditional probability that the EC is greater than 4.0. These results were generated using a 24 × 24 m block. Visible in this figure are zones of high probability which coincide with areas of high EC in Fig. 16.8. Although the shapes of the high probability zones are similar to the shapes of the EC contours in Fig. 16.8, two points with the same estimated value of EC do not always produce the same conditional probability. For example, points (240, 24) and (288, 120) both have estimates of approximately 11 dS/m (i.e., 10.99 and 10.77 dS/m, respectively) but the estimated conditional probability has quite different values of 0.65 and 0.91, respectively. This information suggests that corrective action is necessary in one case but may not be required in the other case. These differences can be explained by considering the PDF at each point.

The conditional probability density and cumulative probability functions at the estimation site can be determined by taking the derivative of Eq. (22)

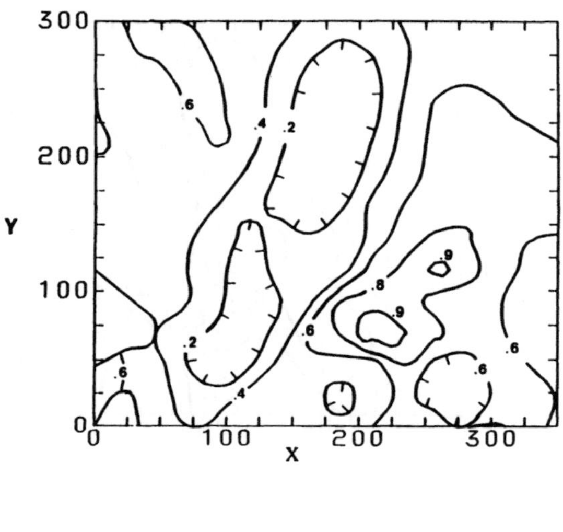

Fig. 16.9. Contour diagram of the conditional probability that the EC is above 4 dS/m using a 24 × 24 m block

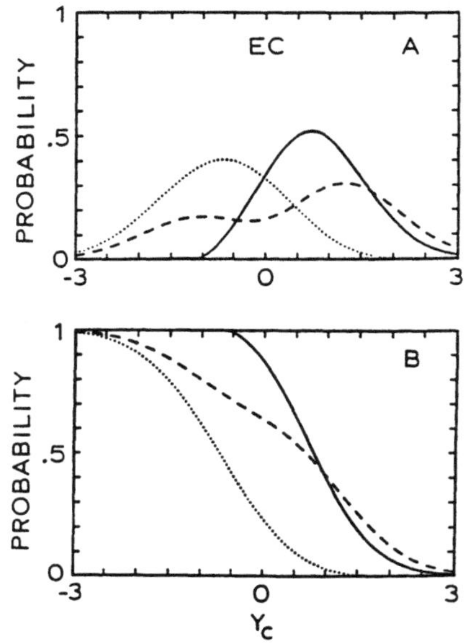

Fig. 16.10. **A** conditional probability density function and **B** probability function for the EC for three points in the field. *Soild line* represents the point (288, 210); *dashed line* represent (240, 24) and *dotted line* represents (96, 96)

and plotting the resulting curve as shown in Fig. 16.10 for the EC. The abscissa in this figure is the transformed value, Y. The three lines represent the PDF for three points in the field: the solid line is for the point (288, 120), the dashed line for (240, 24) and the dotted line for (96, 96). The three points were chosen to show differing behavior of the conditional probability in this field and to demonstrate how the sample values and their associated weights determine the shape of the PDF and, ultimately, the conditional probability.

The importance of the conditional probability can be demonstrated by comparing the behaviour of the PDF for two points with similar mean values (the estimated values at a point) but with differeing PDF values. This is illustrated by the solid (location 288, 120) and the bimodally distributed dashed (location 240, 24) curves in Fig. 16.10. In Table 16.5 the estimated mean value and associated probability for an EC cutoff value of 4.0 dS/m are given for the three points. The bimodal behavior for point (240, 24) is due to clustering about two levels of EC in the samples used to estimate the mean and conditional probability (i.e., Z_1, Z_2 and Z_4 versus Z_3 and Z_5 in Table 16.5). The mean value for each estimate is approximately the same but the conditional probability is vastly different. This is because in one case two distinct levels of EC (with relatively high weights) are used in the estimateion, whereas in the other case the samples are all large and of more uniform value. Information of this type is important when making decisions on how to efficiently manage the subunits.

The situation where the PDF values are similar but the mean value for each PDF is different is illustrated by the solid and dotted curves in Fig. 16.10. For the point (96, 96) the data values used in the calculation of the

Table 16.5. Weights and data values used in estimation of three points for block disjunctive kriging

Point	k	Z^*	P^*	b_{1k}	Z_1	b_{2k}	Z_2	b_{3k}	Z_3	b_{4k}	Z_4	b_{5k}	Z_5
288, 120	1	10.20	0.868	0.113	7.49	0.170	7.29	0.251	12.82	0.1487	7.52	0.144	21.54
	2	10.58	0.866	0.066		0.104		0.170		0.084		0.088	
	3	10.46	0.887	0.030		0.051		0.094		0.038		0.042	
240, 24	1	7.96	0.688	0.093	0.98	0.154	1.22	0.290	32.20	0.182	5.15	0.090	26.58
	2	10.46	0.677	0.026		0.062		0.155		0.085		0.022	
	3	10.63	0.647	0.006		0.019		0.069		0.032		0.004	
96, 96	1	2.27	0.229	0.364	3.77	0.040	2.01	0.188	0.87	0.110	1.03	0.044	4.75
	2	3.12	0.226	0.197		0.018		0.091		0.044		0.018	
	3	3.00	0.247	0.093		0.005		0.034		0.013		0.005	

Z^* and P^* are the estimated mean and conditional probabilities, respectively.
Z^* is within 95% of final value for $k = 2$ and within 99% for $k = 3$.

PDF (i.e., the b_{ik} and $Z(x_i)$ values are of relatively low magnitude (the range for the observed values of EC as from 0.61 to 32.2). Accounting for the effect of the weights shows that the major contribution is from the data values Z_1, Z_2 and Z_4. Since all data are of approximately the same magnitude (i.e., small) this produces a smaller mean value and a strongly peaked PDF. The point (288, 120) behaves in a similar manner, although for this point the magnitude of the data is large (i.e., greater than 7 dS/m); with the exception of Z_5 all the data are between 7 and 13 dS/m and the weights are more uniform.

16.4 Summary and Conclusions

Pseudocross-variograms were used to compute the spatial distribution of NO_3 over a 1000 × 1000 m field. The results of this study suggest that the approach using pseudocross-variograms overcomes several of the shortcomings in more common methods for computing cross-variograms. The main advantage of the pseudocross-variogram approach is that the computation of sample cross-variograms does not require that all variables be measured at the same locations. The method also maximizes the use of all available information.

For our examples, valid cross-variograms of NO_3 and EC, and NO_3 and Ca were obtained from their pseudocross-variograms, which were computed from relatively few NO_3 data, but used all the EC or Ca data. Cokriging using the reduced NO_3 data set gave much more accurate predicted results for NO_3 than did kriging, as evidenced by the smaller MSE and kriging variance. The MSE and mean kriging variance were reduced by as much as 44 and 35%, respectively, when cokriging was used.

The DK method has been shown to be a simple technique for obtaining an estimate of the conditional probability distribution at a point located in space. The method may be used to obtain estimates of the conditional probability density, its mean value (i.e., the estimate) and the conditional probability that an unknown value is greater than a specified cutoff level.

Calculated conditional PDFs at three locations (see Fig. 16.10) showed that this function is strongly dependent upon the samples used in the estimation. Two cases were discussed: conditional probability density functions with approximately the same mean values but different shapes, and vice versa. These two cases explain some of the factors that affect the conditional probability density function produced by the DK estimator.

The conditional probability may be very helpful in making management decisions. The DK method offers a quantitative method for assessing the level of an indicator variable in space, and the probability that the indicator is above an arbitrarily set cutoff level. As illustrated in this chapter, the conditional probability of the estimate is affected by the samples used in the estimation process. When one can manage a given unit area separately from the surrounding units, management efficiency can be improved by utilizing the conditional probability. An example of this was given where, even though the mean values for two blocks were approximately the same, only one case showed a high likelihood that the estimated value was greater than the cutoff level (i.e., p^* of 0.91 vs. 0.65 for the other block). Information of this type cannot be obtained from reading contour maps depicting the estimated values of the EC across a field. Information about the conditional probability provides managers with the means to utilize more efficiently energy, water and various soil amendments, and thus to reduce operating costs.

References

Abramowitz M, Stegun A (1970) Handbook of mathematical functions. Dover Publ., New York

Al-Sanabani M (1982) Spatial variability of salinity and sodium adsorption ratio in a typic Haplarigid soil. M. S. Thesis, University of Arizona, Tucson

ASCE (1990a) Review of geostatistics in geohydrology. 1. Basic concepts. ASCE task committee geostatistical techniques in geohydrology. J Hydraul Eng 116: 612–632

ASCE (1990b) Review of geostatistics in geohydrology. 2. Applications. ASCE task committee geostatistical techniques in geohydrology. J Hydraul Eng 116: 633–658

Burgess TM, Webster R (1980a) Optimal interpolation and isarithmic mapping of soil properties. I. The semivariogram and punctual kriging. Soil Sci 31: 315–331

Burgess TM, Webster R (1980b) Optimal interpolation and isarithmic mapping of soil properties. II. Block kriging. Soil Sci 31: 333–341

Burgess TM, Webster R, McBratney AB (1981) Optimal interporation and isarithmic mapping of soil properties. IV. Sampling strategy. Soil Sci 31: 643–659

Burrough PA (1983a) Multiscale sources of spatial variation in soil. I. The application of fractal concepts to nested levels of soil variation. Soil Sci 34: 577–597

Burrough PA (1983b) Multiscale sources of spatial variation in soil. II. A non-brownian fractal model and its application in soil survey. Soil Sci 34: 599–620

Carr JR, Myers DE (1984) Applications of the theory of regionalized random variables to the spatial analysis of landsat data. In: IEEE computer society proceedings of the Pecora 9 spatial technologies for remote sensing today and tomorrow. IEEE Computer Society Press, Silver Spring, MD, pp 55–61

Clark I, Basinger KL, Harper WV (1989) MUCK – a novel approach to cokriging. In: Buxton BE (ed) Proceedings of conference on geostatistical, sensitivity and uncertainty methods for ground-water flow and radionuclide transport modeling. Battelle, Columbus, pp 473–494

Journel AG, Huijbregts ChJ (1978) Mining geostatistics. Academic Press, New York. 600 pp

Matheron G (1976) A simple substitute for conditional expectation: the disjunctive kriging. In: Reidel D (ed) Proceedings NATO ASI Geostat 75, Hingham, Mass

McBratney AB, Webster R (1983) Optimal interpolation and isarithmic mapping of soil properties. V. Coregionalization and multiple sampling strategy. Soil Sci 34: 137–162

Mulla DJ (1988) Estimating spatial patterns in water content, matric suction, and hydraulic conductivity. Soil Sci Soc Am J 52: 1547–1553

Myers DE (1982) Matrix formation of co-kriging. Math Geol 14: 249–257

Myers DE (1991) Pseudo-cross variograms, positive-definiteness, and co-kriging. Math Geol 23: 805–816

Rao PV, Rao PSC, Davidson JM, Hammond LC (1979) Use of goodness-of-fit tests for characterizing the spatial variability of soil properties. Soil Sci Soc Am J 43: 274–278

Rohlf FJ, Sokal R (1981) Stastistical tables, 2nd edn, WH Freeman, New York, 219 pp

Russo D (1984a) Design of an optimal sampling network for estimating the variogram. Soil Sci Soc Am J 48: 708–716

Russo D (1984b) A geostatistical approach to solute transport in heterogeneous fields and its applications to salinity management. Water Resour Res 20: 1260–1270

Sokal RR, Rohlf FJ (1981) Biometry, 2nd edn. WH Freeman, New York, 859 pp

Stein A, Hoogerwerf M, Bouma J (1988) Use of soil-map delineations to improve (co-kriging of point data on moisture deficits. Geoderma 43: 163–177

Vauclin M, Vieira SR, Vachaud G, Nielson DR (1983) The use of cokriging with limited field soil observations. Soil Sci Soc Am J 47: 175–184

Vieira SR, Nielsen DR, Biggar JW (1981) Spatial variability of field-measured infiltration rate. Soil Sci Soc Am J 45: 1040–1048

Warrick AW, Myers DE (1987) Optimization of sampling locations for variogram calculations. Water Resour Res 23: 496–500

Warrick AW, Myers DE, Nielsen DR (1986) Geostatistical methods applied to soil science. In: Klute A (ed) Methods of soil analysis. Part 1. 2nd edn. Agronomy 9: 53–82

Warrick AW, Musil SA, Artiola JF, Hendricks DM, Myers DE (1991) Sampling strategies for hydrological properties and chemical constituents in the upper vadose zone. Final technical report, Water Resour Res Grand Program, US Dept Interior

Webster R (1991) Local disjunctive kriging of soil properties with change of support. Soil Sci 42: 301–318

Webster R, Burgess TM (1980) Optimal interpolation and isarithmic mapping of soil properties. III. Changing drift and universal kriging. Soil Sci 31: 505–524

Yates SR (1986) Disjunctive kriging. III. Cokriging. Water Resour Res 22: 1371–1376

Yates SR, Warrick AW (1987) Estimating soil water content using cokriging. Soil Sci Soc Am J 51: 23–30

Yates MV, Yates SR, Warrick AW, Gerba CP (1986a) Use of geostatistics to predict virus decay rates for determination of septic tank setback distances. Appl Environ Microbiol 52: 479–483

Yates SR, Warrick AW, Myers DE (1986b) Disjunctive kriging. I. Estimation and conditional probability. Water Resour Res 22: 615–621

Yates SR, Warrick AW, Myers DE (1986c) Disjunctive kriging. II. Examples. Water Resour Res 22: 623–630

Zhang R, Warrick AW, Myers DE (1991) Improvement of soil textural estimates using spectral properties. Geoderma 52: 223–234

Zirschky J (1985) Geostatistics for environmental monitoring and survey design. Environ Int 11: 515–524

Zirschky J, Harris DJ (1986) Geostatistical analysis of hazardous waste site data. J Environ Eng 112: 770–784

Zirschky J, Keary GP, Gilbert RO, Middlebrooks EJ (1985) Spatial estimation of hazardous waste site data. J Environ Eng 111: 777–789

Subject Index

Autocorrelation function 289

Channeling 46, 125
Chemical enrichment 183
Cokriging techniques 284, 286
Convection-dispersion equation 67, 82, 104
Correlation scale 14, 64
Covariance function 64, 74, 289
Crop yield 24, 104, 227, 264
Cross-variograms 284, 286

Decay rate coefficient 104
Dispersion coefficient
  local (pore scale) 18, 65, 88, 104
  field scale 18, 55, 66
Displacement covariance tensor 66, 71
Drainage 240

Effective capillary drive 139
Effective soil hydraulic properties 212
Ergodic hypothesis 64, 250
Erosion 183
Evaporation 109, 217, 236
Evapotranspiration 264

Fingering 125

Hydraulic conductivity 14, 54, 61, 128, 136, 200, 218, 235, 247
Hydrodynamic stability analysis 128
Hysteresis 99
Hillslope 152

Immicible displacement 52, 125
Immobile water 46
Infiltration 20, 29, 99, 125, 152, 175, 217, 247
Infiltrometer 200, 204
Inverse modeling approach 212
Irrigation scheduling 263
  optimization of 266
  in homogeneous field 267
  in heterogeneous field 268

Kriging techniques 284, 288

Leaf water potential 105
Long-slope constant-rainfall infiltration 156

Macropores 34, 42, 47, 245
Mass transfer rate coefficients 35, 56
Micropores 45
Moisture diffusivity 154
Monte-Carlo simulations 15, 51, 213, 247

Overland flow 191

Percolation theory 51
Permeameter 200
  constant head well 203
Ponding 16, 154, 256
Preferential flow 45, 51, 123

Radon barrier 248
Rainfall data 236
Rainfall detachment and redetachment 185
Reactive solute 22, 28
Redistribution 20, 99, 129
Regionalization methods 213
Relative permeability 140
Richards' equation 67, 82, 102, 228, 236
Root density 105
Root water uptake rate 104

Salinity 284
Scaling methods 213
Scanning curves 103
Sediment transport 185
Sediment settling 186
Stimulation
  of flow 69, 99, 217, 233
  of transport 69, 99
  of crop growth 99, 217, 269
Slope-crest infiltration 153
Solute adsorption 101
Solute lifetime 82
Solute travel time 36, 83
Sorbed chemical enrichment 189
Sorptive solute 28, 183
Sorptivity 155
Spatial heterogeneity 14, 33, 61, 247, 263

Spatial moments of concentration distribution 70
Stationary (statistical) 14, 63, 250, 289
Stochastic modeling
of flow and transport 14, 34, 57, 63, 82, 247
of potential evapotranspiration 271
Surface runoff 37, 240

Time domain reflectometry 206
Transient (unsteady) flow 20, 67, 214, 252
Transfer function model 67, 82
Transpiration 105, 227
Two-phase flow 139

Unstable flow 123

Variogram 289

Waste sites 247
Water retention function 99, 218, 235
Water uptake by roots 100, 227